BRETHREN OF THE NET

History of American Science and Technology Series

General Editor, LESTER D. STEPHENS

BRETHREN OF THE NET

American Entomology, 1840–1880

W. CONNER SORENSEN

The University of Alabama Press

Tuscaloosa and London

Copyright © 1995
The University of Alabama Press
Tuscaloosa, Alabama 35487–0380
All rights reserved
Manufactured in the United States of America

∞

The paper on which this book is printed meets the minimum
requirements of American National Standard for Information
Science-Permanence of Paper for Printed Library Materials,
ANSI Z39.48-1984.

Library of Congress Cataloging-in-Publication Data

Sorensen, Willis Conner.
 Brethren of the net : American entomology, 1840–1880 / W. Conner
Sorensen.
  p.   cm.—(History of American science and technology series)
 Includes bibliographical references (p.   ) and index.
 ISBN 0-8173-0755-9
 1. Entomology—United States—History.   2. Entomologists—United
States—History. I. Title. II. Series.
QL474.S67   1995
595.7′00973—dc20   94-5258

British Library Cataloguing-in-Publication Data available

0-8173-1236-6 (pbk: alk. paper)

To Bärbel,
who never lost faith

Contents

Acknowledgments		xi
1.	Entomology in the American Context to 1840	1
2.	"A Few Literary Gentlemen": The Entomological Society of Pennsylvania, 1842–1853	15
3.	Of Cabinets and Collections	33
4.	Agricultural Entomologists and Institutions	60
5.	The Balance of Nature	92
6.	A Weevil, a Fly, a Bug, and a Beetle	107
7.	The Rocky Mountain Locust Plague	127
8.	Profile of the American Entomological Community About 1870	150
9.	Acceptance and Implications of Evolution	197
10.	William Henry Edwards and Polymorphism in Butterflies	214
11.	The Yucca Moth	235
12.	The Debate over Entomological Nomenclature	242
	Conclusion	253
	Appendix 1: Entomological Authors Cited in the *Record of American Entomology*	261
	Appendix 2: Entomological Authors Cited in the *Record of American Entomology* Ranked According to Priority	263
	Abbreviations	265
	Notes	267
	Bibliography	317
	Index	345

Illustrations

Frontispiece by Lesueur for Say's *American Entomology*   9

Plate 9, "Coleoptera," Glover Notebooks   36

Plate 17, "Coleoptera," Glover Notebooks   73

Riley Examining Insect with Hand Lens   76

Plate 8, "Insects Beneficial," Glover Notebooks   98

"Shaking Trees to Release the Plum Curculio"   109

"The Hessian Fly and its Transformations"   113

"The Colorado Beetle"   123

Plate 23, "Coleoptera," Glover Notebooks   195

"Polymorphism in Butterflies"   231

Acknowledgments

It is a pleasure to acknowledge the many people who have helped me at various stages in the research and writing of this book. Morgan B. Sherwood, my dissertation adviser, has helped and encouraged me from the beginning, and I have profited from his conceptual and editorial suggestions. I have benefited from suggestions and criticism from my professors, James Shideler, Wilson Smith, and Daniel Calhoun, in seminars at the University of California, Davis. My wife, Bärbel R. Sorensen, supported me with her steady confidence in the worth of scholarly research in general and this topic in particular in settings that were often inhospitable to such pursuits. She read each draft, suggested changes, and in general persevered when I became fainthearted. This book is dedicated to her.

I had originally intended to write about C. Hart Merriam and American mammalogy only to find that Keir B. Sterling was finishing a dissertation on this topic. Keir helped me switch to the entomologists, a group that had begun to intrigue me during my initial research on Merriam. Margaret Rossiter negotiated the loan of agricultural journals from the library of the University of California, Berkeley, and encouraged me at various stages of the writing. Edward H. Smith helped improve my descriptions of insect forms and behavior, offered material from his research on Charles V. Riley, included my paper entitled "Entomological Societies as Transmitters of Evolutionary Theory" in a session he organized for the History of Science Society, supplied illustrations from the Cornell University Library, and generally provided the companionship only a fellow historian of entomology could offer. Klaus Sander,

Frederick B. Churchill, Mary P. Winsor, Nathan Reingold, Ashley Gurney, John Perkins, Robert Fagan, and Francis Glass commented on various parts of the manuscript. Richard C. Froeschner read the entire manuscript in an early form, and he furnished references from the Library of the U.S. National Museum. Kenelm W. Philip supplied references and comments on Edwards and polymorphism. David C. Smith commented on portions of the manuscript, and he and David B. Danbom edited material that appeared in two special issues of *Agricultural History*. Cyril F. Dos Passos and Joseph Ewan provided important references. Margaret Treat checked references and translated some manuscript letters from the French.

I thank, posthumously, the nineteenth-century American entomologists, librarians, and administrators who took care that the correspondence and other papers of the entomologists would be preserved. As a researcher who has had occasion to search for correspondence on both sides of the Atlantic, I can only express my astonishment at how much more complete the North American records in the history of entomology are than those in Europe. The entomological records in the Museum of Comparative Zoology, the Academy of Natural Sciences, the American Philosophical Society, and the Smithsonian Institution are a unique treasure.

For permission to cite and make quotations from unpublished materials listed in the bibliography, I thank the American Philosophical Society, Philadelphia, Pennsylvania; the Academy of Natural Sciences of Philadelphia, Philadelphia, Pennsylvania; the Museum of Comparative Zoology, Cambridge, Massachusetts; the Museum of Science, Boston, of Boston, Massachusetts; the Smithsonian Institution, Washington, D.C.; the West Virginia University for the West Virginia and Regional History Collection, West Virginia Libraries, Morgantown, West Virginia; the West Virginia Division of Culture and History, Charleston, West Virginia; the New York State Archives, University of the State of New York, State Education Department, Albany, New York; the New York State Library, University of the State of New York, State Education Department, Albany, New York; the Field Museum of Natural History, Chicago, Illinois; and the Bancroft Library, University of California, Berkeley, California.

Librarians, archivists, interlibrary loan specialists (and friends who have helped with interlibrary loans) have assisted me generously in making use of these and other archives and the secondary literature that complements them. I thank Ann Blum, Roxane Coombs, Muriel Conant, Marina Dzidziguri, Mary Brockenbrough, and Katya Fels, all at the

Museum of Comparative Zoology; Mary M. Jenkins, at the West Virginia Division of Culture and History; Barbara Wiseman, Museum of Science, Boston; William P. Gorman, New York State Library; Roland Madany, University of Illinois at Urbana-Champaign; Harold T. Pinkett and James O. Welker, U.S. National Archives; William A. Deiss, William E. Cox, and Pamela M. Henson, all at the Smithsonian Institution Archives; Murphey D. Smith and Carl F. Miller, American Philosophical Society Library; Carol S. Hahn, U.S. National Museum of Natural History Library; and Ruth E. Brown, Academy of Natural Sciences Library. I extend special thanks to the interlibrary loan staff of the Alaska State Library and the University of Alaska Southeast library, who, from Juneau, Alaska, made my research possible. These include Carol H. Ottesen, Kay Rosier, John Ross, Kevin Araaki, Patricia C. Wilson, Sherry Taber, Linda Gowing, Barbara Berg, Kit Stewart, Julie Leary, and Sara Hagen. Gail and Phillip N. Hocker took time from their academic and teaching responsibilities to help me secure many items through interlibrary loan. From Wuppertal, Germany, I have been assisted by the interlibrary loan staff of the Bergische Universität Gesamthochschule Wuppertal. I wish to thank in particular Sabina Böhm, Petra Schonebeck, and Alexandra Erlach.

I wish to thank the editor Malcolm MacDonald, for his encouragement over many years, and his editorial staff; I am grateful for the editorial work of Marcia Brubeck, for her assistance in matters of form and style. I also wish to thank those who helped with various editorial tasks in preparing the manuscript. Trudy Kaliski and Mollie Zrust helped type and format an earlier version of the manuscript. Kathy Hocker researched several items in the Harvard University Libraries and formatted the chapters in a new word processing program. Wendy Selleck and Linda Huff revised the notes and bibliography and formatted the manuscript in its final form.

Several institutions have supported the research for this book. A National Defense Education Act Fellowship at the University of California, Davis, provided time and research funds during the initial stage of research. The History of Science Society (Associate Scholars Grant) and the National Endowment for the Humanities (grant FE-22274-88) supported a research trip to the Museum of Comparative Zoology Archives in Cambridge, Massachusetts.

Chapter 4 appeared in somewhat shorter form in Conner Sorensen, "The Rise of Government Sponsored Applied Entomology, 1840–1880," *Agricultural History* 62 (1988), 98–115, and is reprinted here with permission from *Agricultural History*. Chapters 5, 6, and 7 appeared in part in

Conner Sorensen, "Uses of Weather Data by American Entomologists, 1830–1880," *Agricultural History* 63 (1989), 162–74, and are reprinted here with permission from *Agricultural History*. Chapter 10 appeared in part in Conner Sorensen, "William Henry Edwards, August Weismann, and Polymorphism in Butterflies," in Klaus Sander, ed., *August Weismann (1834–1914) und die theoretische Biologie des neunzehnten Jahrhunderts: Urkunden, Berichte und Analysen,* special issue of *Freiburger Universitätsblätter,* Heft 87/88 (July 1985), 157–65, and is reprinted with permission from the *Freiburger Universitätsblätter.*

I have retained the scientific names of insects as used by the entomologists at the time, except that I have followed the modern practice of using lower case for the species name (*Papilio ajax* instead of *Papilio Ajax*). Scientific names that appear within titles of nineteenth-century works, however, have been retained in their original form with regard to capitalization and italics. In the spelling of scientific names for butterflies, I follow William Henry Edwards, *List of Species of the Diurnal Lepidoptera of America North of Mexico* (Boston: Houghton Mifflin, 1884). As a general guide to scientific names, especially where nineteenth-century names are reasonably close to those now accepted, I follow Manya B. Stoetzel (chair of committee), *Common Names of Insects and Related Organisms* (Beltsville, Md.: Entomological Society of America, 1989).

The final product is my responsibility, and while the individuals and institutions named here share a great deal of the credit, they should not be blamed for any errors of fact or interpretation.

Wuppertal, Germany

BRETHREN OF THE NET

Entomology in the American Context to 1840

The American entomologists who are the subject of the following chapters worked within two distinct yet closely related contexts. They participated in the general enterprise of Western European science with its roots in antiquity, the Renaissance, and the scientific advances of the Enlightenment and the early nineteenth century. This tradition originated in centers in Europe—Bologna, Paris, London, Berlin, and elsewhere—where Western science had experienced its renaissance. At the same time, Americans participated in a growing American scientific enterprise, a much younger community but one that by the early decades of the nineteenth century had developed into a junior partnership with Europe.[1]

In order to interpret the activities and achievements of the generation of American entomologists after 1840, it is necessary to review the development of entomology in Europe and in America up to that time. I will particularly note the activities of those Americans who engaged in entomological investigations between the revolutionary war and 1840.

Entomology as a field of study had its beginnings in the detailed investigation of insects of medical men and others since the Renaissance. From the seventeenth century, European physician-scientists had employed ever more sophisticated microscopes in the study of insects and had acquired an accurate understanding of specialized insect organs such as the compound eye, the antennae, and the organs of smell and hearing. Other investigators took to the field to make important biological and ecological discoveries such as the pollination of plants by insects and the impact of destructive insects on agriculture.[2]

From the mideighteenth century to the early decades of the nineteenth century, natural history enjoyed an exciting period of discovery, expansion, and analysis. At the outset of this period, two eighteenth-century naturalists, Réné Antoine Réaumur and Carl Linnaeus, took decisive steps toward establishing the study of entomology as a special branch of natural history with its own nomenclature, literature, and community of experts. Réaumur was an educated French nobleman with wide-ranging scientific interests who chose the study of insects as his specialty. He insisted that insects deserved to be studied for their own sake, as a matter of pure scientific interest, rather than because of their importance to agriculture or their relevance to anatomical questions. Réaumur's chief work, the six-volume *Mémoires pour servir à l'histoire des insectes* (1734–1742), consisted of small monographs on each insect species, covering all aspects of their anatomy, life history, and habits. His circle of students, fellow investigators, and correspondents in the early to mid-eighteenth century comprised the first self-conscious community of entomologists.[3]

This community expanded exponentially in the later eighteenth century, thanks to Carl Linnaeus and his successors, who developed insect systematics into the main unifying activity of entomological science. Linnaeus was a Swedish naturalist who exercised his genius in classifying wide areas of nature, including plants, animals, and minerals. His *Systema Naturae* (tenth edition, 1758) is still regarded as the starting point for zoological nomenclature. Linnaeus greatly simplified insect classification by choosing the wings as a single set of characters that could serve as a consistent guide to classification. The seven insect orders he constructed according to wing structure were the Coleoptera, Hemiptera, Lepidoptera, Neuroptera, Hymenoptera, Diptera, and Aptera. An equally important contribution was his use of simplified binomial nomenclature. Others had experimented with various forms of shorthand description, but Linnaeus was the first to insist that each designation for genus and species consist of only one word.[4] The seven insect orders,[5] arranged according to wing structure, were outlined by Linnaeus as shown (p. 3).

For the scientist, entomological student, or collector, such an arrangement simplified matters considerably. Thus beetles, with their hard cases over the wings, could be assigned with confidence to the Coleoptera. Butterflies and moths, with their scaly wings, formed a neat group within the Lepidoptera. Flies with two wings, like houseflies, mosquitoes, and gnats, clearly belonged to the Diptera. Insects with membranous wings but no stings, like lacewings, mayflies, and dragonflies, were grouped as the Neuroptera. Bees, wasps, and ants, which had membranous wings and also had stings, formed the Hymenoptera. In the Hemiptera, Linnaeus

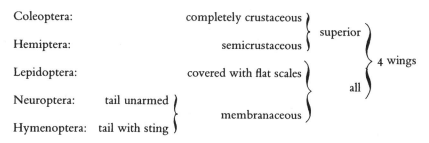

Coleoptera: completely crustaceous ⎫

Hemiptera: semicrustaceous ⎬ superior

Lepidoptera: covered with flat scales   4 wings

Neuroptera: tail unarmed ⎫   all

Hymenoptera: tail with sting ⎭   membranaceous

Diptera: 2 wings, with halteres in place of hind wings
Aptera: 0 wings, without wings and elytra

was not completely consistent with regard to wing pattern. The order contained the cicada, water scorpion, and other "bugs" with semicrustaceous wings, but it also included the aphids and other genera with wings that are totally membranous or completely absent. The real unifying character of the Hemiptera was the mouth structure, a beak that was modified for sucking plant juices or the blood of animals. In this instance Linnaeus followed an affinity other than wing structure; it was his only divergence from the "artificial" system of classifying insects according to this set of characteristics. Nineteenth-century naturalists generally accepted the seven great insect orders established by Linnaeus and worked within these categories.[6]

The generation following Linnaeus modified his artificial system based on the wings to construct the beginnings of a "natural" system of insect classification based on a variety of characteristics. The key figure in this enterprise was Pierre André Latreille. An associate of Georges Cuvier and Jean Lamarck at the Museum National d'Histoire Naturelle, Latreille was regarded as the world's foremost entomologist. Latreille supervised the section on articulata at the museum, and he wrote the sections on the arthropoda for Cuvier's *Règne Animal* (1816 and 1829). In the development of his system, Latreille consulted frequently with his colleague Lamarck and with Johann Christian Fabricius, a student of Linnaeus who had developed a system of classification based on the insect mouthparts. Latreille rejected the use of any single character such as the wings or mouthparts in classification. He also rejected Cuvier's ordering of characters in a hierarchical scheme. Instead, he simply followed pronounced affinities wherever they occurred among insect groups. Because he used various characters and patterns (such as the mouthparts, wings, metamorphosis) to construct his groups, his method has been termed "eclectic." In contrast to Cuvier, who searched for universal ordering principles, Latreille worked from the bottom up, grouping species with marked affinities in genera, grouping genera in families, and so on. Latreille introduced the categories of family and tribe between the genera and orders of Linnaeus. He also promoted Lamarck's definition of "insects," which separated the arachnids and crustaceans from insects, thus restricting the class Insecta to those arthropods with six legs.[7] The American entomologists who were contemporaries of Latreille accepted the genera and families he established, and in fact most of these are still recognized today.[8]

The systematics of Linnaeus and his successors stimulated entomology enormously. Entomologists now had a general map and manageable terms for their discipline. Lists of local insect faunas multiplied, specialists emerged for each of the orders (particularly Coleoptera and Lepidoptera), and naturalists in the Old and New Worlds intensified their efforts to collect and classify the insect faunas of previously unexplored regions.

By the nineteenth century, then, entomology flourished and occupied a niche in European zoological science. The specialists writing about the various orders provided the first guides to identify and classify American insects. In addition to Linnaeus, Fabricius, and Latreille, Americans consulted (among others) works by William Kirby, who with William Spence published an *Introduction to Entomology* (1816–1826), works by the coleopterists Guillaume Antoine Olivier, Compte P. F. M. A. Dejean, and Wilhelm F. Erichson, and works by the lepidopterist J. B. A. Boisduval.

But Europe was distant from America, and the European literature, while outlining currently accepted classification schemes, offered American investigators little help in the identification of North American insect species. Besides, Americans felt empowered by their victory over Great Britain (in the Revolution and again under Andrew Jackson at New Orleans in 1815) to establish American natural history on a level appropriate for a nation that embodied Enlightenment ideals in science and politics. Americans considered it their patriotic duty to discover, name, and classify the natural productions of the New World and to make them known to science.[9]

Though the New World naturalists remained few in relation to their continental undertaking, by the early nineteenth century, Americans had begun to make a creditable showing. The first noteworthy American contributions to natural history were made in zoology and the emerging science of paleontology.[10] Thomas Jefferson, in his *Notes on the State of Virginia* (1782), had defended New World animals against Count George

Louis Leclerk Buffon, who claimed that they represented degenerate Old World species.[11] By the turn of the century, Charles W. Peale was excavating and assembling complete mastodon skeletons, which furnished evidence of New World animals' strength and vitality and supported the conviction of those naturalists who held that significant numbers of species had become extinct. Thus, traditional assumptions embodied in the notion that a Great Chain of Being connected all living forms, past and present, were giving way, partly under pressure from American contributions to international science.[12] By the early decades of the nineteenth century, the American fauna was being ably portrayed by Americans like Peale, Alexander Wilson, Benjamin S. Barton, and Richard Harlan as well as by resident foreigners like Constantine S. Rafinesque, Charles A. Lesueur, and Charles L. Bonaparte.[13]

To a great extent these American contributions to zoology were possible because of the development of institutions devoted to natural history in regional scientific centers located primarily in Philadelphia, Boston, and New York. These local communities of naturalists and the institutions they built provided the support necessary for the early growth of American entomology.

Philadelphia, home of Benjamin Franklin, the American Philosophical Society, and the botanical gardens of John and William Bartram, was the first important center of American natural history. Philadelphia's leadership in natural history continued with Peale's museum, which flourished from about 1800 to 1816, and with the founding of the Academy of Natural Sciences of Philadelphia in 1812. The academy was supported generously by William Maclure, the industrialist, geologist, and social reformer, who served as its president from 1817 to 1840. Its collections and its published proceedings placed vital resources in the hands of Philadelphia naturalists and established the academy's reputation abroad. The academy's leadership in natural history remained unchallenged until around midcentury, when the Smithsonian Institution, Louis Agassiz's museum, and other centers of natural history appeared. As will be seen, the Academy of Natural Sciences played a decisive role in nurturing entomological activities during the formative midcentury decades.[14]

The Boston–Cambridge area, second only to Philadelphia as a regional center of American science, enjoyed the unique advantage of Harvard College, with its loyal and generous alumni, intent on making their alma mater superior in science as well as in other areas. Harvard was a magnet for New Englanders and others eager to pursue natural history. Another focus of science in Boston was the American Academy of Arts and Sciences, founded in 1785 at the urging of John Adams as Boston's equivalent of the American Philosophical Society. Though never a serious challenge to the American Philosophical Society and the Academy of Natural Sciences in natural history studies, the American Academy gave birth to the Massachusetts Society for Promoting Agriculture, which proved important to entomology. Originally formed as a committee of the American Academy and incorporated as a separate institution in 1792, the agricultural society awarded a series of prizes for studies of insects injurious to agriculture that helped stimulate interest in this subject. In 1805, the agricultural society also helped fund a natural history professorship and a botanical garden at Harvard College. The post was filled by William Dandridge Peck, a Harvard graduate and self-taught naturalist whose primary interest was insects.[15] In 1814–1815, naturalists in the Boston area organized the Linnaean Society of New England, which flourished until the mid-1820s, when a proposed transfer of its collections to Harvard aborted and the society fell into disarray. In 1830, however, a core group from the Linnaean Society reorganized as the Boston Society of Natural History. Beginning by reassembling what was left of the natural history collections of the Linnaean society, the Boston Society prospered in the 1830s, and its collections have grown steadily to the present.[16] The Boston Society of Natural History, with its strong emphasis on the study of insects, played a significant role in the early careers of many American entomologists in the midcentury decades.[17]

New York, though outstripping its rival Philadelphia as the center of trade and commerce in the United States, remained behind both Philadelphia and Boston as a center of regional science. Of most importance to entomology was the Lyceum of Natural History of New York. Organized in 1817 by professors and students of the New York College of Physicians and Surgeons, it attracted able naturalists like Amos Eaton and John Eatton LeConte. The Lyceum's cabinet of natural history specialized in reptiles, fishes, and minerals, with relatively little attention given to insects, but its journal, first published in 1823, featured some early papers on entomology by LeConte. All in all, New York lagged well behind Philadelphia and Boston as a center of entomological activity.[18]

American activities in entomological science in the years up to 1840 combined various strands of European science with the growing American scientific tradition. Because systematics flourished above all other entomological activities during this era, Americans who wished their work to be recognized internationally had to prove themselves primarily in this area. In doing so, they had certain advantages and disadvantages. On the one hand the Americans had ready access to the insect fauna of a continent that was of interest to international science.[19] On the other hand, the Americans suffered from lack of ready access to the printed literature and to the great insect collections in Europe, which were necessary for reference and comparison. Likewise, they suffered from a shortage of scientific publications in which to announce their discoveries. They were handicapped further by the need to secure patronage for their scientific pursuits, a need only partly met through the growth of regional scientific institutions.

Rejecting the role of collector, the Americans used their favored geographical position to concentrate on the description of new species. While American attention to the practical concerns of agriculturists emerged at an early date, these studies of agricultural pests were initially less effective than the description of new species in establishing a reputation for American entomology within the context of international science.

These various advantages, limitations, and problems served to define the relations among American entomologists as well as between them and the Europeans in the generation prior to 1840. These relationships may be traced in the careers of six Americans who devoted extensive attention to entomology: John Abbot, William D. Peck, Thomas Say, John Eatton LeConte, Frederick Valentine Melsheimer, his son, John F. Melsheimer, and Thaddeus William Harris.

John Abbot, born in London in 1751, was encouraged by his father, an attorney, to pursue his interest in natural history. His talent for drawing brought him to the attention of Dru Drury, a former president of the Linnaean Society and owner of one of the best collections of insects in England, who became Abbot's entomological mentor. Following the example of Mark Catesby, who in his *Natural History of Carolina, Florida and the Bahama Islands* (1754) had portrayed American animals in striking illustrations, Abbot set out in 1773 to paint American birds and insects. Abbot joined the American revolutionary forces and for his military service was later awarded a substantial acreage in Georgia. He married, settled down as a planter, and took up the painting of insects and birds in earnest. He was the first artist in the New World to produce an extensive series of insect drawings, an estimated three thousand illustrations that showed insects in all their stages. In order to render accurate reproductions of the early stages, he raised thousands of insects from the egg through their transformation to adults. Abbot kept in close contact with naturalists and art dealers in England, to whom he sold his paintings, his pinned insect specimens, and his stuffed birds. In 1797, James E. Smith, president of the Linnaean Society of London, published (without Abbot's prior knowledge) a two-volume collection of Abbot's butterflies entitled *The Rarer Lepidopterous Insects of Georgia*. It became Abbot's best-known publication, though he received no payment beyond the modest amounts for which he had sold the original drawings to London dealers.

Abbot had little direct contact with American naturalists, the primary exception being the ornithologist Alexander Wilson, who visited him in Savannah in 1809. Abbot supplied Wilson with substantial material on southern birds, which were featured in Wilson's *American Ornithology* (1808–1814). Although Abbot remained mostly removed from the centers of natural history in America, his work was known to American entomologists. John E. LeConte reportedly collected two or three thousand of his original watercolors of insects. Abbot continued to collect insects and birds, which he illustrated and sold mostly to London customers, until his death around 1840. Although he was relatively isolated from the developing taxonomic tradition that guided entomological science in those years, he made hundreds of American insects known for the first time through his beautiful and accurate illustrations. He also established a tradition of rearing insects for study, which became a characteristic strength of American entomology.[20]

William D. Peck was born in Boston in 1763 and educated at Harvard, but he left Boston soon after graduation to join his father on a farm near Kittery, Maine. For twenty years he devoted his leisure hours to the study of natural history, becoming an expert in botany, entomology, ornithology, ichthyology, and other sciences. In 1793 he won the prize awarded by the Massachusetts Society for Promoting Agriculture for his essay on the canker worm, the first of several such essays to appear in the published papers of the society. He also wrote articles on insects injurious to agriculture for the *Massachusetts Agricultural Repository and Journal* and other periodicals. His growing reputation as an entomologist and botanist led to his appointment, in 1805, to be the first professor of natural history and keeper of the botanical garden established at Harvard, the position created with financial support from the agricultural society. Peck was sent to Europe for three years to prepare himself further for his new duties. While in Europe, he contacted naturalists in England, Sweden, Denmark, Germany, and France, including James E. Smith (who had published Abbot's butterflies) and Latreille. Peck established a firm reputation at Harvard as a quiet but effective teacher and administrator of the botanical garden. He sent many insect specimens to European correspondents, in particular to William Kirby, who published them in the *Transactions* of the Linnaean Society of London. His investigations of injurious

The frontispiece drawn by Charles Lesueur for Thomas Say, *American Entomology; or, Descriptions of the Insects of North America* (3 vols., Philadelphia: S. A. Mitchell, 1824, 1825, 1828)

insects marked another characteristic of American entomology, a tradition carried on after his death in 1822 by his student Harris.[21]

Thomas Say, the grandnephew of the venerable American naturalist William Bartram, grew up among the circle of Philadelphia naturalists whose work brought American zoology and paleontology to prominence in the early nineteenth century. An eager learner, he had access to Peale's museum, which contained many thousands of insect specimens, and to the cabinets and libraries of Philadelphia naturalists. He was one of the founders of the Academy of Natural Sciences of Philadelphia in 1812 and in the course of his lifetime became the most prolific writer associated with the academy. Say had wide-ranging interests in the natural sciences, but his most important publications were in entomology and conchology. The title of his best-known work, *American Entomology; or, Descriptions of the Insects of North America,* published in three volumes (in 1824, 1825, and 1828), stated his intention to extend the systematic description of American insects. Say's Quaker leanings led him to endorse the educational and social reform efforts of William Maclure and Robert Owen, who founded a utopian colony at New Harmony, Indiana. In 1825, Say joined the colony as superintendent of literature, science, and education, and he remained at New Harmony after most others had abandoned the project. Say's remote location, his ideal of improving society through education, and his commitment to the diffusion of scientific knowledge by means of elaborately illustrated publications, caused his estrangement from the more narrowly focused systematists who assumed leadership in the Academy of Natural Sciences in the mid-1820s. Only the first two volumes of his *American Entomology* were published through the Academy of Sciences, the third volume being published at New Harmony. Nevertheless, when Say died in 1834, his *American Entomology* and other papers contained descriptions of over fifteen hundred species, including a thousand that had not been previously described. Say established his reputation as the foremost American entomologist of his generation. He was the first American whose systematic work in entomology gained widespread European recognition.[22]

John Eatton LeConte, an army engineer and a wealthy landed proprietor in New York state, followed his primary career as an engineer until he retired in 1831, but he is better known for his contributions in various branches of natural history. Of French Huguenot descent, LeConte collaborated with French experts in the publication of systematic works on the Lepidoptera and the Coleoptera. With the lepidopterist Boisduval, he coauthored the *Histoire générale et iconographie des lépidoptères et des chenilles de l'Amérique septentrionale* (1829–1834). He also assisted the French coleopterist Dejean in the publication of the American portion of the *Spécies général des coléoptères* (1825–1838). LeConte thus linked American entomologists with two of Europe's leading specialists.[23]

Frederick Valentine Melsheimer studied with the German naturalist August Wilhelm Knoch prior to coming to America in 1776 as a chaplain with the Braunschweiger Dragoons. Having been captured and later released by the American forces, Melsheimer settled in Lancaster County, Pennsylvania. There he began to collect insects of the region, and with the help of Knoch, he published a *Catalogue of the Insects of Pennsylvania* (1806), which was the earliest work devoted exclusively to insects published in America. Following Melsheimer's death in 1814, his son, John Melsheimer, maintained and enlarged his father's collection, and he corresponded with Say, who used the Melsheimer names in his published descriptions. After John's death, the younger son, Frederick Ernst, inherited the family collection. He continued to add specimens and to expand the manuscript catalog of descriptions of Pennsylvania insects. In the 1840s, Frederick Ernst played a central role in the Entomological Society of Pennsylvania, America's first entomological society.[24]

Thaddeus William Harris was a medical doctor who had acquired a taste for entomology from his father, Thaddeus Mason Harris, and from his Harvard professor Peck. In the 1820s, while practicing medicine in Massachusetts, he assembled a large collection of insects, and he published notes on insect life histories and descriptions of insects in the agricultural press. Harris sought the aid of Say and LeConte in the arrangement of his collection and the description of new species. In 1831, he became Harvard librarian. At Harvard, in addition to his regular duties, he compiled a list of Massachusetts insects for the Massachusetts Geological and Natural History Survey. When in 1836 the survey was extended, Harris received a commission to provide a comprehensive report on the insects in relation to agriculture. Published in 1841 as *A Report on the Insects of Massachusetts Injurious to Vegetation,* Harris's report earned him a reputation as America's outstanding agricultural entomologist.[25]

These early entomologists frequently voiced their frustration at trying to pursue entomological studies in an American setting. Say wrote to John Melsheimer concerning the shortage of entomological literature in America: "I am . . . sending to Europe for entomological books, and shall be very happy to receive your opinion as to such as I ought to order . . . it is certainly of the first importance to a naturalist to know what has been done by others in his particular science."[26]

While Say had access to relatively good libraries at the Academy of

Natural Sciences and the American Philosophical Society, Harris for many years was removed from such libraries. In 1823, Harris sought Say's help in the identification of the insects in his expanding collection, noting that he had been able to identify only about one-tenth of his specimens.[27] Harris confided to Say that he considered himself "a traveller in a strange land, anxious to obtain . . . the best of guides."[28] Later, when Say moved to New Harmony, Harris turned to LeConte for assistance.[29] Harris explained to LeConte that the reason his species were not listed in systematic order was because "in commencing Entomology without books I merely occupied myself with collecting and studying the habits of insects. The first preserved was numbered and the others in the order in which they were captured."[30] The shortage of European literature on insects, coupled with the expense of obtaining those books that could be purchased, led to almost unbelievable exertions on the part of American entomologists. Asa Fitch, an aspiring entomologist in New York state, borrowed standard European reference works on entomology from Say, Harris, and others and copied them by hand into manuscript volumes for his own use.[31] The lack of access to entomological literature remained a major limitation to American entomologists until at least the 1850s.[32]

Americans also faced a shortage of scientific periodicals through which to make their findings known. Since the priority of description was in many cases the entomologist's only reward, competition for this recognition was keen. In the 1820s and 1830s, publication in Benjamin Silliman's *American Journal of Science* was the surest means of reaching an extensive audience in America and Europe, but its pages were often too crowded to serve the entomologists' needs. Harris, wishing to reach agriculturists with reports of his investigations of injurious insects, published many descriptions of new insects in the *New England Farmer.* Incensed that his species names were not recognized in Dejean's catalog, Harris insisted that his descriptions still had priority, though they had not appeared in "scientific" journals. LeConte came to Dejean's defense, explaining that if the descriptions had appeared in a work with a circulation in Paris, the French author would have respected Harris's prior claim. To avoid similar disputes, LeConte urged Harris to republish his descriptions in the *American Journal of Science* or the *Transactions* of the New York Lyceum of Natural History, either of which he said would come to the attention of European authors.[33]

Already handicapped by their lack of entomological reference works and by inadequate publication outlets, the Americans were further frustrated by the European practice of claiming priority for names assigned to specimens in European collections. The assumption that reference collections were generally available put the Americans at a distinct disadvantage. Say voiced his objection to those who gave "a very laconic description [and] a reference to a cabinet in which the specimen may be inspected, by the comparatively few persons who have the opportunity."[34] As LeConte pointed out, "Even if all the books . . . were available, we still should not [know] what insects are known to naturalists in Europe."[35] One solution was for the Americans to send their specimens to Europe for comparison. Both LeConte and Harris tried this, with the result that their claims for priority were reduced. In 1830, LeConte reported that half of the two thousand specimens he sent Dejean had already been described.[36] Another solution was for Americans to visit the European collections themselves. This LeConte did in 1828 when he traveled to Paris with an extensive collection to compare with those of Boisduval and Dejean. Such travel, however, was costly and normally impractical.[37] Finally, Harris, exasperated at hearing that Dejean had described four species that he (Harris) had previously described, proposed a rule governing priority in scientific nomenclature: "*The right of priority in imposing a specific name* does not . . . depend upon this name having been given in a cabinet, or a bare catalogue, but in its being designated to the world by a *description*. . . . the *first describer* alone can claim this right."[38] The position taken by Harris, that the first describer should have the absolute right of priority regardless of the place of publication, was debated extensively in the nineteenth century. Finally, in 1901, the International Zoological Congress adopted a position reflecting Harris's view that the first description, rather than the first catalog designation, should have priority.[39] For nineteenth-century American entomologists, however, the practice of assigning names to specimens in distant European collections remained a troublesome obstacle to the advancement of their science. Only by establishing and maintaining permanent reference collections in America could they overcome this obstacle.

Although limited by a shortage of books, publication outlets, and collections, the Americans held an important advantage in their access to unnamed species. As the elder Melsheimer noted, the "ardent desire" of European naturalists for American insects provided a powerful incentive for Americans to become acquainted with the products of their own country.[40] Like Adam and Eve in the Garden of Eden, American entomologists had only to look around them and name the creatures within their realm.

Say pressed this advantage of access to specimens to the fullest. Leaving the formation of new genera and families to Latreille and other Europeans, Say concentrated on the description of new American insect species. With his extensive knowledge of accepted genera, Say was ideally positioned to seek out, describe, and arrange the thousands of American species according to the accepted classifications.

Say had unparalleled access to undescribed material. In a relatively short period, he covered a large portion of North America at a time when this area was unknown entomologically. In 1817 Say, Maclure, George Ord, and Titian R. Peale undertook a collecting expedition through Georgia, Florida, and the Sea Islands. In 1819, as zoologist with Major Stephen H. Long, Say traveled across the Great Plains west to the Rocky Mountains. In 1823, he accompanied Long across the northern plains to the source of the Mississippi River. In addition, he received specimens from Thomas Nuttall, his colleague at the Academy of Natural Sciences, and other widely traveled naturalist-explorers. From 1825 to 1834, while residing in New Harmony, Say published insect descriptions based on specimens from his wide travels through the American interior. His success in the description of American species inspired a generation of American entomologists.[41]

The descriptive work of Say, Harris, LeConte, and the Melsheimers in the years prior to 1840 altered the traditional status of American colonial science in some important ways.[42] First, Americans participated on a more equal footing with Europeans in their degree of specialization. Commencing with the elder Melsheimer, these Americans specialized according to current European fashion. In contrast to general natural history narratives and artistic renditions of New World fauna like those of Peter Kalm, Catesby, the Bartrams, or Abbot, Melsheimer's *Catalogue* dealt with only one order of insects, and it used binomial nomenclature.[43] The writings of Say, LeConte, and Harris carried this pattern to new levels of exactness. Second, Americans no longer contented themselves with collecting specimens for European scientists. LeConte was an active collaborator and author with European specialists on American material, and Say was recognized on both sides of the Atlantic for his original descriptions. Third, American entomologists were coming into closer contact with one another. The figure of Abbot working in relative isolation was being replaced by the first signs of an American entomological community, apparent in the correspondence among Say, John Melsheimer, LeConte, and Harris. What this generation of American entomologists lacked, however, were the institutions necessary to practice scientific entomology fully on a level with Europeans. In the 1840s, American entomologists turned to the task of constructing these institutions.

"A Few Literary Gentlemen": The Entomological Society of Pennsylvania, 1842–1853

2

In July 1839, the Reverend John G. Morris, pastor of the English Lutheran Church in Baltimore, Maryland, wrote to Thaddeus William Harris suggesting the formation of a society devoted to the study of entomology in America. Introducing himself as one "who has paid as much attention to Entomology as my parochial engagements would allow" and noting that "the science of Entomology has but few students in this section of the country," Morris congratulated Harris for maintaining an extensive entomological correspondence among their countrymen. Yet the time had come, thought Morris, when something more than correspondence among members of America's entomological brethren was needed:

> I have sometimes thought that the science [of entomology] would be greatly promoted, if its friends could hold a meeting at some central place for mutual consultation, and I have no doubt that if the call were made, a convention would assemble. It would reflect honor on its members, as being perhaps the first to begin a course, which would doubtless be followed by students of other sciences, and the whole might eventuate in an American Association of Naturalists.[1]

Morris's suggestion for a more formal organization among the American entomological fraternity reflected a growing call for scientific organization in America. In fact, the most significant developments in American science in the 1840s were the organization of the American Association for the Advancement of Science and the founding of the Smithsonian Institution.[2]

Morris's hopes for an entomological society materialized in August

1842, when he and four other "literary gentlemen" met at the home of the Reverend Daniel Ziegler in York, Pennsylvania, to found the Entomological Society of Pennsylvania.[3] This was the first society formed exclusively for the pursuit of entomology in America. It marked a new departure in the way American entomologists viewed themselves and in the way they pursued their science. The story of its formation, its membership, activities, accomplishments, and shortcomings during the decade of its existence documents an important transition from an era characterized by the strictly personal contact between Thomas Say, John E. LeConte, and Harris and their circles to an era characterized by entomological institutions designed to ensure order and permanence in American entomology.

The entomologists' decision to organize at that particular time seems to have been influenced by three circumstances. First, they faced a crisis in their discipline that arose from the fact that just at the time when American entomologists felt a new urgency to organize and advance their science, three of the leading descriptive entomologists of the preceding decade were no longer active. Thomas Say, the American whose entomological work was most widely known in Europe, had died in 1834. John Eatton LeConte, who had collaborated with leading French coleopterists and lepidopterists, suffered a physical collapse in 1836 and never fully recovered.[4] Harris, who had hoped that his pathbreaking work on agricultural entomology would be the start of an academic career at Harvard, saw these hopes fade in 1842 when the chair of natural history went to the botanist Asa Gray rather than to him.[5] Surveying the outlook for American entomology in light of these circumstances, Harris concluded that American entomology might slip from its promising start to a state of relative stagnation. With Say dead, and with Harris and LeConte hoping only to finish the projects they had begun, Americans might be unable to compete with Europeans in describing the insects of their own country.[6] Given this situation, it was natural that the Entomological Society of Pennsylvania adopt as its primary task the unfinished program of descriptive entomology begun by Say.

A second circumstance that prompted the organization of the society was the rapid improvement in transportation and communication by means of roads, canals, and railroads that linked urban centers, like Philadelphia, New York, and Baltimore, with each other and with their hinterlands.[7] Improved transportation and communications helped the society in many ways. Better service made possible the safe, speedy transfer of packages containing insect specimens by means of carriers like the Adams Express Company. It also facilitated the correspondence among members, the exchange of specimens, and the gathering of members at meetings.[8]

A third circumstance behind the organization of the society was the strident nationalism of the 1840s. In that decade, Americans reasserted national claims in diplomacy, in territorial expansion, and also in the arts and sciences. Entomologists shared the nationalistic sentiments that propelled the American republic on its course of Manifest Destiny. They joined with spokespersons of American culture in various fields in a general effort to elevate American science and letters to full equality with their counterparts in Europe, thus achieving a cultural independence worthy of America's status as a powerful, independent country.[9]

The original membership of the society comprised men who lived in or near the Susquehanna valley in south central Pennsylvania.[10] Present at the first meeting were Morris, Daniel Ziegler, Samuel S. Haldeman, Dr. Frederick Ernst Melsheimer, and Reverend Solomon Oswald. Oswald, a retired Lutheran minister, operated a bookstore in York, Pennsylvania. Ziegler lived at Kreutz Creek, six miles from York, and served as pastor to a circuit of German-speaking congregations in the area. Melsheimer was a country physician, now mostly retired, who lived about nine miles outside York. Morris lived in Baltimore, Maryland, about fifty miles from York, and Haldeman lived at Chickies near Columbia, Pennsylvania, about twelve miles from York. Corresponding members of the society included Harris of Cambridge, Massachusetts; Karl [Christian] Zimmermann, who lived near Charleston, South Carolina; and Nicholas Marcellus Hentz, who lived in Georgia. John L. LeConte, in New York, although not listed as an original member, played a key role in the affairs of the society.

Melsheimer was elected president of the society, Haldeman vice president, Morris corresponding secretary, Ziegler recording secretary, and Oswald treasurer. The society approved a "simple constitution," of which apparently no copy has survived.[11] Meeting several times a year at the homes of members in York, Baltimore, and Chickies, the members compared and exchanged specimens, discussed entomology, and shared dinner. Regular meetings were apparently held from 1842 to 1847 and perhaps as late as 1850 or 1851. From the outset, one of the main goals of the society was the publication of a definitive catalog of American beetles.[12]

The society was in some ways a formalization of personal ties already established. Haldeman, the prime mover in the formation of the society, had studied natural science under Henry D. Rogers at Dickinson College in Carlisle, Pennsylvania, and at the University of Pennsylvania.[13] He preferred independent study over the school environment, however, and he spent most of his time in the library of his home near Columbia, Pennsylvania. There he supported his scholarly interests by joining two brothers as a silent partner in an iron smelting business. In 1842, the year the society was formed, Haldeman taught zoology at the Franklin Institute in Philadelphia, one of many short-term professorships he accepted.[14] During the 1840s he was interested primarily in the natural history of invertebrates, and he published on the systematics of the Coleoptera and freshwater mollusks. In the 1850s he turned to linguistics and later to archaeology.[15]

Haldeman first contacted Melsheimer with the specific proposal for an entomological society in Pennsylvania.[16] At that time, Melsheimer had the largest insect collection in America, devoted primarily to the Coleoptera. The second son of Frederick V. Melsheimer, Frederick Ernst had inherited the Melsheimer collection from his older brother, John, in 1830. Though not in close contact with naturalists in Philadelphia, Melsheimer had continued to add new specimens to the collection and to record new species in a manuscript catalog. On the basis of his distinguished entomological lineage and the importance of the Melsheimer collection, Frederick Ernst was named president of the new society.[17]

Morris's entomological interests brought him into contact with Haldeman and Melsheimer, and the discussions among the three apparently prompted Haldeman to suggest the formation of the society. Morris maintained a wide circle of contacts that included leaders in the Academy of Natural Sciences of Philadelphia and in the growing scientific community of Washington, D.C. In addition to his pastoral duties, he found time to collect and study the American Lepidoptera, to teach natural history at the University of Pennsylvania, and to advise in the planning and staffing of the Smithsonian Institution.[18] Morris actively promoted the advancement of American science, and the emerging entomological fraternity profited from his wide contacts and his advocacy of their interests. Morris publicized the efforts of the entomological society in the *American Journal of Science,* and he served as liaison between the Entomological Society of Pennsylvania, the Academy of Natural Sciences of Philadelphia, and, later, the Smithsonian Institution.[19] In 1846, Morris toured Europe, where he met many European entomologists and visited entomological collections in the major museums. He reported to his American colleagues on the grand size and superior facilities of these European collections.[20]

Daniel Ziegler was probably introduced to entomology by his neighbor Melsheimer.[21] By 1842, he had assembled a large collection that like other American collections contained primarily Coleoptera and Lepidoptera. At the first meeting he exhibited his "larvae nursery."[22] His fellow entomologists praised his ability to identify insect specimens accurately. Although he published only one paper, Ziegler contributed substantially through his knowledge, his correspondence with American and foreign entomologists, and his large collection, which was eventually purchased by the Museum of Comparative Zoology.[23] Reverend Oswald, on the other hand, was not the entomological equal of the others in the society, and apparently his contributions were limited mostly to his role as treasurer.[24]

John Lawrence LeConte, though not named as one of the original members of the society, soon became a key participant. LeConte graduated from St. Mary's College, in Georgetown, Maryland, in the year the society was formed.[25] The only child of John Eatton LeConte, who guided his instruction in natural history, John Lawrence came into contact with Harris through his father's entomological work, and he may have met Morris while a student at Georgetown. From 1842 to 1846, LeConte studied medicine at the New York College of Physicians and Surgeons, earning an M.D. in the latter year, but he practiced medicine for only a short time during the Civil War. While still a medical student, LeConte began a series of explorations to the Mississippi Valley, the Great Lakes, and the far West, during which he gathered the material for his descriptions and systematic revisions of the American Coleoptera. In 1844, 1846, and 1848, LeConte collected insects in the Mississippi Valley, from St. Louis north to the headwaters of the Mississippi, and around Lake Superior. In 1849, he traveled to California via the Isthmus of Panama. He spent the next year and a half collecting insects around San Francisco, San Diego, and the Gila River country in Arizona.[26]

When the Entomological Society was formed, the LeContes lived in New York, which may explain why LeConte apparently did not attend a meeting until 1847. In 1852, the LeContes moved to Philadelphia, prompted, one may speculate, by the desire to be near the center of entomological activity in America. Even when he was located some distance away, LeConte participated actively in the affairs of the society. In 1844, for example, Morris wrote to Melsheimer that LeConte had gone to Wisconsin on a collection expedition and would bring back many treasures for them.[27] LeConte's first publication, in 1844, was submitted as a communication from the entomological society. In 1848 he published a major revision of the Coleoptera, the first of thirty such monographic essays to appear during the next thirty years. By that time, he was well on the way to becoming the leading authority on the American Coleoptera.

Harris, who following Say's death was America's most prominent entomologist, was elected as the first corresponding member. Harris had been in contact with Melsheimer since 1835 and with Morris since 1839.[28] Haldeman was apparently in contact with Harris prior to the formation of the society, and he visited Harris in Cambridge during the fall of 1842.[29] Harris exerted considerable influence on the early activities of the society through his correspondence with Haldeman, Melsheimer, and Morris, but this was cut short in 1844, when Harris suffered a severe eye inflammation and had to restrict his correspondence.[30]

Other corresponding members included Karl Zimmermann and Nicholas Marcellus Hentz. Zimmermann, who immigrated to the United States from Germany in 1832, was a university graduate and an accomplished entomologist with a wide acquaintance among European naturalists.[31] Hoping to earn his living by collecting insects for sale and by teaching school, he settled in South Carolina and divided his time between extensive collecting trips along the East Coast from Massachusetts to Cuba and working up his collections at home. On his frequent trips, Zimmermann visited others of the entomological brotherhood. Zimmermann visited Melsheimer en route to South Carolina in 1832.[32] He became acquainted with Harris in 1836 and with Morris shortly thereafter.[33] In 1839, in a planned move back to Germany, Zimmermann's extensive collection and library were lost at sea. Following that disaster, Zimmermann resided with Harris and Morris for several months during the winters of 1840 and 1841 while reestablishing his collection and library through purchase and exchanges.[34] Prior to his association with the society, Zimmermann maintained closer contact with leading European entomologists like Johann C. Klug (with whom he had worked at the Berlin Museum) and Wilhelm Ferdinand Erichson than he did with Americans.[35] Morris described Zimmermann as "a strange genius with many most excellent traits of character."[36] John L. LeConte, who visited him in 1849, reported that Zimmermann "does not seem disposed to associate much with the townspeople. . . . He says he finds no congenial spirits among them. He engages principally in studies of classification, but is not content with anything he writes . . . unless his system is absolutely perfect."[37] The German entomologist Hermann Schaum, who also visited Zimmermann, considered him "a first rate entomologist . . . equal to any in discriminating species."[38]

Hentz, another correspondent from the South, had immigrated with his parents from France in 1816.[39] In Philadelphia and Boston he became acquainted with many American naturalists, in particular Harris, with whom he continued to correspond.[40] Hentz made his living teaching in finishing schools for young ladies. In 1824, he taught in North Carolina and later in Kentucky, Alabama, and Georgia. Hentz's first publications were descriptions of new species of beetles from Massachusetts and Pennsylvania that appeared in the *Journal of the Academy of Natural Sciences*. Later, he specialized in spiders, becoming America's first authority on arachnology. By 1842, when Haldeman contacted him, Hentz had described and illustrated 141 species of spiders, and he was preparing a monograph on the subject.[41] Hentz eagerly sought the new contact with Haldeman, Melsheimer, and others of the society. He expressed the hope that he might eventually be able to move away from what he called "this benighted region" and return to a place "nearer the seats of science," but his wish was not fulfilled.[42] Hentz participated little in the society, even by correspondence, but even this limited contact with those interested in entomology in the mid-Atlantic and New England regions was a source of encouragement to him.

The membership and gaps in the membership furnish some clues to the way the society members regarded themselves and their work. Although Morris claimed (forty years later) that the society included all Americans with an interest in entomology as "honorary members," the invitation to membership seems in fact to have been more selective.[43] The society clearly did not include all those who could have been members. Morris recalled that there were about ten individuals "actively pursuing entomology" in the 1840s, which comes reasonably close to the five or six regular members plus three corresponding members whom he and others named.[44] In 1846, however, Morris listed two American collections (Titian Peale's in Philadelphia and Count Castleman's at the National Institute in Washington, D.C.) belonging to nonmembers.[45] Selective membership also applied to corresponding members. While Harris, Zimmermann, and Hentz were elected as corresponding members, the society made no effort to contact or elect individuals in Harris's wide circle of entomological correspondents, some of whom eventually became entomologists of renown. William LeBaron, a future state entomologist of Illinois who corresponded with Harris in the early 1840s, for example, was not offered membership.[46] Another significant omission in the society's membership was Asa Fitch, a physician-naturalist and farmer near Salem, New York, who in the 1840s began publishing on the insect pests in his region and who began an insect collection for the New York State Cabinet of Natural History.[47]

As such omissions indicate, though there were no formal requirements for membership, there was an understanding that membership would be restricted, at least in the sense that there was no public call for all interested people to join.[48] Morris indicated a possible basis for including some and excluding others when he referred to "Gentlemen of taste" who collected insects "merely as objects of beauty." Although Morris acknowledged that those persons did occasionally discover a new species and "thus contribute to science," the implication was that their interests were not primarily scientific, by which he meant "systematic." Again, when describing Peale's collection, Morris noted that it was arranged geographically, not systematically.[49] In other words, the society was composed of a select group whose efforts were dedicated to the improvement of insect systematics rather than to displaying insects as objects of beauty or demonstrating the habits of injurious species.[50] From the outset the society was dedicated to the advancement of systematic entomology rather than to the diffusion of entomological information or its application to agriculture, and there was little interest in contacting those who did not share this primary goal. In keeping with this emphasis on the advancement of science rather than its diffusion, the society chose not to develop a central collection or library that might attract a host of casual amateurs and thus dilute efforts to improve systematics.[51]

The members of the society, though representing different backgrounds and geographic regions, thus shared a common orientation and a common purpose. They all aspired to advance American entomology in the area of insect systematics.[52] A good indicator of their orientation is Haldeman's own career. In the 1840s, Haldeman, Spencer Fullerton Baird (Haldeman's former classmate at Dickinson College), and John L. LeConte were leaders in the movement to reform American zoology.[53] Whereas the founders of the Academy of Natural Sciences had emphasized the diffusion of knowledge through the publication of beautiful portfolio volumes like those on ornithology compiled by George Ord, Haldeman, Baird, and LeConte strove for the advancement of scientific knowledge through the publication of taxonomic monographs devoted to selected groups of animals.[54] Their approach required an exclusive scientific elite working toward exact taxonomic precision in the description and classification of animal species rather than an inclusive group of amateurs with varying degrees of competence in taxonomy.[55] Guided by their belief that authoritative descriptions could be made only after assembling the most extensive collections and researching the European and American zoological literature, these aspiring reformers of American zoology emphasized precise anatomical description over accounts of life histories and the habits of animals. Haldeman, for example, shared the Philadelphia zoologists' determination to mark a clear distinction between what they considered to be "scientific" writers and popular nature writers like John James Audubon, whom they regarded as an artist who depicted animals with human characteristics.[56] Though Baird did not share the Philadelphians' animosity toward Audubon, he did agree that writing taxonomic monographs was the best means to advance American zoology to a level worthy of the respect of European zoologists.[57]

In order to carry out their program, the entomologists agreed that it was necessary for Americans to publish descriptions of American insects rather than leaving this task to the Europeans. Here they echoed Say's complaint that Europeans collected American insects and placed them in collections without giving full descriptions. Such complaints had of course been voiced before, but in the 1840s this traditional complaint of inadequate descriptions was transformed into the patriotic claim that Americans had the right to first priority in the naming of American insects. Harris had formulated this position initially when he urged the elder LeConte to follow his own practice of never sending an unnamed or undescribed specimen abroad.[58] Whereas in the 1830s J. E. LeConte and others declined to follow Harris's example, by the 1840s the entomological society heartily endorsed his position. Morris knew of several industrious Germans in his vicinity who regularly spent days in the field collecting rare specimens that they sent to Europe. Because of such practices, Morris warned the society members that "our rule" of not sending unnamed specimens abroad might not be very effective.[59] Nevertheless, the society members considered that their call for a protective embargo on undescribed insects was justified on the same grounds as were protective tariffs for infant industries. In 1845, young John LeConte announced "America's Entomological Declaration of Independence": "I trust the day is past, when our insects must be sent to Europe for determination. Are we to be bound by the mere dictum of some European entomologist . . . who chooses to name the insects which we have discovered?"[60]

The entomologists recognized that in order to carry out effectively their program of describing American species, they needed a regular, speedy, and authoritative means of publishing their findings. Publication seemed especially critical in the 1840s, because just at the time American nationalism was reaching a new intensity, the older entomologists responsible for the lion's share of published descriptions were no longer active. As Melsheimer explained the situation to Haldeman: "More of our insects have been described by Europeans since 1834 than . . . for many years before. . . . The reason . . . is obvious. Our Say was then at work, and he alone kept [them] at bay. Now no opposition is made to these entomological invaders."[61] Publication, Melsheimer believed, would place American entomologists in the position they did not now occupy—the position of the consulted.[62]

In the urgency to publish new material, the Melsheimer collection and catalog assumed critical importance, for they were the most readily available sources of new descriptions of species. Melsheimer, Haldeman, and Morris urged the rapid publication of the Melsheimer material.[63] Their initial plan called for Melsheimer to edit the catalog for publication in 1843, within a year following the organization of the society.[64] Each day they waited represented lost opportunities to publish their material. Melsheimer cited a European catalog published in 1844 that contained descriptions of eight or nine species in his collection. If his descriptions had been published a year earlier, Melsheimer said, the catalog would have contained his names. Now he had to buy the book to find the names of insects in his own cabinet![65] Harris and LeConte preferred to postpone publication until the European literature could be surveyed.[66] The majority insisted, on patriotic grounds, however, that every delay ran the risk that European descriptions would take priority.

In their haste to publish, the entomologists faced a choice between their patriotism and their desire for the approval of European experts. Haldeman conceded that the society's publications "will doubtless inflict many synonyms on the science, and meet with much approbation from abroad," but, he concluded "we must either pursue this independent course or quit the study."[67] Their ambivalent position with respect to European science endangered their credibility with the very people they hoped to impress. Facing this dilemma, the entomologists reached a compromise. They would publish immediately in the *Proceedings of the Academy of Natural Sciences* all the descriptions of insects determined by them since the death of Say. These descriptions, mostly from the Melsheimer catalog, were published as "Communications from the Entomological Society" in 1844 and 1845, and additional descriptions were published (though not credited directly to the Entomological Society) in 1846 and 1847.[68] With respect to the descriptions, Haldeman explained to officials at the academy that the society was "more anxious to secure an early publication than to be very particular about the mode."[69] At the same time, they arranged for LeConte and Haldeman to edit the Melsheimer catalog as a monograph according to rigorous scientific standards.

As they expected, the preliminary descriptions were criticized by European entomologists. Schaum, who visited America in 1847, just after the series of new descriptions was published in the *Proceedings,* objected to the notion that "American insects ought to be described by American entomologists."[70] The result, he said, had been the proliferation of isolated descriptions, a practice that had been characterized by Schaum's colleague, Erichson, as the "nuisance of science."[71] Schaum regretted that the American entomologists had added to this nuisance. Despite Schaum's criticism, Haldeman and Melsheimer agreed that publication in the *Proceedings* had served to establish priority for the names in Melsheimer's catalog.[72] Obviously, they hoped that Europeans would, as a result, be deterred from describing the insects in question until the publication of the definitive monograph.

For the publication of the Melsheimer catalog as a monograph, the entomologists turned to Joseph Henry, secretary of the recently established Smithsonian Institution. Henry approved of their plan because he too considered the sponsorship of original scientific monographs, worthy of notice by European authorities, to be the best means of advancing American science.[73] Melsheimer's *Catalogue of the Described Coleoptera of the United States,* revised by Haldeman and LeConte, was published by the Smithsonian Institution in 1853.[74] It was one of the earliest Smithsonian monographs, and it received the approval of European entomologists.[75] This publication set a new standard of thoroughness and precision in the taxonomy of American Coleoptera. Its appearance accelerated the collection and study of Coleoptera in America, and it served as a model for later monographs covering other insect orders.[76]

With the publication of the Melsheimer catalog, the last vestige of the Entomological Society faded away. Its demise may be attributed to a combination of circumstances. Haldeman, who from the beginning had served as the key organizer, turned from entomology to linguistics, thus depriving the society of effective leadership. More important, the society apparently neither aspired to the publication of regular proceedings nor sought the support of wealthy patrons for such an endeavor, both of which were prerequisites for permanent scientific organizations in that era.[77] The society relied instead on the Academy of Natural Sciences and the Smithsonian Institution to report its meetings and to publish the researches of its members. With the publication of the descriptions and the catalog, the essential patriotic, scientific, and personal goals of those who had formed the society had been achieved. Melsheimer, for example, expressed satisfaction that these publications had secured priority for the names in his collection.[78] By publishing the material in the manner they had, the Americans felt they had effectively asserted their claim to practice scientific entomology on an equal footing with the Europeans. Haldeman later wrote to Baird about his own role in the publication of the catalog: "It was very useful, and I think it was about the best thing I did for Entomology to suggest, plan, and push it through."[79]

The historical significance of the society may be suggested by noting first what it was not, second what it was, and third how it served an important transitional function in the development of American entomology. First, those who formed the society were clearly not interested in uplifting or educating the general public. They expressed no interest, for example, in giving public lectures or establishing a central collection or library, though all these activities had been important in promoting the early growth of the Academy of Natural Sciences, and science topics were current favorites on the American lyceum circuit.[80]

More surprising was the lack of any significant attempt to promote agricultural entomology. The Pennsylvania-based entomologists were slow to recognize the potential importance of agriculture to their discipline, despite the solid foundation already laid by Harris and Peck. The correspondence of the members contains few references to agricultural applications, and these are contradictory. On the one hand, they apparently did not share the concerns of Pennsylvania farmers who in 1842 and 1847 suffered from serious outbreaks of the Hessian fly that destroyed an estimated 20 percent of the wheat crop. It was left to Asa Fitch, outside the society, to publish an account of these events.[81] On the other hand, Morris wrote of a three volume illustrated work on noxious insects he had seen in Germany, and he proposed that the society members produce a similar work for American farmers. Arguing that Harris's report lacked popular appeal because it lacked illustrations, Morris proposed the publication of a well-illustrated work on noxious insects in America. He urged Haldeman, Melsheimer, and Ziegler, all of whom lived in or near the countryside, to take up the study of noxious insects. He also proposed a chapter by Baird on noxious birds.[82] Yet Morris's proposal was not taken up. More indicative of the attitude of the society members was Melsheimer's apology, in a letter to Harris, for including "extraneous material" about insect damage to Pennsylvania crops.[83]

The society members knew of the attention to entomology in the Natural History Survey of New York, the leading sponsor of applied science. In 1845, the New York legislature commissioned Ebenezer Emmons, a geologist with the survey, to conduct an agricultural survey as a continuation of its natural history survey. Both Haldeman and LeConte expressed interest in reporting on the insects of New York, but one suspects that their interest was more in systematics than in agricultural applications. In the end, Emmons chose to compile the five-volume series on soils, climate, fruits, and insects himself.[84] Haldeman and LeConte reviewed Emmons's first volume, which included some discussion of insects, and they criticized it for numerous errors in insect identification.[85] Despite some society members' awareness of agricultural entomology as practiced by Harris and by the New York survey, the society as an organization remained generally silent, or ambivalent, with regard to agricultural entomology.

The society's attitude most likely arose from the example and close working relationship with the Academy of Natural Sciences, whose leaders had consistently declined any official connection with other institutions. The academy limited its contacts strictly to the personal exchange of information by members. Whereas the natural history societies in New York and Massachusetts promoted geological and natural history surveys in those states, the academy declined to take part in planning the Pennsylvania Geological Survey.[86] The academy's reluctance to enter into close working relationships with other organizations sprang from two sources. One was the desire of key leaders, particularly its president, William Maclure, to maintain an informal gathering for the sharing of information rather than to allow the proceedings to be preempted by experts. A second source was the unique role of Philadelphia's ruling patricians in the support of science in their city. During the first half of the nineteenth century, Philadelphia's elite compensated for their city's decline in politics and trade, typified by the loss of the national and state capitals to Washington and Lancaster and the diversion of trade to New York following the completion of the Erie Canal, by lavishing support on Philadelphia's cultural organizations. The prosperity of institutions like the Academy of Natural Sciences did ensure that Philadelphia remained dominant as a cultural and scientific center through the first half of the nineteenth century, but patrician support was limited to institutions that brought renown to Philadelphia and did not extend to institutions in competing cities. The state geological survey, for example, was administered from the new state capital at Harrisburg (the capital had moved in 1812). One Philadelphia organization that was supported by the patriciate, the Philadelphia Society for Promoting Agriculture, might have joined with the academy in the promotion of agricultural entomology, but its urban-based leadership had been weakened by decades of infighting and was apparently uninterested in new endeavors.[87]

A final reason that many in the academy and in the entomological society were reluctant to establish formal ties was because they feared that other organizations, especially agricultural groups, might fall under the influence of charlatans and that this association would harm their own reputation. The reaction of LeConte and Haldeman to Emmons's performance as an agricultural entomologist indicated that this was an important reason for their reluctance to support agricultural entomology.[88] At that stage in the development of their science, ambitious naturalists like LeConte and Haldeman hesitated to encourage new institutions or organizations that were outside well-established traditions, because undirected "amateur" science might harm their efforts to gain European recognition.[89] Such aloofness may help explain why the entomological society apparently had no direct ties to the Pennsylvania Horticultural Society, which by the early 1850s had appointed its own entomological committee to report on noxious insects and to propose remedies.[90]

A final reason why the society members did not rush to embrace agricultural entomology was because many of them were financially independent. LeConte and Haldeman, two obvious candidates for careers in agricultural science, had ample financial resources independent of their scientific work. Neither was motivated by personal financial need to develop a career in applied entomology. LeConte in particular deplored the preoccupation of his countrymen with moneymaking, which he described as "the great motive power of modern times."[91] Others supported themselves from their professions in the ministry or medicine. Typical of the attitude of the members toward outside support was Harris's suggestion that entomologists seek backers for their science among wealthy amateurs.[92] This profile of entomologists' finances contrasts sharply with the agricultural chemists and the geologists of the period, most of whom faced "formidable obstacles in acquiring an education and in establishing a scholarly career."[93] Whatever its basis, the reluctance of Philadelphia-oriented societies to undertake projects with other organizations created an anomalous situation in which Pennsylvania had the best naturalists but lagged behind Massachusetts and New York in the public support of applied sciences like agricultural entomology.[94]

What most interested the society was what A. Hunter Dupree has called the creation of an information system.[95] The local society served to link American entomologists with each other and with their entomological counterparts in Europe as one component in a global information system.[96] It did so by facilitating the exchange of both verbal and non-verbal information through meetings, correspondence, and the exchange of specimens. The members of the society considered themselves a select group of experts who were advancing a specific discipline.[97] They rightly regarded the establishment of an information system geared to their needs as the best means to advance their science.

The system operated most efficiently in the contact of members at the meetings. These gatherings allowed the entomologists to assess the current state of their discipline and to take steps to advance entomological knowledge. Such contact was directly responsible for the most important publications of society members, of which the Melsheimer catalog is the best example. Until 1847, LeConte and Melsheimer had never met in person. Their contact had been solely through Haldeman and by correspondence. Melsheimer had learned that LeConte was planning an extensive study of the Coleoptera, his own specialty, and he expressed concern that their "respective performances do not clash."[98] Haldeman suggested a division of labor between the two (the details are not clear) that satisfied both parties.[99] When LeConte finally met Melsheimer, apparently at a meeting of the society in 1847, it was agreed that LeConte would help Melsheimer prepare the catalog for publication.[100] The contact was mutually helpful, because LeConte found the catalog useful in his preparation of a monograph that appeared in 1848.[101] Similarly, Haldeman acknowledged the aid of other members of the society in the preparation of his own monograph, which was published in 1853.[102]

The contact between members of the society also allowed the Americans to derive greater benefit from visiting foreign scientists. When Schaum came to the United States, in 1847, he stayed with the LeContes, Haldeman, Ziegler, Morris, and Zimmermann.[103] Schaum was reputed to have a greater working knowledge of the Coleoptera than any other living person. His presence in the United States was particularly important to young LeConte, with whom he stayed for several weeks, going over the LeConte collection and adding to it from specimens collected on his journey from New Orleans to Canada.[104] Louis Agassiz, who had recently moved from Switzerland to the United States, also visited LeConte and Haldeman in 1847, offering encouragement and advice to them in their efforts to promote American entomology. Agassiz was pleased with the zeal of the Americans, and he urged them to do "more than has ever been done in Europe."[105]

The exchange of nonverbal information in the form of insect specimens was a central part of the information system.[106] One important reason for holding meetings was the opportunity this afforded members to examine the insect cabinets of other members.[107] For example, Morris, who lived at a transfer point on the sea and land routes, regularly conveyed packages between Harris and Melsheimer.[108] Between meetings, members continued to exchange specimens by mail, personal delivery, or delivery by another member. The letters that accompanied such shipments carried lists of specimens shipped, desiderata, and comparisons of specimens from different collections (e.g., "Your no. 319 is my no. 984").[109]

The members frequently testified to the improved flow of information and to the encouragement increased contact provided. Haldeman admitted that if it had not been for the society, he would not have turned his attention so exclusively to entomology.[110] Hentz testified that "the correspondence between us [Haldeman, Melsheimer, and others] have [*sic*] awakened a new ardour in me for the study of entomology."[111] Schaum noted improvements in the organization of Melsheimer's collection following the loan of books and the advice he had received from Morris and Zimmermann.[112]

The intent of the society to establish an information system became most rigorous in the control over publication, an area that Sally Kohlstedt has identified as the critical sign of professionalization in American science.[113] The entomologists' desire to establish control over publication is indicated by the appointment, in 1843, of the Entomological Committee to the American Association of Geologists and Naturalists, which comprised Haldeman, LeConte, and Melsheimer, who certified descriptions for the association.[114] Likewise, when Haldeman sent descriptions for publication in the *Proceedings of the Academy of Natural Sciences,* he informed the editor that the descriptions had been examined by the members of the society and had "passed."[115] When it came to publication of the Melsheimer catalog, the entomologists chose the Smithsonian Institution because it had established the practice of critical review of material for publication, a precedent that was soon followed by the American Association for the Advancement of Science.[116] The entomologists, like other American scientists, viewed the control over publication as vital to establishing American credibility with European scientists. Only through the strict enforcement of internationally accepted rules of priority and the prohibition of synonymy could they demonstrate their responsible participation in the global information system.[117]

Finally, the society served as a vital transition institution, bridging the time when entomology in America was characterized by a few isolated investigators to the time when it was characterized by leading specialists supported by extensive scientific institutions. One of the prime legacies of the society may be the influence its members had on the subsequent career of young John LeConte. Although it is impossible to say exactly how much influence derived from the society and how much contact there would have been in the absence of the society, the connections are worth noting. LeConte's early contact with Haldeman and the others may have influenced his decision to make the study of American Coleoptera his life's work. His contact with Melsheimer, and his subsequent editing, with Haldeman, of the catalog clearly benefited his career. Le-Conte acknowledged his own indebtedness to Zimmermann for "many valuable hints on classification,"[118] and after Zimmermann's death, Le-Conte edited and published several unfinished memoirs left by Zimmermann, citing their improvements over previous systems.[119]

Whatever the extent of the society members' influence on LeConte, he developed into the ideal systematic specialist called for by Haldeman, Baird, and others of the emerging scientific elite at midcentury. Beginning with his association with the entomological society in the 1840s, LeConte's knowledge of the systematics of the American Coleoptera steadily increased. By 1848, his publications were replacing European standard references in the field of American Coleoptera.[120] By the 1860s, he was regarded as America's leading coleopterist, and by the time of his death in 1883 he had increased the number of described American Coleoptera by nine times what it had been when he began. In all, he was the author of nearly half of the approximately 10,000 descriptions of North American Coleoptera then known to science.[121]

The contributions made by the society in the transition from individual to institutional science may be seen in other areas. In the 1850s, many of the functions of the society were assumed by other organizations. The Smithsonian assumed the authority to review and publish entomological monographs. It also coordinated entomological exploration in the West, and it facilitated correspondence and exchanges with European scientists. In the 1850s, the annual meetings of the American Association for the Advancement of Science substituted for the periodic meetings of the society. When the entomologists organized again, in 1859, as the Entomological Society of Philadelphia, there was some direct carryover in personnel. LeConte served as the first president of the Philadelphia society, and Morris continued to represent the entomologists' interests to the Smithsonian.[122] In the 1870s, when American entomologists organized the Entomological Club (later a subsection) within the AAAS, Morris was chosen as president. In 1881, he delivered a presidential address in which he recalled the entomologists' initial organization in Pennsylvania forty years earlier and noted the advances in American entomology since that time.[123]

Another important link between the Pennsylvania Entomological Society and successor institutions can be found in the disposition of the insect collections of the members. The collections of Melsheimer, Ziegler, LeConte, and others were eventually donated or sold to the Museum of Comparative Zoology, in Cambridge, Massachusetts, where they formed the nucleus of the first great public insect collection in North America.[124]

Because of these developments, American entomologists in the 1850s had a new confidence in their competence and their institutions. To a considerable extent, the basis for this change in attitude related directly to the activities and the achievements of the Entomological Society of Pennsylvania. Through the society, the entomologists had established effective and continuing communications among themselves and with the Europeans, they had published descriptions and systematic revisions according to the best scientific standards, and they had assembled, standardized, and preserved the insect collections that gave a permanent basis for the advancement of entomology in North America. In one significant area, agricultural entomology, the society had failed to provide leadership. With this exception, the members of the Entomological Society of Pennsylvania had initiated the institutional structure necessary for the development of their science.

Of Cabinets and Collections   3

For American entomologists at midcentury, the central institutional requirement for the advancement of their science was the establishment of permanent insect collections where specimens could be safely stored and retrieved for study. Europeans had led the way in the generation following Linnaeus by amassing vast insect collections in Paris, London, Berlin, Vienna, and elsewhere. Thousands of insect specimens gathered from Europe, America, Africa, and Asia comprised the basic data of systematic entomology that flourished in Europe and promised to flourish in America.

While the Entomological Society of Pennsylvania had elected to concentrate its efforts on the publication of descriptions from Melsheimer's catalog of Coleoptera rather than to assemble a central insect collection, the society members recognized the need for Americans to match European institutional models in the construction of permanent insect collections. This awareness is clear in the letters of John G. Morris, who in 1846 embarked on a reconnaissance of European entomological collections. Writing home to his American colleagues from the University of Berlin Museum, Morris described the collections with awe: "Everything here is on a grand scale. The insects alone occupy over 5 rooms each larger than Haldeman's parlor. . . . there are over 40,000 species of Coleoptera alone. . . . All the other orders are equally well represented. There are 4 or 5 men constantly at work on them, cleaning, arranging, classifying, describing, etc. etc."[1] The challenge was clear. Americans must find the means to erect and maintain similar institutions that could act as central depositories and support paid curators. Only then could entomological science mature in America.

Within forty years of Morris's admiring report, American entomologists had achieved this goal. By the 1880s, America had three central collections that equaled and in certain respects surpassed the European collections. By that decade the collections at the Smithsonian Institution, the Academy of Natural Sciences of Philadelphia, and above all the Museum of Comparative Zoology at Harvard University compared favorably with those in Europe in quantity, quality, arrangement, curatorship, and permanence.

The story behind this institutional development involved all segments of the entomological fraternity plus a host of supporting individuals and institutions with various related interests. The cast of characters behind the growth of major insect collections involved a range of personalities, motivations, and interests almost as varied and colorful as the insects collected for display. Those who captured, exchanged, bought, sold, raised, arranged, and stored insect specimens were prompted by aesthetic appreciation, acquisitiveness, competitiveness, jealousy, and class consciousness as well as scientific curiosity.

The entomologist's collecting activities contributed to the stereotype of eccentric "bug hunter" so common among nineteenth-century Americans. After all, the capture, pinning, and arrangement of specimens in cabinets were the most conspicuous "activities" in which entomologists engaged, and these activities reinforced the impression of eccentricity that was associated with the collecting and hoarding of esoteric items of nonutilitarian value. Those who ridiculed entomologists of course failed to recognize the worth of the scientific information derived from specimens and collections. On the other hand, entomologists could at times employ means as ruthless and unethical as those of any robber baron when they wanted to increase their collections. Charges of theft of specimens circulated frequently among the entomological fraternity. In 1877, William Henry Edwards wrote to his fellow lepidopterist Joseph A. Lintner: "What is there about our science that makes one set of men so inflammable, and another rascals and thieves, like [Tryon] Reakirt, and 3 or 4 still left in the country, whom one cannot leave two minutes amongst your butterflies without knowing that they have stolen something!"[2] While the popular stereotype of "bug hunter" was unfair, entomologists are human and subject to the same strengths and weaknesses of character as other human beings. The human side of the entomological community emerges in collecting and collection building perhaps more than anywhere else.

Oblivious to the scoffing of their fellow countrymen, and motivated by visions of scientific and personal advancement, American entomologists busied themselves with additions and improvements to their insect collections. As they did so, they profited from three related developments that ensured the rapid growth of central reference collections during the midnineteenth century. First, as a result of territorial expansion and scientific exploration on the North American continent, Americans could count on a burgeoning supply of insect specimens. Second, the entomological fraternity at midcentury increased rapidly in numbers and in competence within a context of increasing urbanization and specialization of American society. The entomologists were thus better prepared to convert the wealth of insect specimens into permanent scientific collections. Third, the agricultural and commercial communities provided solid financial support for entomological collections.

The territorial expansion of the American republic in the nineteenth century produced a quantum leap in the numbers of insects collected. The most important sources of specimens were the state and federal surveys designed to identify exploitable minerals, transportation routes, boundaries, and other features of the continent. Privately financed entomological expeditions supplemented the government surveys, often producing more specimens per dollar invested than the official surveys. Entomological collections also profited from the spread of railroads across the continent and from the opening of new zoological provinces like Texas, California, and Alaska.

Entomologists participated in various capacities in the reconnaissance of the West. Some explored in person, either attached to a government expedition or traveling privately. LeConte, for example, explored much of the Mississippi Valley, California, and Arizona on various private and government expeditions from the 1840s to the 1860s. On other occasions, entomologists made special arrangements with those traveling to unexplored regions. In 1849, LeConte and Haldeman arranged for Haldeman's brother to accompany a private trading expedition to Santa Fe and California.[3] At other times, entomologists pooled their resources to sponsor entomological exploring expeditions, such as when LeConte, Louis Agassiz, and others raised $600 to send a collector named Pease to Oregon in 1849.[4]

Entomological interest in the American West and other peripheral regions gave rise to a special kind of explorer: the insect collector who supported himself through the sale of specimens to Americans and Europeans. Gustav Wilhelm Belfrage, who made his headquarters in Texas in the 1860s and 1870s, was one of the best known of these explorer-

Plate 9, "Coleoptera," Townend Glover Notebooks, ca. 1850–1878, box 9, Record Unit 7126, Smithsonian Institution Archives, Washington, D.C.

collectors. LeConte, who considered Belfrage the most expert collector of small insects, suggested Belfrage as naturalist for the Telegraph Expedition to Alaska in 1865, but unfortunately for Belfrage the application came too late.[5] The following year Belfrage received a commission from the Swedish Academy of Sciences to collect insects in Texas. Over the next two decades Belfrage sold specimens collected in Texas to every important entomologist in the East, as well as to his many European contacts. The list of Belfrage's correspondents reads like a who's who of the American and European entomological establishment. It included Alexander Agassiz, Henry Edwards, Herman A. Hagen, John L. LeConte, Alpheus Spring Packard, Jr., J. D. Putnam, James Ridings, Samuel H. Scudder, and Philip Reese Uhler as well as entomologists in Austria, England, France, Germany, Sweden, Russia, and Mexico.[6] Many of these correspondents received annual shipments of insects, while others placed orders for exotic specimens that they desired for their collections.

Another professional collector, Herbert K. Morrison, made his career choice in 1874 while on an excursion to the White Mountains in New Hampshire with the Cambridge Entomological Club. During the next decade his travels took him to Georgia, North Carolina, Colorado, the Black Hills, Nevada, Washington, Oregon, California, Arizona, Florida, and Key West. Morrison often walked forty miles a day in pursuit of insects, then mounted and sorted them at night, meanwhile capturing moths that flew to his camp light.[7] The few references to actual terms of payment indicate that Belfrage, Morrison, and other professional collectors must have been extraordinarily efficient to support themselves in this trade. Packard paid Belfrage five cents per specimen, which was apparently a standard rate.[8]

These explorers, most of whom were familiar only to those within the entomological fraternity, deserve wider recognition in the saga of westward expansion. The insect traders, like the fur traders, miners, ranchers, and settlers later eulogized by Frederick Jackson Turner, played a role in the dynamic relationship between the romantic and distant West and the centers of culture and learning in the East that defined the frontier in the American imagination.

The most important result of the Great Reconnaissance of the West for entomological collections was the veritable flood tide of new insect specimens flowing into the hands of American entomologists. LeConte, upon his return from the far West in 1851, wrote Haldeman that he brought with him about 20,000 duplicate specimens containing a total of about 1,000 new species.[9] To place LeConte's 1,000 new species in perspective, it may be noted that all four of the Pacific Railroad surveys of the 1850s resulted in a total of only 52 new species of mammals and 210 species of birds.[10]

No less important than the escalating supply of insects was the growing capacity of American entomologists to preserve and classify the specimens in collections. This competence was partly a function of increasing numbers of experienced entomologists. Whereas in the 1840s the working entomologists in America numbered only about a dozen, by the 1850s this number had risen to about thirty, and by the late 1860s, the number of publishing American entomologists had risen to over forty.[11] These publishing authors represented the elite of the entomological fraternity in the 1860s and 1870s. During the same period, the total number of American naturalists who were serious enough in their study of insects to identify themselves as entomologists increased to somewhere around eight hundred in the 1870s.[12]

Whatever the exact numbers, the sheer increase in serious entomological collectors in America contributed to the rapid growth of collections. When more individuals collected more specimens, this material ultimately found its way, through trade, exchange, and sale, to the holdings of experts in each insect order, like LeConte for the Coleoptera. Accordingly, collecting was encouraged by all members of the fraternity. Printed instructions for the collection of insects were supplied by LeConte and others interested in procuring. Similar instructions, to assist the amateur in collecting, labeling, and shipping specimens, were printed on the letterheads of federal and state entomologists and in entomological journals. Experienced entomologists encouraged neophytes to work up specialties, particularly in less-known orders, by holding out the possibility of their discovering a new species or other new information. In 1870, for example, the editors of the *American Entomologist* explained how an amateur, even with limited books and knowledge, could still make significant discoveries: "Should [the beginner] care only to acquaint himself with the nomenclature of some limited group or order, and wish to increase his cabinet in that speciality, he will find that he has the powers of a capitalist to invest his miscellaneous collection of specimens . . . in such a manner as he may prefer. . . . There is always some eager specialist . . . standing ready to give full value for . . . such material."[13]

The rapidly expanding cadre of American entomologists proved to be effective in the collection, classification, and preservation of insects in cabinets because of a shared vision of the comparative and descriptive functions of insect specimens as components of collections. The study of entomology as it had developed primarily in Europe down to the midnineteenth century consisted chiefly in the comparison of specimens.[14] As heirs to this European entomological tradition, Americans were guided in their collecting, comparing, and arranging by the systematics of Linnaeus and his heirs, the open-ended search for affinities developed by Pierre André Latreille, and the development of the "type concept" as an aid in organizing and labeling specimens in large permanent collections. With every new acquisition of specimens, Americans acquired new experience as practitioners of this tradition in the arranging, classifying, and rearranging of specimens. Out of this combination of new material and applied systematics, individual collections of leading American entomologists grew rapidly in size and sophistication.

The Americans' use of the type concept illustrates how they incorporated an important aspect of the European tradition in building their collections. By the early nineteenth century, "type specimens" were recognized as vital components of large natural history collections, primarily because they functioned as name carriers of species and as authoritative sources in the resolution of difficulties in classification.[15] As the number of specimens in collections grew, and as the consequent splitting of genera and species became more frequent, museum professionals began the practice of carrying the original name of the species or genera with the group that retained the type specimen. Type specimens in standard collections thus served as references for classification. As the collections of individual American entomologists grew in size and complexity, and as the disposition of these collections became pressing with the aging of America's first entomological elite, the preservation of types in these collections emerged as a matter of grave concern. American entomologists sought to preserve intact the material in the early comprehensive collections made by Say, Melsheimer, and Harris. Their efforts met with varying degrees of success. The Say collection, donated upon Say's death in 1834 to the Academy of Natural Sciences, was almost completely destroyed by shipping and by insect pests before it was sent to Harris for evaluation in 1836.[16] With the loss of the Say collection, Harris and Melsheimer, who had exchanged specimens with Say, thus held the majority of original specimens collected by Say. Baron Charles Robert Osten-Sacken, North America's resident authority on the Diptera, recommended that the Harris collection, which was purchased by the Boston Society of Natural History following Harris's death, should be kept "sacred," without changing its arrangement, because it contained some of the few original specimens [types] of Say.[17] The presence of Say's types in the collections of Harris and Melsheimer added a further incentive for American entomologists in the 1850s and 1860s to preserve these collections, with Say's types, intact.[18] With the growth of American collections from the 1840s on, the preservation of types became increasingly important.[19] This concern demonstrates Americans' practice of an important aspect of the European natural history tradition with respect to collection building and administration. The value they placed on types and the means they used to preserve them ensured the scientific value of large reference collections as repositories of information about American insects.

In addition to the well-established classification and type concepts, Americans also profited from the initial advances made by Europeans in the preservation of collection specimens through the use of standard-sized boxes, standard-sized pins to secure specimens in the boxes, and toxic preservatives to control museum pests. By the 1840s, many American entomologists had adopted the method of constructing boxes of uniform size in which to house their specimens. As new specimens were added, cabinets could be expanded by rearranging the boxes to accommodate the new material. William Henry Edwards related to Joseph A. Lintner the construction of the boxes he used in his cabinet and the costs: frame, $0.50; glass, $0.25; tinfoil (bought by the pound; no price given), binding $1.50. Lintner was so impressed with the design and the modest cost that he ordered six boxes to be constructed by Edwards's carpenter.[20] Through the period under consideration, American entomologists continued to share ideas on improved design of specimen cases, pins, and labels, and this sharing contributed to steady improvement in the physical design and arrangement of cabinets.

The matter of most importance in the curation of collections, however, was their protection against live insect vermin, in particular moths, mites, and the larvae of *Anthrenus,* and other Dermestidae. By the early nineteenth century, European naturalists had made great strides toward solving the problem of preserving museum specimens indefinitely through the application of mercury, arsenic, cyanide, and other poisons to cabinets and specimens.[21] Of all natural history collections, insect collections were the most fragile and vulnerable to destruction by pests. The application of poisonous compounds to collections, which came into general use in the early nineteenth century, served to preserve great natural history collections of all species, but such compounds were probably most crucial to the preservation of insect collections. In 1846, Morris wrote about the preservatives he observed in the Berlin Museum, where each case contained three or four drops of mercury.[22] By the 1860s, Americans were using arsenic, cyanide, camphor, benzene, turpentine, and other lethal compounds in their cabinets.[23] Packard recommended plunging each specimen in a solution of alcohol containing fragments of arsenic, followed by regular maintenance of cabinets, which consisted of placing pieces of camphor or bottles with sponges soaked with benzene into the bottom of cabinet drawers.[24] During the midnineteenth century, American entomologists, along with their European colleagues, continued to experiment and share ideas about how to eradicate museum pests, particularly the destructive *Dermestes*.[25] This experimentation and sharing brought about marked improvements in the preservation of specimens in the individual cabinets.

Advances in the use of preservatives were necessary, though not sufficient, safeguards to preserve insect collections. Even the most lethal poison lost its strength over time. The single indispensable element in the maintenance of insect collections was constant vigilance and attention. LeConte recommended a thorough inspection and replacement of preservatives twice annually.[26] Packard recommended a careful examination every month.[27] In the formative years of entomological collection building, from the 1830s to the 1860s, individuals proved more capable than institutions in the sustained attention necessary to preserve large insect collections. Those American collections that survived intact owed their longevity to the care of individuals and families, like the LeContes and Melsheimers. Only in the post–Civil War era did stable scientific institutions prove capable of assuming the curation of large, pest-free insect collections.

In addition to the expanding supply of insects and the increasing competence of the American entomological community in the assembling and curation of insect cabinets, American entomological collections depended on the support of the agricultural and commercial communities. These communities funded the scientific institutions where insect collections could be housed and preserved.

American agriculture contributed to both the quantity and the richness of American entomological collections through the support of entomological scientists and institutions. Agriculturists, seeking professional assistance in combating harmful insects, readily agreed to the necessity of insect collections devoted to insects of economic importance. Entomologists easily translated this support for economcally important species into support for general collections. The collections of agricultural entomologists emphasized the biology of insects, including all stages in the life of injurious species, and they displayed specimens in ways that showed their economic and ecological importance as well as their systematic classification.

The earliest important agricultural collection originated with the agricultural survey sponsored by the New York Agricultural Society in the 1840s. In 1852, the insect collection was transferred to the New York State Cabinet of Natural History in the statehouse. In 1847, Asa Fitch was hired by the regents of the state university to supplement and arrange the insects in the state cabinet, which were then displayed in the capitol. When Fitch was named state entomologist in 1854, he continued to attend to the cabinet, though the state entomologist's office was administered separately from the state cabinet. In 1865–1866 the New York regents proposed the consolidation of all the state's natural history collections in a reorganized state museum of natural history to be housed in a new museum building. James Hall was placed in charge of the reorganization, and in 1870 he was named director of the state museum. In 1868, Hall named the entomologist Lintner as his assistant. From 1868 until 1898, Lintner, first as assistant to Hall and later as state entomologist, administered this important entomological collection as part of the New York State Museum of Natural History.[28] The collection suffered a major setback following Fitch's death in 1879, when Lintner failed in his attempt to persuade the state to purchase Fitch's personal collection, which was then sold in Europe.[29] Despite this setback, the entomological collection of the New York state museum continued to grow and to serve the entomological and agricultural communities.

Other significant collections begun by early agricultural entomologists were those of Harris, which went to the Boston Society of Natural History, Benjamin D. Walsh, which went to the Chicago Academy of Sciences (where it was subsequently destroyed in the great Chicago fire of 1871),[30] and Charles Valentine Riley.

In addition to the support provided by agriculture, American entomological collections profited from the philanthropy of Americans (and some foreigners) who concentrated the wealth from commerce and later returned a portion of this wealth to the public by funding scientific institutions where American productions and knowledge of American natural history would be displayed. William Maclure, a wealthy Scots merchant and geologist immigrated to the United States early in the nineteenth century. There he became associated with the Academy of Natural Sciences of Philadelphia and gave lavishly of his fortune to the Academy of Natural Sciences and to the utopian community of New Harmony, Indiana. Thomas B. Wilson, an executive of the Pennsylvania Central Railroad, succeeded Maclure in 1840 as the main financial benefactor of the Academy of Natural Sciences. Beginning in 1859 Wilson underwrote a large share of the expenses of the Entomological Society of Philadelphia. Francis C. Gray, heir to his father's Salem, Massachusetts, shipping fortune, contributed substantially to Harvard University and, through his heirs, to Louis Agassiz's Museum of Comparative Zoology. James Smithson, a British aristocrat and scientist, provided in his will for the founding of an institution for the "increase and diffusion of knowledge" to be located in Washington, D.C., and thus initiated the scientific institution that bears his name. Other institutions where philanthropy promoted the assembling of entomological collections were the Boston Society of Natural History, the Peabody Academy of Science, and the Essex Institute.

Support from agricultural organizations and private and public philanthropy helped establish a secure niche for entomological collections within the edifices of American scientific institutions around midcentury. The accelerated pace in the growth of central reference collections centered on three institutions: the Smithsonian Institution, the Academy of Natural Sciences in conjunction with its daughter institution, the Entomological Society of Philadelphia, and the Museum of Comparative Zoology of Harvard University.

The Smithsonian Institution, named as the official recipient of specimens from government expeditions, became critical to the development of all American natural history collections, including entomology. Spencer Fullerton Baird, a vertebrate zoologist from Dickinson College, was appointed in 1850 as assistant secretary in charge of collections, publications, and foreign exchanges. Baird presided over Smithsonian natural history during an extraordinarily active period in the reconnaissance of the West, which included the Pacific Railroad surveys, the boundary surveys, the fortieth parallel survey, the Geological and Geographical Survey of the Territories, the Wheeler survey, and the Alaskan telegraph survey. Recognizing the potential for natural history (especially zoology), Baird assumed a major role in the planning and staffing of the surveys. To him fell the task of reporting the scientific results of the surveys and ordering, preserving, and distributing the wealth of material that came to the Smithsonian.[31]

Like Baird, the entomologists were eager to use the federal surveys as a new source of specimens. In 1849, LeConte pointed to the example of botanists like John Torrey, Asa Gray, and George Engelmann, who had already mapped out botany in the American West, and he urged that future explorations devote more attention to zoology.[32] For entomologists, as for other naturalists, the American West offered untold wealth in the form of thousands of specimens inhabiting scientifically unmapped territory now rapidly coming within the scientists' reach. The transfer of this profusion of freely flying and crawling insects from their home territory in the West to securely pinned specimens arranged neatly in Eastern collections, however, posed special problems for the entomologists. While the personnel recruited for the early government expeditions, like the Pacific Railroad surveys, normally knew the rudiments of killing and skinning vertebrate specimens like birds, mammals, or reptiles, they rarely possessed the knowledge or (apparently) inclination, to collect small, fragile invertebrate specimens. As a result, the shipments from survey leaders in the 1850s and 1860s typically included an abundance of vertebrate specimens with only a paltry selection of insects.[33] The Pacific Railroad surveys of the early 1850s, though resulting in the most comprehensive reporting of the natural history of the American West up to that time, produced only one significant report on insects, LeConte's report on the insects of the northern route explored by the I. I. Stevens party.[34] In 1859 LeConte complained to Baird that "the Vertebrata of the interior seem to monopolize attention, to the entire exclusion of the Articulata."[35] This state of affairs was particularly irksome to LeConte, who confided to Haldeman his belief that insects held a greater potential than vertebrates for compiling impressive reports, since "they are more numerous, and a few hundred insects will make a much greater display (on paper) than three or four birds and mice."[36]

To remedy this problem of haphazard and ineffective collection of insects, LeConte endeavored to place naturalists who were knowledgeable and reliable insect collectors on government expeditions. In 1849, LeConte and Haldeman asked Baird's advice on how to place Haldeman's brother (a good insect collector) on a government expedition. Baird replied that he should get recommendations from superiors stressing "energy and excellence," then find some congressman to "bore the President" with these recommendations.[37] They succeeded in placing their candidate on the boundary survey, and he in turn provided LeConte with bountiful specimens from the Southwest.[38] LeConte noted that collectors had little patience or aptitude in the art of pinning insects, so he simply supplied collectors with small bottles filled with alcohol in which they could place specimens for shipment.[39] LeConte also prepared printed circulars for distribution by the Smithsonian, explaining how to collect, preserve, and ship insect specimens.[40] He also corresponded directly with expedition commanders.[41] These practices, LeConte noted, brought good results in the form of abundant specimens arriving in bottles every month or two.[42]

The greatest improvement in the collection of insect specimens from government expeditions came with the appointment of survey leaders who were sympathetic to the entomologists' needs and who cooperated in assigning trained entomologists to their expeditions. Of all the survey leaders, Ferdinand Vandiveer Hayden, director of the U.S. Geological and Geographical Survey of the Territories (1867–1879), developed the closest working relationship with the entomologists. His survey published many important reports on insects and provided entomologists with the opportunity to gather a wealth of insect specimens. Three entomologists, Packard, Scudder, and Cyrus Thomas, served as regular members of Hayden's expeditions, and Thomas eventually became Hayden's administrative assistant. These and other entomologists either collected insects themselves or received regular shipments of insects from Hayden's collectors, and they contributed entomological monographs to Hayden's annual reports.[43]

Hayden was a naturalist with catholic interests. His encouragement of collection and reporting in all branches served naturalists, including entomologists, who were eager to collect and learn about the western fauna.[44] This symbiotic relationship served both Hayden and the entomologists well during the life of Hayden's survey, but in 1879, in the struggle over the consolidation of the surveys, Hayden's practice of collecting natural history specimens was challenged by Major John Wesley Powell, who took the narrower stance of a geological and topographical specialist. Powell made a virtue of not collecting natural history material because, he said, this wasted government funds in pursuits that could be handled by individuals and private institutions.[45] LeConte came to Hayden's defense on behalf of the entomologists when he rejected Powell's "narrow views regarding zoology and botany." To Hayden he wrote, "You did good work in the West . . . before . . . the other surveys. . . . you have procured the collateral material without impairing or diluting the results of the geological investigation."[46] Hayden's retirement in 1879 thus ended an era of fruitful collaboration between the entomological fraternity and federal surveys of the trans-Mississippi West. This collaboration had benefited the entomologists by providing them with a wealth of specimens for their own collections, new outlets for entomological publications, and the institutional beginnings of great national collections. The growth of permanent collections came about primarily through Baird and the Smithsonian Institution.

The growth of the Smithsonian's own entomological collection proceeded with a certain amount of circumlocution. Joseph Henry, the first secretary of the Smithsonian, whose earlier experiments in electricity and magnetism led to the development of the telegraph, believed the Smithsonian should function primarily as a supporter and disseminator of research and knowledge gained from research and not as a storehouse of material objects. Nor should the Smithsonian compete with existing institutions either as a library or as a museum. While Baird agreed with Henry on the primacy of original work, as a working naturalist he stressed the need for comprehensive collections upon which to base zoological research. In 1855, following a shakeup in the Smithsonian that involved the firing of Charles Coffin Jewett, the assistant secretary in charge of the library, who opposed what he considered Henry's narrow views regarding collections, Henry and Baird apparently reached an agreement regarding the natural history collections. Baird was given free rein to build the Smithsonian collections provided that this activity remain subordinate to research and publication and provided that the curation of the collections be funded through congressional appropriation rather than directly from the Smithsonian fund. By 1878, when Baird succeeded Henry as secretary, Baird had built a world-class natural history collection with the tacit approval of the secretary, who publicly disavowed this goal.[47]

The circumlocution went one step further in the case of entomology, which lay outside Baird's main area of expertise. Baird at first distributed the specimens to entomological experts, who arranged them and wrote the reports. Although Baird retained the Smithsonian's claim to specimens collected under its auspices, the distribution amounted in practice to permanent loans to the collaborators. This practice served the interests of both the Smithsonian and the entomological community because the Smithsonian was not yet equipped to handle entomological specimens, whereas individual entomologists had already established what were becoming standard reference collections. LeConte and Baird had no difficulty with the informal understanding because they were related (John Eatton LeConte was a cousin of Baird's mother) and had known each other at least since 1841, when Baird stayed with the LeContes in New York. They addressed their letters "Dear Spence," and "Dear John." LeConte was concerned, however, that some future Smithsonian administrator might want specimens from his cabinet returned.[48] In that event, property rights would be extremely complicated, because LeConte had made arrangements with many collectors who worked at times for government expeditions. He considered specimens that were collected for him by army personnel during off-duty time to be his own property.[49] Further complications arose from Baird's practice of soliciting funds from entomologists to help finance collectors who accompanied government expeditions. LeConte regularly contributed amounts ranging from twenty-five dollars to fifty dollars per season, for which he expected to have, and received, the first choice of the Coleoptera for his own collection.[50] One example of a way in which Baird's system of distribution might lead to difficulties began in 1859, when Baird sought funds from LeConte and from the Entomological Society of Philadelphia to help meet expenses of Robert Kennicott, who initiated a collecting expedition to Northwest Canada and Alaska. The expedition was financed through contributions from individuals and institutions including the Chicago Academy of Sciences and the Smithsonian Institution. The distribution of Kennicott's specimens soon provoked contention among the entomologists.[51]

While the Smithsonian policy of collaboration and distribution in the field of entomology enriched important private cabinets like LeConte's, it did not address the need for a central entomological reference collection. This need became the paramount concern of the Entomological Society of Philadelphia, which was organized in March 1859. With the appearance of the Entomological Society of Philadelphia, the Smithsonian's own modest insect collection receded to the background for a time during the 1860s and 1870s, while the Entomological Society and the Museum of Comparative Zoology (discussed below) forged ahead in the establishment of central reference collections for North American insects.

The Entomological Society of Philadelphia was organized by four charter members: James Ridings, Ezra Townsend Cresson, George Newman, and Thomas B. Wilson.[52] The constitution they adopted emphasized the gathering of accurate information about American insects: "In our opinion the cause of science would be materially advanced by the establishment of a society whose members would make it incumbent upon themselves to ascertain the *Name, locality, Habits, Time,* etc., *of Insects* taken within the United States of America, and communicate the same to the Society."[53] Though it paid patriotic attention to American insects much as its predecessor, the Entomological Society of Pennsylvania, had done, the new society differed from the earlier one in the class consciousness of its membership, its emphasis on collections, and its permanence as an institution affiliated with the Academy of Natural Sciences.

The members described themselves variously as "active collectors," "practical collectors," those "interested in entomological science," or those "interested in entomology." Cresson, its longtime corresponding secretary and historian, has characterized the society members as "Franklinesque," meaning that they pursued knowledge for its own sake.[54] The bonds that drew them together, however, originated in their social and economic class as much as their scientific interests. Most were petite bourgeoisie, the clerks, shopkeepers, and small entrepreneurs who sought in science a camaraderie, participation, and sense of achievement in an interesting and important pursuit. Like many groups in Philadelphia at midcentury, the society was responding to rapid population growth, immigration, and economic expansion, which were transforming the old-style Quaker city, with its merchant leadership and its neighborhood life in the central city, into a sprawling industrial metropolis marked by obvious economic disparities in income. Philadelphians responded by organizing a multiplicity of clubs, societies, associations, and lodges. Those who formed the Entomological Society of Philadelphia represented the middle-class, and they faced new opportunities and new uncertainties in the emerging industrial order. The middle-class orientation of the membership militated against the kind of elitism implied by the designation of a scientific authority on an insect order, like LeConte for the Coleoptera. At the same time these middle-class origins promoted the kind of values involved in building a reference collection. Such an enterprise would be the joint accomplishment of a society comprising dedicated, efficient, thrifty, resourceful, and persevering members.[55] Such scientific organizations for the middle classes typically appeared earlier in Republican America than in class-conscious Europe. By contrast, the membership of the East London Entomological Society, organized in 1873, was restricted to eight "gentlemen."[56] The members of the Entomological Society found an able spokesman in Cresson. They also found an influential backer in Wilson, a railroad magnate from the new industrial elite.

Cresson longed for a career in science, but he was acutely aware of his limited education, having completed only eight years in public school. His interest in natural history had been encouraged by his future father-in-law, Ridings.[57] In 1857, Cresson and some fellow naturalists, apparently not feeling at ease at the meetings of the Academy of Natural Sciences, formed the Henry Institute (named for the secretary of the Smithsonian), which was devoted to the study of zoology. Cresson's deference to his scientific superiors is indicated by the use of Henry's portrait for the Henry Institute (which Cresson later sent to Baird, saying, "I have never had the honor to look at his countenance in reality") and Cresson's reluctance to approach Baird in person when Baird visited the academy in 1858.[58] Cresson requested Baird's advice on the best of two proposed plans of animal classification, so that the committees could be arranged accordingly, and he asked that the Smithsonian donate extra zoological specimens to the institute's cabinet.[59] The Henry Institute failed to gain the necessary support and disbanded in 1858. A few months later Cresson helped organize the Entomological Society, and he was elected corresponding secretary. Cresson resigned in 1859 to go with his mother and family to New Braunsfels, Texas, to start a cattle ranch. The project failed, and Cresson returned to Philadelphia within the year, where he resumed the office of corresponding secretary, which he retained for the remainder of his life.[60] Cresson also returned to secure employment as a clerk at the Pennsylvania Railroad Company. In 1859, he was offered the curatorship of the academy's collections, but he declined because he could not support himself from it, and he was advised that previous mismanagement of the collections had made them the subject of a factional dispute among the academy's members. He then applied to Baird for a position at the Smithsonian: "It has been my ardent desire to obtain a situation in some scientific institution where I would command a small salary and where I could advance my knowledge of science. . . . Altho' I am not a scholar, yet I may be of use in the Entomological line and am handy at writing."[61] Cresson soon found the kind of "situation" he sought when he was promoted to a position as Wilson's private secretary at the Pennsylvania Railroad, a formal title that allowed Cresson to spend all his time at the room of the Entomological Society.

For their first president, the members of the society chose the city's most eminent entomologist, LeConte. At first, the society held its meetings in LeConte's home, where LeConte lectured on entomological classification.[62] Within a few months, however, the members voted to rent a room for their meetings because some members felt "hesitancy" at coming to a private residence.[63] Their sensitivity to differences in social standing was indicated by Cresson when he wrote to Baird: "I find it very objectionable [to go to LeConte's house] for it is not to be expected that any man would like to have entomologists, or those interested in the science, of all grades in life, to be running and tramping through his private residence in all sorts of weather to examine . . . specimens."[64] Cresson added that "the working members[,] some very poor (dues only 6 cts. a month[,] are all zealous and practical collectors."[65] Over LeConte's strong objections, the meeting place was moved.

The society's steady progress during the next five years came from the combination of volunteer effort and the financial patronage of Wilson. In addition to paying Cresson's salary, Wilson paid the rent for the society's new building (which was constructed by Ridings), and he contributed generously to the society's library, its cabinet, its collecting fund, and its publication fund. The publication fund ensured the success of the society's *Proceedings,* which were begun in 1861. To save money, the members bought printing equipment, set the type, and edited and printed the *Proceedings* themselves. By 1864, the membership had increased to seventy-one resident and eighty-four corresponding members; in 1867 the society acknowledged its leading role by changing its name to the American Entomological Society.[66]

Wilson's patronage of science stemmed from something akin to religious appreciation for the wonders of nature. As a young man he had traveled throughout Pennsylvania and the Old Northwest, studying the region's geology and natural history. For him, the patterns in nature revealed the Design of the Creator. Though he wrote almost nothing himself, he continued his own studies and supported the research of naturalists in the Academy of Natural Sciences, which he financed from his earnings as an executive of the Pennsylvania Railroad.[67] After the death of William Maclure in 1840, Wilson became the academy's primary patron.[68] His decision to switch his primary support from the academy to the Entomological Society apparently originated in a dispute with LeConte over the care of the academy's insect collection.

Until 1857, relations between the two men had been harmonious. Both Wilson and LeConte shared the hope that the academy might serve as a permanent repository for Smithsonian insect specimens. LeConte reported to Baird that Wilson had purchased "bug proof" boxes where specimens "would be safe and ready for use . . . for all time."[69] His suggestion that Philadelphia was better suited than Washington for the Smithsonian insects included the significant proviso that the Coleoptera should go to his collection while the other "optera" should go to the academy.[70]

Soon thereafter, for reasons not entirely clear, LeConte expressed his dissatisfaction with the management of the academy's insect collection. Several months prior to the organization of the Entomological Society, certain "losses" in the academy's cabinet provoked a heated discussion between LeConte and Wilson, whereupon Wilson threatened to sever all ties with the academy. Wilson wrote of the encounter to Joseph Leidy:

Dr. LeConte's remarks in regard to the losses in the entomological room, had rendered it impossible for me to continue my studies in connection with the Academy:—we mutually laid the blame of the losses on each other and from that time my separation from the Academy was inevitable. . . . I cannot help regarding Dr. LeConte's election on the Entomological Committee as a decision by the Academy that he was the innocent party and I the guilty one.[71]

Wilson did not break completely with the academy; he was elected president in 1864 and resigned six months later, probably due to failing health.[72] Still, Wilson's sense of betrayal by fellow academy members, many of whom also opposed the formation of a separate entomological society, probably strengthened his decision to support the new society.

With Wilson's support, the society began a cabinet of its own. Soon the members decided to establish the cabinet as a standard reference collection with the intention of securing representatives (preferably types) of all American insects.[73] As with the Henry Institute, the society hoped for substantial donations of specimens from the Smithsonian.[74] LeConte's selection of a set of insects for the cabinet from the Pearsall collection, on which he was reporting for the Smithsonian, provided the occasion for renewed hostilities with Wilson, this time involving the Entomological Society and the Smithsonian as well.[75]

John Pearsall had been hired by Baird, in 1859, to travel to the upper Missouri and to collect for the Smithsonian. According to Pearsall's written statement to the Entomological Society, he collected birds, mammals, and insects that were sent to the Smithsonian. Baird was evidently satisfied with his work. Pearsall also collected insects along the Missouri both before and after he was employed by the Smithsonian. Soon after he returned to Philadelphia, the Entomological Society was organized; Pearsall became a charter member, and he donated his own insects to the society. Some portions of the Pearsall insects, which were turned over to LeConte for cataloging, were thus claimed by the society and some portions by the Smithsonian.[76] Meanwhile, the society had adopted a rule, apparently at Wilson's suggestion, which forbade the loaning of any named specimens from its cabinet. LeConte wanted to take the remainder of the Pearsall insects to his home for study, but his request was refused.[77]

Both Baird and Henry objected to the new rule of the society. They insisted that all specimens from the Smithsonian be available for loan to scholars.[78] Cresson then explained to Baird, at some length, why the members had made the ruling, and why they remained steadfast in their decision. Prominent in the explanation was his charge that LeConte had appropriated specimens without returning them: "It was . . . the decided opinion of . . . the members that a standard collection could never be made if . . . specimens were allowed to be loaned because as soon as a unique species . . . is received, a Certain Entomologist would immediately loan [borrow] it and *(as is generally the case, so far) it will never be returned!*"[79] Cresson recounted that he had loaned specimens of his own, only to find them later "quietly and snugly labeled and placed in his [LeConte's] own extensive collection."[80] For the sake of "peace" and in order to retain the privilege of using LeConte's cabinet, Cresson refrained from pressing the matter; but he vowed never again to loan specimens to

LeConte.[81] Cresson ascribed LeConte's behavior to his desire to retain his prestige as the preeminent describer of Coleoptera. "I may say," wrote Cresson, "that the Doctor is not at all favorably disposed towards the Society because he knows that he cannot wear the crown of laurels (tho' gained by many selfish motives) many years longer."[82] In contrast to LeConte, many members preferred that the credit for a new species go to the discoverer (not the describer) or to the society as a whole.

Under these circumstances, Cresson argued, the rule helped, rather than hindered scholars, because it ensured a standard reference collection:

> Now please consider what good the collection [would be] as reference, if specimens are to be loaned. Here a member presents a . . . unique specimen, and it is published to the world by the Society. . . . Soon thereafter someone applies for the loan of it. . . . the . . . borrower either keeps it, mislays it, or sends it to Europe, and . . . the specimen is never returned. Now, someone who may be very anxious to examine the specimen . . . comes a great distance. . . . he is disappointed and the Society is in hot water.[83]

Cresson's explanation failed to change Henry's decision that Smithsonian duplicates would be placed in the society's collection only if they could be loaned for study.[84] Cresson was certain that LeConte was behind Henry's refusal. As evidence, he pointed to the continued donation of specimens to the academy, which (he said) had the same restrictions.[85]

LeConte refused to accept a second term as president, citing his objection to the society's restrictive policy.[86] The society members, determined to build their reference collection even without the help of LeConte or the Smithsonian, distributed their own printed pamphlet asking for specimens and offered free subscriptions to the society's publications as inducement to collectors.[87] The circulars were sent by Cresson to LeConte's correspondents with the request that they send specimens to the Entomological Society rather than to LeConte.[88] The ensuing competition between LeConte and the society accelerated the growing competition for entomological specimens from the trans-Mississippi West. In 1869, for example, LeConte wrote to Belfrage: "Mr. Ridings [of the Entomological Society], for some reason unknown to us will neither sell to us, nor permit us to see what he received from you. Perhaps if your collections passed through our hands, after taking what we need . . . we could obtain for the remainder as much as you now get for them. . . . we would be glad to make any arrangement that is agreeable to you to secure first choice."[89] LeConte attempted, apparently with little success, to organize a boycott of the Entomological Society. He wrote to Edwards, "As there are no students in this city competent to name any insects except Coleoptera a refusal to name specimens on the part of those engaged in the study of other orders (vis. yourself, Clemens, Norton, Osten-Sacken, and Uhler) would necessitate at once a change of policy."[90]

The deteriorating relationship between LeConte and the society broke out in open hostility over a remark that LeConte placed in a footnote in his report on the Pearsall collection. Referring to the insects he was not allowed to remove from the cabinet, LeConte wrote that he "regretted that the most valuable portion of the [Pearsall] collection . . . has been rendered . . . unavailable for scientific research, in the restrictions placed . . . by the Entomological Society."[91] On January 13, 1862, Newman, the society's president, offered a resolution expelling LeConte on the ground that this note constituted a "flagrant act of disrespect" to the society. The resolution passed (over Wilson's plea against hasty action) with only three votes in opposition.[92] LeConte wrote to Baird of the action:

Dr. Wilson and parasites have gotten into a funk about . . . the Pearsall Collection . . . and . . . a long series of cloudy sentences were presented to expel me. . . . if Dr. Wilson wants fight, I am ready for it, but it appears to me that the action . . . is very childish. They propose to insult everyone who will not act in accordance with Dr. Wilson's monomania for collecting stuff which no one is prepared to study and which no one could study if he wished owing to the restrictions placed in his way.[93]

LeConte was ostracized from the society for ten years following his expulsion. In 1872, when altered conditions permitted a reconciliation among the Philadelphians, LeConte was again elected president of the society.[94]

Meanwhile, Cresson led the society's continued effort to have Smithsonian insects deposited with it. In 1864, Henry agreed to a proposition made by the society whereby two series (consisting of a male and female of each species) would be selected from those forwarded by the Smithsonian. One series would be labeled and returned to the Smithsonian, and the other would remain with the Entomological Society, subject to its own restrictions. Credit would be given for all Smithsonian contributions, and all extra insects would be returned to the Smithsonian.[95]

For a time the society's collection prospered through donations from members and from the Smithsonian and through the success of the society's collecting expeditions, like the one to the Rocky Mountains in 1863. Donations included types from Uhler, Cuban insects purchased by Wilson, specimens collected by Ridings in Colorado Territory and "presented" by Wilson, and larvae of Diptera from Osten-Sacken.[96] The society's hope of establishing a national reference collection collapsed when Wilson died in 1865, however, leaving the society (and Cresson) in serious financial difficulties. Wilson's death came just as he was preparing to endow the society with shares in the Pennsylvania Railroad, which would have made the society financially independent.[97] Cresson, who took a position with the Philadelphia Fire Department, continued as secretary, but he could no longer serve as a full-time curator for its collection. In 1876, the Entomological Society was reorganized as a section of the Academy of Natural Sciences, and thereafter its collections were housed in the academy building.[98] Despite these difficulties, the society took the first tentative steps toward a national reference collection, and through its efforts the Academy of Natural Sciences became the recipient of an important scientific collection that has served entomologists from the 1860s to the present. Because of the conflicts and because of the loss of Wilson's patronage, however, the society failed to attract the majority of private collections of American entomologists during the critical decades of the 1860s and 1870s. The opportunity for assembling the first comprehensive reference collection of American insects thus passed from Philadelphia to Boston.

The Philadelphians marveled at Boston's lavish support of science. In 1865, Cresson, seeking new sources of funding following Wilson's death, wrote to Scudder, "If it is not an impertinent question, how did the Boston Soc. Nat. Hist. go about getting money from the merchants of Boston and also the Community at large? . . . The Merchants of Philadelphia are much more 'close fisted' than those of Boston, and don't seem to like the idea of 'bugs' at all."[99] Boston's generous support of science was amply displayed in Louis Agassiz's Museum of Comparative Zoology at Cambridge. Agassiz, the protégé of Georges Cuvier and Wilhelm von Humbolt, had earned fame and fortune with his studies of fossil fishes, his theory of the ice age, and the founding of his scientific research institute at Neuchâtel, Switzerland.[100] From the time of his arrival in America and his triumphant Lowell Lectures in 1846, Agassiz had charmed and impressed Boston's social and financial elite, appearing to them the personification of the dedicated scientist, cosmopolitan European, educator of the masses, and hardheaded, efficient scientific administrator.[101] Material evidence of Agassiz's ability as a fund raiser was evident in an endowed chair at Harvard's Lawrence Scientific School, $300,000 in subscriptions (mostly from Boston industrialists) for the publication of a projected ten volumes of "Contributions to Natural History," and finally the $50,000 willed by Francis C. Gray and matched by $100,000 from the state and $71,000 in private contributions for the Museum of Comparative Zoology.[102]

"My great object is to have a museum . . . which will equal the great museums of the Old World," announced Agassiz upon the founding of the museum in 1859.[103] Within a decade his dream had been fulfilled. Beginning with his own collections, Agassiz set about building the museum's holdings through purchase, exchange, and the solicitation of donations. The best means to raise the museum to first rank, said Agassiz, was to "purchase . . . private collections which contain the labors of distinguished naturalists[104] . . . [and] to obtain well authenticated specimens of species."[105] The lack of reliable standards, he said "has been a serious obstacle to . . . American naturalists."[106]

From the outset, the Museum of Comparative Zoology had a series of able curators for the insects: Scudder (1859–1863), Packard (1863), Uhler (1864–1866), and Herman August Hagen (1867–1892). Until 1866, the curators were occupied with the acquisition, inventory, and storage of specimens. That year Uhler completed the arrangement of the entomological collections (having compared them with the specimens in Philadelphia) so that they could be displayed and used for reference.[107]

In 1867, the ascendance of the museum's entomological collection over that of the Entomological Society of Philadelphia was assured when Agassiz hired Hagen, a professor at Königsberg, Prussia, and one of Europe's foremost entomologists, to take charge of the museum's entomological department.[108] Under Hagen's direction, new boxes and cabinets modeled after the Berlin Museum were built, and the entire entomological collection was reorganized, cleaned, and repacked to secure it "for all time."[109] With Agassiz's reputation and financial backing and the presence of a permanent entomological curator like Hagen, the museum attracted the majority of the entomological collections from American entomologists between 1864 and 1885. The most important were those of Melsheimer and Ziegler (1864), Osten-Sacken (1871), and LeConte (1875).[110] In addition to these complete collections, the museum acquired specialized collections like the gall insects of Benjamin D. Walsh and Osten-Sacken, the Texas Lepidoptera of Jacob Böll, and the Curculionidae beetles of the French entomologist Lacordaire. By the mid-1870s, the museum was the outstanding repository in such areas as North American Coleoptera, Diptera, microlepidoptera, gall insects, and biological material.[111]

The museum's reputation and stability served to reverse the flow of specimens from America to Europe. One striking example was the return of F. Hermann Loew's American Diptera from Germany to the

United States. Osten-Sacken, a member of the Russian delegation in Washington, D.C., and New York City from 1856 to 1877 and a leading specialist in Diptera, collaborated with Loew, of Germany, the world authority on Diptera. Osten-Sacken sent Loew American specimens under an agreement that they would be returned to the United States when a suitable repository was found. In 1877, having become satisfied that the Museum of Comparative Zoology was the most suitable place, Osten-Sacken returned to Europe and supervised the shipment of Loew's American Diptera to the museum.[112]

The presence of an expanding entomological collection at the Museum of Comparative Zoology served as an important catalyst in the formation of the Cambridge Entomological Club, which was organized at a meeting in Hagen's house on January 9, 1874. Scudder, who studied under Agassiz and worked as his assistant until 1864, was elected as the first president of the club, and he was a seminal influence in its activities until his death in 1911. In the 1860s and 1870s, he specialized in the study of Lepidoptera, Coleoptera, and cave insects, and he reported regularly on his studies at the meetings of the club.[113] The club's publication, *Psyche,* featured biological contributions on Arthropoda, lists of captures, and a bibliographical record of all writings pertaining to North American entomology.[114] From its founding in 1874 to the present, the Cambridge Entomological Club and the personnel associated with the entomological collection of the Museum of Comparative Zoology have maintained a close working relationship.[115]

Despite the growing success of the Museum of Comparative Zoology, many entomologists and others, like Baird, still felt there was a need for a great national entomological collection located in Washington, D.C. Apparently Baird's decision about 1860 to retain for the Smithsonian a series of all insects (rather than distribute all specimens to the Academy of Natural Sciences and to Smithsonian collaborators) was made in response to the requests of the Entomological Society of Philadelphia for specimens from the Smithsonian. Up to 1859, the year when the Entomological Society of Philadelphia and the Museum of Comparative Zoology were founded, Baird followed a policy of placing entomological specimens on permanent loan with the Smithsonian collaborators who wrote the reports. That year Baird wrote to Scudder: "At present we do not ask insects for ourselves. Your contributions to Osten-Sacken and others who are preparing works for us will answer all our immediate purposes."[116] Four years later, while the dispute with the Entomological Society was in progress, he informed Scudder that a full series of each species, including sex, age, and geographical distribution, must remain at the Smithsonian.[117] It is not clear, however, who selected the specimens, who was in charge of them, where they were stored, and how many of the specimens from the 1860s and 1870s "survived."[118] In 1874, Baird, after many years of unsuccessful effort in maintaining an insect collection at the Smithsonian, began depositing the Smithsonian specimens in the Agricultural Department collection.[119] Baird considered this a temporary expedient, however, and vowed at some point to return the entomological collection to the Smithsonian. He wrote to LeConte, in 1881, saying "What you have already turned into the Agricultural Department may have gone to dust. I hope to make a new collection in the future."[120] Baird's goal of a national entomological collection was eventually realized through agricultural support for entomology, which fostered the growth of such a collection first in the Department of Agriculture and eventually in the Smithsonian Institution.

The entomological collection in the Department of Agriculture originated with the work of Townend Glover, an English immigrant turned gentleman farmer, who was appointed to the patent office in 1854 for the purpose of "collecting statistics and other information on seeds, fruits, and insects."[121] Having trained as an artist in England and on the Continent, Glover took little interest in specimens as such, believing that drawings were as good as specimens. Much of his energy was channeled into the production of an encyclopedic set of illustrations of American insects, which he hoped to see published for distribution to agricultural societies and colleges. The distance between Glover, the artist and applied entomologist, and the entomological systematists who treasured type specimens can be gauged by Glover's disdain for "species grinders" and by the fact that during his entire career he did not describe a single new species. As a result of Glover's attitude toward specimens, the departmental collection was haphazardly arranged with boxes of uneven sizes containing broken pins and indistinct labels. Despite his neglect of entomological specimens, Glover envisioned a "Grand Agricultural Museum," with national, state, and economic sections. Though he spoke often of the museum, Glover was not adept at the politics necessary to gain congressional appropriations. Charles R. Dodge, an assistant who later wrote an admiring biography of his chief, said that the museum's most interesting exhibit was Glover himself. When a congressman or someone with influence happened through the door, the slightly built man would hop up from his desk and lead the visitor to the corner of the room where the plan for the museum was laid out. After carefully explaining each part, Glover would step back and ask, in a phrase familiar to those in the department, "There, how do you like the Plan?"[122]

With the appointment of Dodge as Glover's assistant, in 1867, the entomological collection of the department began to improve. The department's holdings steadily swelled with specimens received from agricultural correspondents. Following Baird's decision in 1874 to place the Smithsonian insects in the Department of Agriculture collection, the department collection grew even faster. By 1876, the collection was substantial enough for Glover and Dodge to prepare an impressive insect exhibit for the department's display at the Centennial Exposition at Philadelphia.[123]

Though Glover, Dodge, and Baird had made a substantial beginning by the late 1870s, the real impetus for a national insect collection came from Charles Valentine Riley, the former state entomologist of Missouri, whose work with the Rocky Mountain locust and other injurious insects brought him to Washington to head the U.S. Entomological Commission and made him a driving force within the Department of Agriculture.[124] Arriving in Washington in 1877, Riley intended, among other things, to found a national insect collection that would be unsurpassed in the country.[125] In 1881, upon completion of a new building for the Smithsonian Institution, Baird reorganized the Smithsonian museum, and this work gave Riley his opportunity.[126] Baird made several rooms available to Riley for his own extensive collection and appointed him honorary curator of the Smithsonian insects.[127] For a while, Riley was hesitant to trust his collection to the Smithsonian, but his initial doubts were overcome by his belief that the Smithsonian must now take its place as the central repository of the nation's insect treasures. In 1885, he wrote to Baird: "While the future of any institution dependent on congressional support may not be as certain as one supported by endowment, I make this donation in the . . . belief . . . that the National Museum . . . must inevitably grow until it shall . . . surpass other institutions . . . as a repository of natural history collections."[128] That year Riley formally donated his collection to the Smithsonian.[129]

The collection, which had been assembled over the previous forty years, was estimated at the time to be the largest general collection in the country (though several collections surpassed it in single orders), and the most extensive in biological material, including the larval and pupae stages of 3,000 species of economic importance. The collection, containing 30,000 species and 150,000 specimens, was housed in 800 boxes (as compared to 623 boxes already in the Department of Agriculture collection).[130] Riley proceeded to purchase other collections for the U.S. National Museum, some with his own money and some with funds made available from the Department of Agriculture.[131] The nucleus formed with the departmental collection, Riley's collection, and subsequent purchases began to attract the hoped-for donations. By the close of the 1880s, the dream of a truly national insect collection at the Smithsonian was well on the way to realization.

The establishment of permanent insect collections in Philadelphia, Cambridge, Boston, Washington, D.C., and elsewhere marked a quiet but vital transition in American entomological science. These collections, cared for by paid curators, ensured American entomologists of access to comprehensive, accurate information, which was necessary for their science. In an important sense these collections culminated the era of institution building. Following the example of European entomologists, the Americans constructed a system to store, retrieve, and compare information about insects. These collections represented the handiwork of a major community of American scientists, with institutions now comparable in every way to their European counterparts.

4 Agricultural Entomologists and Institutions

While American entomology owed much to the European example and influence, the most distinctive characteristics of the emerging community derived from its close association with American agriculture. To some admirers, the breathtaking expansion and dynamic transformation of American agriculture in the nineteenth century seemed proof that a Divine Providence was assisting American farmers in the fulfillment of the American mission of progress and plenty in the New World. While American faith in agrarian progress may have been overstated, what does seem clear is that the dynamics and requirements of agriculture in the context of the North American environment propelled American applied entomology to the front rank of world science.

Three key aspects marked this transformation of the American entomological community. The first was a shift in personnel away from self-supported "amateur" investigators to paid professionals. The second was an increased emphasis on public service and practical applications of entomological knowledge. The third was an increase in the level of funding for agricultural entomology that placed the American community in a dominant position in the world. Taken together, the most distinctive aspect was the development of agricultural entomology primarily as a North American institution rather than as an extension of European science. Though Europeans developed similar practices, the American profession predominated in terms of total financial support, numbers of people employed, numbers of publications, range of investigations, and successful applications. This transformation of American entomology from amateur to professional status was part of the most distinctive development in nineteenth-century western science, the replacement of gentlemen scholars by professional scientists as the dominant members of the scientific community.[1]

American predominance in agricultural entomology came about through the interaction of two features within the social context that might be termed "supply" and "demand."[2] As has been seen, on the "supply side" there had emerged, around midcentury, a community of entomologists in North America who were dedicated to advancing their science according to the standards of post-Linnaean systematics. These entomologists developed their own information systems, societies, and institutional structures based largely on European models. The "demand" side came primarily from a new leadership among agriculturists and horticulturists, who presided over the commercialization and mechanization of agriculture and horticulture throughout the northern states and Canada. Their goal was to increase production through expansion of acreage and the mechanization of agricultural methods while at the same time controlling production costs. Collaboration between the two groups developed where their interests coincided, that is, in the investigation of insects that damaged crops and orchards (thus adding to production costs) and in the search for methods to control this damage.

Entomological leadership in this collaboration was assumed by fourteen individuals who sought and eventually gained support for their applied entomological investigations. These were Albert John Cook, Asa Fitch, Stephen Alfred Forbes, Townend Glover, Thaddeus William Harris, William LeBaron, Joseph Albert Lintner, Manley Miles, Alpheus Spring Packard, Jr., Charles Valentine Riley, Francis G. Sanborn, William Saunders, Cyrus Thomas, and Benjamin D. Walsh. Although they lived in widely separated locations and belonged to distinctly different age groups, they developed a common sense of mission in the service to agriculture and horticulture.

Like others aspiring to become professional scientists in midcentury America, the agricultural entomologists came mostly from families in the traditional professions (medicine, law, the church, and college teaching) or in business. Most of them studied for a profession like medicine but practiced it only in transition to a career in agricultural entomology. Their financial resources were typically sufficient to allow them the leisure to devote some months or even years to the collection and study of insects; however, none had sufficient financial means to continue these pursuits indefinitely.[3]

One-third of the agricultural entomologists were born and educated in England (Glover, Riley, Saunders, and Walsh). While many of these men acquired early interests in entomology in Europe, their expertise in agricultural applications developed after they arrived in America. Of the American born and educated, the majority took degrees in medicine, the usual training for antebellum Americans in the geological and biological sciences.[4]

The most influential mentor and role model for the agricultural entomologists was Thaddeus William Harris. He exerted his influence primarily outside the formal classroom, a circumstance not of his choosing. Born into a minister's family with strong ties to Harvard College, Thaddeus William studied medicine and practiced for a few years but soon returned to Cambridge, where he became the Harvard librarian. In 1842, having served as librarian and lecturer in natural history at Harvard College, and having published his influential *Report on the Insects of Massachusetts,* Harris competed unsuccessfully with the botanist, Asa Gray, for the professorship of natural history. Thereafter, until his death in 1856, Harris served as the Harvard librarian, while continuing his entomological pursuits during evenings and weekends. It is interesting to speculate how agricultural entomology might have developed had Harris won the appointment at Harvard and had he had the opportunity to train a generation of students in the same way as Asa Gray trained botanists, as Louis Agassiz trained zoologists, and as Justus Liebig trained agricultural chemists.[5]

Restricted as he was by his library duties, Harris still continued to exert a decisive influence. He set the scientific standard in his 1841 *Report,* corresponded widely, loaned reference works, and provided access to his collection to those who visited him in Cambridge. Among those who benefited directly from his encouragement, aid, and advice were Fitch, LeBaron, and Glover.[6] Through his writing, primarily the *Report on Insects* (which was reprinted as a "treatise" in 1842, 1852, and 1860), Harris influenced the career decisions of many other agricultural entomologists, including Packard, John Henry Comstock, and Leland O. Howard.

The agricultural entomologists came from three distinct generations. Harris, who was born in 1795 and began publishing in agricultural entomology in 1828, stands apart as the original prototype of the profession. The first generation (excluding Harris) included Walsh, Fitch, Glover, and LeBaron, all of whom were born around 1810 and began publishing in the 1840s. The second generation consisted of Lintner, Thomas, and Miles, who were born in the 1820s and who began publishing in the 1850s and 1860s. Members of the third generation, which included Riley, Packard, Saunders, Sanborn, Cook, and Forbes were born around 1840 and began publishing in the 1860s and 1870s. Although coming from different generations, these individuals as a rule paid increasing attention to agricultural entomology after 1840.

The daily round of activity developed by the agricultural entomologists developed into a typical pattern, one marked by compulsive, organized activity. Everything had to be done at once and done correctly: specimens collected at the right time and season, broods of insects raised from eggs and the results recorded, insect cabinets built, specimens arranged and safeguarded, specimens examined by microscope, specimens exchanged with correspondents in America and Europe, reference books obtained and studied, publications prepared, and so on. The agricultural entomologists not only maintained scientific contacts, but also established relations with agricultural and horticultural groups, with editors of farm journals, with farmers, and with state and federal legislators, all of whom ultimately had some say in the support of entomology.[7] Fitch maintained that an entomologist needed "as many limbs as a centipede" to carry out these myriad tasks.[8] He coped with this problem by cutting his sleep to five hours a night and even fewer hours when observing insect broods as they hatched in his cages. Fitch also developed the practice of borrowing major reference works and copying them by hand, tasks that required months of unremitting labor.[9] The strain of such activity often took its toll. Fitch spent so many hours at the microscope that his left eye was permanently distorted. Riley worked so compulsively that he suffered a nervous collapse early in his career. Later in life, Riley was plagued by insomnia, and he found that sound sleep came only when he sat in a barber's chair (he often paid his barber extra to allow him to sleep for hours in the shop!). Yet the general impression one gains from their correspondence is that the agricultural entomologists viewed this kind of sustained energy as the price they willingly paid to further their chosen specialty.

Given the importance of religion in nineteenth-century America, one might reasonably expect that such intensity derived from religious piety. Asa Fitch took religion seriously throughout his life, and Walsh and Thomas both studied for the ministry.[10] Yet with the exception of Fitch, the public and private writings of the agricultural entomologists are almost devoid of religious references, in striking contrast to the religious terminology that permeates the correspondence of agricultural chemists like Evan Pugh and Samuel William Johnson.[11] In general, the attitude of the entomologists seems closer to the rationality of the European Enlightenment. Riley, who was European born and educated, scoffed when his fellow entomologist Francis G. Sanborn tried to convert him to "religion." In his view, Sanborn's religiosity rendered him an object of pity.[12] Walsh once wrote to Herman A. Hagen, "in science we want to eliminate faith as much as possible."[13] Likewise, the writings of the entomologists carry relatively few references to "design," the idea that order in the natural world proves the existence of a Creator. Only Harris (whose father was a minister) used the terminology, and these occasional references seem techniques of style rather than intellectual arguments.[14] It is noteworthy that when Darwin's *Origin of Species* appeared in 1859 Walsh, Riley, and other agricultural entomologists became early and vigorous champions of the theory of evolution.

The agricultural entomologists' basic self-identification was with the American "entomological fraternity," a term by which American entomologists were starting to designate their own community.[15] While developing their special agricultural emphasis, they played an active role in the development of major entomological collections in Philadelphia, Cambridge, and Washington, D.C. They also took the lead in developing new societies, notably the Entomological Society of Canada and the Entomological Club of the American Association for the Advancement of Science but always within the context of the larger entomological fraternity. Not until the passage of the Hatch Act of 1887, which "created" a demand for agricultural entomologists serving at land grant experiment stations, did agricultural entomologists consider themselves distinct from nonagricultural entomologists. The new self-image called forth a new society, the American Association of Agricultural Entomologists (1889).[16]

While maintaining the systematic rigor and pure science ideals of the "entomological fraternity," agricultural entomologists defined their special mission as assisting farmers and horticulturists. Packard summarized the distinction in these words: "the work of classification is by no means the highest. . . . He who studies . . . the habits and structure of one insect . . . is a true benefactor to agriculture, more than he who describes hundreds of new species."[17] Choosing as their subject matter those species that were harmful to cultivated crops and orchards, agricultural entomologists investigated their systematic classification and the natural conditions and biological mechanisms that regulated their populations as the basis for their control.[18] In contrast to systematists and collectors, who concerned themselves primarily with adult insect specimens, agricultural entomologists often paid more attention to insect larvae, which were at the stage when insects often inflicted the most damage.[19] As agents of agricultural improvement, they linked the venerable tradition of studying insect life histories to the scientific and material progress of their respective states, provinces, and nations. Harris (and his mentor William

Dandridge Peck) had pioneered this blending of traditions in the early years of the century. As agricultural change intensified after 1840, agricultural entomologists identified their work more closely with material and scientific progress. By placing their studies at the strategic intersection of science and the material progress of farmers and horticulturists, the agricultural entomologists eventually succeeded in establishing an institutional structure within which they could carry out their investigations.

The agricultural entomologists spoke in concert with the advocates of agricultural improvement who represented the spread of commercialized and mechanized agriculture in the northern states and Canada. The agricultural revolution began in the 1840s, when railroads and canals connected the eastern urban markets with the new states and territories of the West. Under the stimulus of larger markets and greater productive capacity, northern agriculture shifted quickly from subsistence to commercial farming, accompanied by heavy reliance on investment capital, machinery, and substantial increases in production per farm laborer. The Civil War accelerated the commercialization and mechanization of northern agriculture by drawing off much of the farm labor force and at the same time increasing the wartime demand for products like pork, wheat, and wool.[20]

The spread of commercial wheat production on successive frontiers provided the cutting edge of the new system. In 1840, the main wheat-growing region was located in western New York state, Pennsylvania, Virginia, and eastern Ohio. By 1860, the center of wheat production had moved west to Illinois, Indiana, and Wisconsin, which together produced 50 percent more than the four eastern states combined. After 1865 the wheat belt moved farther west to Minnesota, the Dakotas, Kansas, and Nebraska.[21]

Eastern agriculture meanwhile shifted to the production of hay, fruit, and garden produce for nearby urban markets. By 1859, the leading states in horticultural production were New York, Pennsylvania, Ohio, Indiana, Illinois, and Michigan. As the wheat belt moved farther west, the older prairie region centering in Illinois, Iowa, Wisconsin, and Michigan shifted the emphasis to a corn–cattle–hog production complex and to dairying and horticulture.[22]

Each aspect of agricultural change intensified the problem of insect pests. As farmers turned from subsistence to market-oriented crops, they increased the practice of monoculture, which created a favorable environment for the multiplication of insects that fed on those crops. The continuous production of a commercial crop like wheat depleted soil resources and weakened the crop resistance to disease and insect pests.[23] Producers in frontier regions typically witnessed a drop in production about ten years after an area was first farmed.[24]

Mechanization contributed to the problem of insect pests by increasing the cultivated acreage per person, thus rendering traditional hand control methods impracticable. In central Europe, where farms were small and labor plentiful, workers could be hired to "uncaterpillar" plants by hand, a term that existed in German *(abraupen)* and in French *(décheniller)* but never entered the American lexicon.[25] Machinery represented greater capital outlay that made American farmers more dependent on credit and therefore less capable of sustaining the losses of commercial crops to insects. Farmers with a heavy investment in equipment were among the most active supporters of applied entomology as a way of protecting these investments.[26] The mechanization and commercialization of agriculture meant that insects were now serious competitors for producers' pocketbooks.[27]

Agricultural change prepared the way for the support of applied entomology in diverse ways, by stimulating the growth of agricultural societies, the agricultural press, special interest lobbies, and agricultural colleges and experiment stations.[28]

Institutional support for agricultural entomology came initially from the agricultural societies. The early agricultural societies, organized by gentlemen farmers and urban landowners of the eastern seaboard cities in the late eighteenth and early nineteenth centuries, were clearly out of place in the world of rapid agricultural change that prevailed after 1840. The New York State Agricultural Society, founded in 1833, was the first to found its philosophy and practice on agricultural change, and it became the prototype for successful agricultural organizations after 1840. The movement for agricultural reform began around Albany, where farmers were threatened by a combination of western competition, soil exhaustion, and insect pests. The Albany farmers demonstrated a willingness to experiment with techniques like the use of fertilizers and diversified farming. The movement for agricultural reform profited from the leadership of able agricultural editors like Jesse Buel of the *Cultivator* and Luther Tucker of the *Genesee Farmer,* whose papers had the widest circulation of any agricultural press in the world. In 1841, the New York State Agricultural Society secured annual state funding to finance its *Transactions,* its county agricultural societies, and its county fairs. Other state agricultural societies soon organized along the New York pattern. Led by editors of popular farm journals and progressive farmers, these societies established strong political lobbies in the state capitals and secured state funding for their activities, including support for trained agricultural experts like chemists and entomologists.[29]

The most effective lobbyists for applied entomology were the commercial fruit growers, or pomologists. The diversification of American agriculture stimulated commercial horticulture and imparted renewed vigor to horticultural organizations under leaders like Andrew J. Downing and Marshall P. Wilder. In 1850, horticulturists in New York and on the eastern seaboard, responding to the need for standard nomenclature for commercial varieties and for communication and coordination among the commercial growers, organized the American Pomological Society, the largest such organization in the world. As horticultural specialty crops reached substantial volume in the Midwest, state horticultural societies were organized there. The Ohio Pomological Society was organized in 1847, the Illinois State Horticultural Society in 1856, and the Missouri Fruit Growers Association (later the Missouri State Horticultural Society) in 1859.[30] Of the horticulturalists, Benjamin D. Walsh, the state entomologist of Illinois, remarked, "The Fruit-growers, as a class, are a remarkably intelligent and well educated set of men. One fruit-grower would cut up into a hundred grain-farmers."[31]

The horticulturists were understandably concerned with damage from insects. They had to deal with a host of specialized pests, like weevils, borers, bark lice, tent caterpillars, and codling moths. They also had more to lose. The capital and labor invested in an orchard represented a long-term financial commitment. Unlike a field crop, an orchard could not be plowed up and replanted, nor could the acreage be expanded within a season to take advantage of favorable weather conditions or high crop prices.[32]

By the 1850s, both the entomologists and the agricultural and horticultural groups had developed an effective leadership and institutional structure. Leaders of each group called for more effective application of entomological science to agricultural improvement. The effectiveness of the combined groups was aided by their common outlook and by the journalistic and oratorical skills employed by key leaders.

The common outlook developed among leaders of both groups drew strength from the farming and horticultural experience of several key entomologists. Fitch and Glover owned large orchards in New York state, Cook managed a fruit farm in California, Saunders raised fruit and berry crops in the province of Ontario, and Walsh and Riley farmed in Illinois. Furthermore, there was no educational barrier between expert and client, since the entomologists employed observational methods of investigation and reporting and relied heavily on the observations of nonspecialists in their work.

The effectiveness of the two groups was increased also by the journalistic and oratorical abilities of the agricultural entomologists. Harris, Fitch, Walsh, Riley, LeBaron, Lintner, Thomas, and Packard all contributed regularly to the agricultural press. Walsh began his career as a journalist in England. In 1865–1867, he applied his skill as editor of the *Practical Entomologist,* published by the Entomological Society of Philadelphia. In 1868–1870, with Riley as a partner, he edited and published the *American Entomologist.*

Oratorical skills were a particularly important means of persuasion in rural nineteenth-century America. Walsh and other entomological speakers perfected the art of exhibiting insect collections and delivering popular lectures (as Walsh said) like "itinerant showmen" at state and county fairs and other gatherings.[33] In 1855, Fitch had spoken before the New York State Agricultural Society and helped persuade the society to press for continued funding for his investigations.[34] Glover entertained audiences with his model fruit display, which included exhibits of insects that attacked each fruit. Walsh was such an entertaining speaker that he held the attention of audiences at fairs and horticultural meetings for hours at a time.[35] Riley drew large colored drawings to illustrate his talks.[36] When teaching in a classroom where a blackboard was available, he often drew insect illustrations with both hands as he lectured. In Walsh's opinion, a willingness to enter into public controversy was necessary if entomology were to move from its sheltered seclusion to a public service function.[37] Fitch shared that willingness, even to the extent of running for state office himself if necessary.[38]

The entomological, agricultural, and horticultural leaders agreed on the seriousness of insect depredations. Beginning in the 1840s, various farm sectors suffered from major insect infestations. Wheat crops in some regions were entirely destroyed by the Hessian fly or the wheat midge, and orchards of soft-skinned fruits like plums and pears suffered increasing damage from weevils of the curculio family. In the 1840s, the chinch bug began ravaging western corn and grain crops, and in the 1860s, the Colorado potato beetle began its destructive march to the East Coast. Estimates of the economic losses due to insects lacked the sophistication of a later day, when such concepts as the "sacrificed opportunity" entered the agricultural economists' vocabulary. In 1868, Walsh and Riley estimated the annual losses in the United States at $300 million.[39] There was agreement, however, that the losses were (1) much greater than generally realized, (2) above acceptable levels considering the economic investment and expected rate of return, and (3) subject to substantial reductions through the support of entomological investigations.

Beginning in the 1840s, entomologists and agriculturists proposed three institutional changes to provide support for those who could conduct these investigations. The first was an effort to link entomological research to the state surveys. The second was the appointment of state (or federal) entomologists by the agricultural and horticultural societies independent of the surveys. The third was the establishment of positions for entomological teaching and research in agricultural colleges.

The state and federal surveys were the focus of what one historian has termed an "anthill of investigatory activity" under way in Jacksonian America.[40] Sustained by faith in the Baconian scientific tradition that all knowledge would lead ultimately to material and scientific progress, the states (and increasingly) the federal government compiled inventories of available resources. In the mid-1830s, the scientists and assistants employed in the surveys coalesced into a distinct new group of professional scientists in America. Geologists associated with the state surveys led the way, developing into a distinct professional group about 1836.[41] The state surveys thus seemed to offer fertile ground for the growth of agricultural entomology as well, and in some cases these hopes were fulfilled. The entomologists soon came to view the surveys with suspicion, however, because they emphasized the search for mineral resources. Whereas mineral surveys offered quick returns with a minimum of expense, entomological research required long-term commitments with no promise of immediate, spectacular results.[42]

Harris's experience with the Massachusetts survey in the 1830s illustrates the difficulty of bringing about institutional change in connection with the surveys at a time when, despite entomological leadership, agricultural change had not yet proceeded sufficiently for the agricultural community to require professional appointments. Edward Hitchcock, a geologist from Amherst College who directed the survey, appointed Harris as a zoological commissioner to report on the state's insects. The zoological work of the survey, which was strictly subordinate to the search for mineral resources, was so limited that Harris and the other two zoological commissioners had only enough funds to compile lists of species, without investigating their economic importance. Harris's catalog, published in 1833, advanced the systematics of Massachusetts insects, but it satisfied neither him nor the agriculturists, who had hoped his work would expand his investigations of economically important insects.[43] When the Massachusetts legislature extended the zoological and botanical work, with the stipulation that the survey personnel "keep . . . in view the economic relations of every subject," Harris went on to produce his *Report on the Insects of Massachusetts, Injurious to Vegetation* (1841), a monograph that detailed the taxonomy, life histories, and habits of the most economically important species and suggested appropriate control measures. Though still wedded to the natural history tradition in its organization, with insects discussed sequentially, according to their taxonomic order, Harris's work broke new ground by emphasizing the practical control of insects. Harris's *Report* was the first book on agricultural entomology in America and one of the first monographs on agricultural entomology in the world.[44] It exerted an enormous influence on the midcentury generation of American entomologists. Harris and the Massachusetts survey may thus be considered transitional. The commitment to applied entomology clearly set Harris apart from the natural history–taxonomic tradition that predominated in the Entomological Society of Pennsylvania, yet despite Harris's contribution, neither the needs of the agricultural community nor its leadership was sufficient to secure his continued services either as a full-time professor (as he hoped) or as a state entomologist.[45]

The New York Natural History Survey, established in 1836 as the most ambitious of the state surveys, demonstrates the growing influence of the agricultural and horticultural interests. The New York State Agricultural Society supported the bill calling for the survey but soon found these concerns subordinated to the industrialists' search for coal deposits. The report by the survey zoologists, Ebenezer Emmons and James DeKay, omitted the insects altogether.[46] Upon completion of the survey in 1842, the agricultural society joined with other groups to secure funding for a continuing agricultural and paleontological survey and the creation of a state cabinet of natural history. The agricultural society favored either Fitch or Haldeman to conduct the agricultural survey; however, Emmons (one of four survey geologists) summoned sufficient political influence to win the appointment as agricultural commissioner.[47] Emmons produced a work on the agriculture of New York, with volumes on soils, climate, and fruits and a final one on insects, published in 1854. When the volume on insects appeared, it was criticized by both the agriculturists and the entomologists. The agriculturists criticized it because it lacked attention to life histories and control measures. The secretary of the Agricultural Society charged that the volume was "too purely scientific [and that it would] be of little consequence to know that a particular kind of moth or fly is an inhabitant of this state, unless we are also informed of its history and habits, and whether it is a depredator upon any substance which is of value to man."[48] The primary criticism from entomologists like Haldeman and LeConte was the report's incorrect identification of insect species and its poorly drawn insect illustrations.[49]

The entomologists who criticized the report included Asa Fitch, a gentleman farmer and naturalist living at Fitch's Point, near Salem, New York. Fitch had studied with Amos Eaton and Emmons at Rensselaer College in the 1820s, had then gone on to earn an M.D., and had practiced medicine for a few years before returning to Fitch's Point to run the family farm. Fitch was fascinated with natural history. By the early 1840s he had decided to make entomology his life's work. Following the example of his father, Fitch assumed an active role in the growing activities of the county and state agricultural societies in the 1840s.[50] Although Fitch had assisted Emmons by supplying him with some information for the report on insects, he joined other entomologists in criticizing the final version of the report.[51] He clearly considered Emmons, whose primary interests lay in geology, as an outsider to the "entomological fraternity" and therefore unqualified for the exacting work required in agricultural entomology.

Meanwhile, in 1847, the agricultural society secured state aid to sponsor a series of county agricultural surveys. The survey of Washington County, written by Fitch, served as a model for other surveys.[52] The report featured accounts of insect depredations on crops and orchards and discussion of the controls available.[53] The agricultural society considered Fitch's investigations so important that, in 1854, its leaders introduced a bill in the New York legislature to appropriate one thousand dollars per year to support his investigations on a continuing basis. The bill as passed into law authorized the agricultural society (rather than the survey commissioners) to name the entomologist. Fitch was appointed immediately, and the comprehensive instructions for the entomologist probably reflected his own hand. The entomologist was instructed to investigate insect pests, to make annual reports, and to assemble an insect collection for the State Cabinet of Natural History.[54] Fitch's first report, in obvious deference to the horticulturists who assisted in this appointment, was a comprehensive account of the insects infesting fruit trees.[55] When Fitch assumed his new position, he became the first full-time professional entomologist in the United States.[56]

Fitch's fourteen annual reports, published from 1855 through 1872, differed in some important respects from Harris's *Report*. Following in the public service tradition of Amos Eaton, his teacher, Fitch endeavored to make the reports eminently useful to farmers. In a break with Harris's style of organization by taxonomic order, Fitch arranged his accounts of insect pests according to the visual damage to plant or crop. For example, he described insects that attacked the roots of apple trees, then insects that attacked the apple tree trunk, then the blossom and the fruit. First he described the symptoms an orchardist might observe, then the insect that caused the condition, and then the remedies available to rid the orchard of the pest. This method of organizing reports was generally followed by later agricultural entomologists.[57] Fitch's early reports, building on Harris's pioneering report, placed the United States in the position of world leadership in applied entomology.[58]

During the 1840s, while Fitch developed his program of agricultural entomology in conjunction with horticulturists and agricultural reformers around Albany, Townend Glover was developing a similar program in conjunction with horticulturists along the Hudson River Valley. An immigrant from England in 1836, Glover roamed the country for a few years before marrying Sarah T. Byrnes and settling down as the manager of her father's large estate on the Hudson River. Glover divided his time between farming, fruit culture, the study of insects, and (his special avocation) the modeling of fruits. Over a period of half a dozen years, assisted with funding from the New York State Agricultural Society, Glover prepared a display of over two thousand model fruits, accompanied by etchings of insects that destroyed fruit orchards. Through his model fruits and etchings, Glover became well known among New York pomologists. It is curious that, although Glover and Fitch lived in the same state and knew many people in common, the two apparently had little or no direct contact with each other. A. J. Downing, editor of the *Horticulturist,* proposed that Glover prepare a complete volume of engravings of insects for publication, but the project was cut short by Downing's death. Glover displayed his model fruits at the New York State Agricultural Society museum in Albany, the American Institute (a group based in New York City that promoted science and technology in America), and the meetings of the American Pomological Society in Philadelphia. His exhibit in Washington, D.C., in 1853–1854, led to his appointment in 1854 as special agent in the recently created agricultural division of the U.S. Office of Patents. He remained with the patent office until 1859, when he resigned to teach at the Maryland Agricultural College, but returned to Washington in 1863 as the first U.S. entomologist in the new Department of Agriculture, a position he held until forced by illness to resign in 1878.[59]

Meanwhile, in Massachusetts, where Harris had tried and failed to win sustained public funding for applied entomology, the agricultural community took steps to separate entomological work from the geological

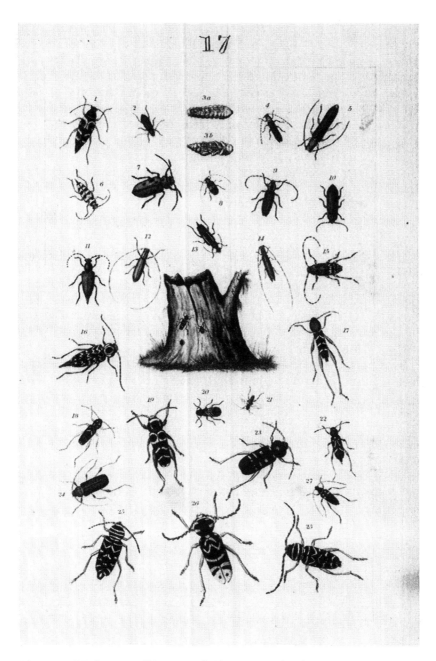

Plate 17, "Coleoptera," Townend Glover Notebooks, ca. 1850–1878, box 9, Record Unit 7126, Smithsonian Institution Archives, Washington, D.C.

survey. In 1858, two years after Harris's death, C. L. Flint, secretary of the Massachusetts Agricultural Society, hired Francis G. Sanborn to improve the society's entomological collection. Sanborn, a youth of twenty, had come to Flint's attention the previous year when he displayed a collection of one thousand insect specimens at various agricultural fairs. From 1858 to 1865, Sanborn worked for the Massachusetts Agricultural Society, scouring the state for specimens to be used as the basis for illustrations for a new edition of Harris's *Report,* adding specimens to the cabinet of insects, and preparing a lengthy report entitled "Insects of Massachusetts Beneficial to Agriculture." Sanborn left the Massachusetts Agricultural Society in 1865 to take charge of the insect collection at the Boston Society of Natural History.[60]

Fitch and Glover began their long tenures as professional entomologists in 1854. Sanborn served, in effect, as state entomologist of Massachusetts from 1858 to 1865. Applied entomology had thus made considerable progress before the outbreak of the Civil War. The Civil War accelerated agricultural change. The losses to destructive insects reached crisis proportions, particularly in the West. In Illinois, which by 1860 was the leading agricultural state, the problem was addressed by agricultural leaders and by the growing community of western agricultural entomologists, including Walsh, Thomas, LeBaron, and Riley.

In 1862, the president of the Illinois State Horticultural Society announced that "an almost innumerable multitude of the insect tribes seem to bid defiance to our skill. . . . these enemies must be met and vanquished."[61] By 1868, John P. Reynolds, president of the Horticultural Society, was advising his fellows that the great question facing them was no longer when to plant orchards, or what varieties to plant, but "how to protect what has been already planted from the ravages of . . . insect enemies."[62] Citing the example of Fitch in New York, Reynolds termed the appointment of a state entomologist for Illinois the matter of first importance. Fortunately, he added, it would not be necessary to go outside Illinois to find a competent candidate. He referred to Benjamin D. Walsh, who like Fitch, had earned a reputation for his ability to apply entomological science to the problem of agricultural pests.

Born in Frome, Worcestershire, England, in 1808, Walsh was trained for the ministry at Trinity College but later followed a career in journalism. At age thirty, he moved to America, where he bought a three-hundred-acre farm in Henry County, Illinois. He eventually settled in Rock Island, Illinois, where he ran a successful lumber business.[63] While still in England, Walsh had become an expert entomologist, but during his first years in America he found little time to pursue this interest.

Upon his retirement from the lumber business in 1857, he took up the study of agricultural entomology, to which he applied his journalistic and managerial skills. By 1859 he was speaking at the Illinois State Fair, where he displayed 1,650 species of insects that he had collected in the vicinity of Rock Island during the preceding two years.[64] Walsh's articles on injurious insects were singled out for awards by the Illinois State Agricultural Society.[65] In 1865, Walsh served as western editor for the *Practical Entomologist,* a publication sponsored by the Entomological Society of Philadelphia, and in 1866 he became editor-in-chief. In 1868, the *Practical Entomologist* having ceased publication, Walsh and Charles Valentine Riley started their own journal, the *American Entomologist.* Both publications were popular among farmers and entomologists because of their plain, clear style and their accurate, detailed illustrations of insects.

In 1866 both the horticultural society and the agricultural society called for the appointment of Walsh as state entomologist.[66] The bill passed the legislature, but soon political opposition developed to the governor's appointments that year. The horticultural society then offered to pay a portion of Walsh's salary on the expectation that it would be reimbursed. Walsh agreed to this arrangement and started his work in 1867. Like Fitch in New York, Walsh devoted his first report to insects that damaged fruit orchards and gardens.[67] In November 1869, Walsh was struck by a train and killed. His death ended a career in agricultural entomology barely a decade after he had taken up the study of injurious species. Fortunately, Walsh was succeeded in Illinois by a series of capable state entomologists.

Missouri was the third state to establish the office of state entomologist, and there the horticulturists also played the leading role. In 1867, a spokesman for the Missouri State Horticultural Society declared, "the annual losses by insects are becoming more and more serious. They filch from us a large part of our earnings [and] there is an earnest demand for information on this subject."[68] The president of the horticultural society estimated the annual losses to insects to be $60 million.[69] That year, the horticultural society petitioned the Missouri General Assembly to create an office of state entomologist.[70] Though unsuccessful, the horticulturists were joined in 1868 by the Missouri State Board of Agriculture and the St. Louis Academy of Science, and with this combined support the proposal was passed by the General Assembly. Upon recommendations from Walsh and Norman J. Colman, the editor of *Colman's Rural World* and future secretary of agriculture, the Missouri lawmakers chose Charles Valentine Riley as state entomologist.[71]

Riley's career in some ways paralleled Walsh's. Born in Chelsea, London, in 1843, Riley was educated in art and natural history in France

Charles V. Riley examining insect with hand lens (Courtesy of the
National Agricultural Library Special Collections, Beltsville, Maryland)

and Germany. In 1860, at age seventeen, he immigrated to the United States. Riley eventually joined G. H. Edwards, who owned a stock farm near Kankaka, Illinois, about fifty miles from Chicago. Like Walsh, Riley had studied entomology in Europe and then, influenced by his American farm experience, had turned to applied entomology. Riley worked and studied so incessantly that his health broke, whereupon he moved to Chicago to try his hand at journalism. In 1862, he became a reporter for the *Evening Journal* and soon thereafter began writing an entomological column in the *Prairie Farmer,* the leading agricultural paper of the West. Walsh, whom Riley first met in 1864, had a decisive influence on his development as a scientist, writer, and promoter of agricultural entomology.[72] At his new position in Missouri, Riley assumed the role of a flamboyant promoter of applied entomology. His nine *Annual Reports,* published from 1869 to 1877, set new standards in the comprehensive treatment of insect pests and in their superb insect illustrations. In the judgment of Leland O. Howard, who succeeded Riley as entomologist in the U.S. Department of Agriculture, the *Annual Reports* produced by Riley "formed the basis for the new economic entomology of the world."[73] Riley's appointment as Missouri state entomologist began his meteoric career first as state entomologist, then as head of the U.S. Entomological Commission, and finally as chief entomologist for the U.S. Department of Agriculture.[74]

The appointment of state entomologists in New York, Illinois, and Missouri served to catalyze efforts among Canadian entomologists and agriculturists to establish similar institutions in Canada. The leadership was taken by William Saunders and Reverend Charles James Bethune, who with a group of entomological friends formed the Entomological Society of Canada.

Saunders had immigrated with his family from England to Canada as a child. He became a druggist in London, Ontario, later expanding the business to include the sale of medical extracts of plants. Still later, he was appointed professor of materia medica at the University of Western Ontario. Saunders owned a farm, where he experimented in the development of new varieties of crops, fruits, and berries. A knowledgeable collector of plants and insects, Saunders had a special interest in the Lepidoptera. He raised many species of butterflies from the egg or larva, of which he published careful descriptions.[75]

In 1861, Saunders met the Reverend Bethune, an Anglican priest, in Credit, Ontario. Bethune shared his enthusiasm for the study of entomology. In 1862, these two with some like-minded friends met in Kingston, Ontario, and formed the Entomological Society of Canada (later Ontario). The following year the society moved its central meeting to Toronto, where the majority of the membership lived, but maintained branch societies in Kingston and London, Ontario, and in Quebec Province. Among the first activities of the society was the publication of lists of the known Canadian butterflies and moths by Saunders and Bethune, which appeared in 1864 and were periodically enlarged and revised.[76]

From the outset, the Entomological Society of Canada developed a balance between systematic and economic concerns. In 1868, the society began to publish the *Canadian Entomologist*. Bethune, the editor and leading systematist of the society, stated that the new journal would provide those "wielders of the Butterfly net and Beetle bottle in Canada [with] some mode of telling one another what they have taken, how and where they have taken it, and what they are in want of."[77] In addition to notices of captures and offers of exchanges, the journal contained original articles on the classification, description, habits, and history of insects and on the transactions of the society. The Canadians had a cordial relationship with the American Entomological Society; many members of each society became honorary or corresponding members of the other, and each society's publication contained articles and notices from members of the other.

Though similar to the American Entomological Society in many ways, the Canadian society laid less emphasis on developing a centralized collection and more on economic entomology. Early issues of the *Canadian Entomologist* carried extensive material on applied entomology, including a series of articles by Walsh on insects of economic importance in Canada and a series by Saunders giving a chronicle of the insect pests in his orchard and his experiments with various control measures.[78] In 1868, upon Riley's appointment as Missouri state entomologist, Bethune urged the Canadian government to appoint official entomologists following the precedent set in New York, Illinois, and Missouri.[79]

Canadian economic entomology took on institutional form in 1870 when the Council of Agriculture and Arts Association of Ontario offered to help fund the activities of the entomological society on the condition that it publish an annual report devoted to economic entomology and that it form a permanent cabinet devoted to insects of economic importance. In 1871, the society accepted the offer of four hundred dollars per year for support of its publication, changed its name to the Entomological Society of Ontario, began publishing the annual report on economic entomology (in addition to the *Canadian Entomologist*), and established the collection of economic entomology in Toronto. From that time on, the Entomological Society of Ontario combined the functions of a scientific society with those of a governmental agency. In its new public service role, the society in 1871 issued an emergency report on the spread of the Colorado potato beetle in Canada. Later, the society expanded its work with the Fruit Growers Association of Ontario and the agricultural officials of the provincial government who were concerned with insect problems.[80]

By 1870, the agricultural and entomological leaders had succeeded in placing professional entomologists in three states and one Canadian province and in the U.S. Office of Patents (which after 1862 was in the Department of Agriculture). They had not succeeded everywhere, however, and their failures indicate critical shortages in leadership, support, or finances.

In Pennsylvania, failures in leadership and shortages of funds blocked the appointment of professional entomologists at critical points. At the time the Pennsylvania Geological Survey was established under Henry D. Rogers, from 1836 to 1842, the Philadelphia entomologists emphasized systematics to the conscious exclusion of applied science. Lacking support for agricultural concerns from this pivotal group, the survey geologists devoted exclusive attention to minerals. In 1842, in part because of a state financial crisis, funding for the survey was withdrawn, and in the period from 1851 to 1858, only enough funds were provided to Rogers to finish his report. By the latter year, some Pennsylvania entomologists had taken up agricultural investigations. Jacob Stauffer, a druggist and self-taught botanist and entomologist from Mountjoy, Pennsylvania, petitioned the state legislature to support him as a state entomologist for a period of three years. Stauffer apparently did not attract sufficient support among the Pennsylvania agriculturalists and horticulturalists. At any rate, continuing shortages in state finances precluded even the modest expenditures he proposed.[81]

In Michigan, the early failure to establish agricultural entomology as part of the state survey or as an independent office seems to have been due to a lack of effective leadership. When the first state survey began in 1837, agriculturists and horticulturists called for the investigation of destructive insects. Their concern failed to impress some prominent state legislators, who ridiculed the work of the botanical and zoological assistants. These assistants apparently lacked the promotional and public relations skills of Fitch or Walsh, for they resigned in 1840.[82]

When a second survey under Alexander Winchell was authorized in 1860, horticulturists in the Grand Traverse region asked for a special survey of the fruit-growing potential of their area. Despite this interest, the zoological and botanical portions of the survey were turned over to systematists whose compensation was the right to retain rare specimens for their cabinets. Still later, in 1869, Mark W. Harrington was hired to continue the natural history work of the survey. His investigations in botany, entomology, and ornithology had barely begun when, in 1871, Winchell and the entire survey staff came under fire for studying weather patterns and other subjects that the politicians charged were unnecessary. Winchell resigned and the natural history portions of the survey were dropped.[83] By the 1860s, entomological studies were making headway at Michigan Agricultural College, and this progress may have weakened the effort to establish entomology in the survey.

In the state of Maine, Packard served as entomologist with the Geological Survey in 1860 and was hired by the Maine Board of Agriculture to write a report on destructive insects. A reduction in state finances, however, cut off appropriations for the entomologist and zoologist after only one year.[84] Later, Sidney Irving Smith, a professor of zoology and comparative anatomy at Yale University, served for several years as state entomologist in Maine and Connecticut.[85]

There were other short-lived attempts to establish applied entomology as a state responsibility. Scudder wrote one report on the distribution of insects for the New Hampshire Geological Survey (1874).[86] Georgia was apparently the only southern state to establish an office of state entomologist. J. T. Humphreys was appointed to the position in 1877, but nothing further is known of Humphreys or his work.[87] In North Carolina, Ebenezer Emmons was hired in 1852 to make a geological and agricultural survey; however, following the public disparagement of his New York report, Emmons apparently did not attempt any further entomological work.[88]

By the late 1870s, the effort to establish applied entomology in permanent state offices was losing momentum, as indicated by the failure in Missouri in 1877 to replace Riley, who resigned to head the U.S. Entomological Commission. By that time, however, there had developed a promising new area of institutional change, the establishment of positions for applied entomologists in colleges and universities. By the 1880s, the universities would provide the most stable institutional base for the discipline. This development was related to wider reforms in the American system of higher education that transformed the classical liberal arts colleges into centers of teaching, scholarship, and research.[89] Entomologists and agriculturists contributed to these educational reforms by promoting applied entomology and other agricultural related disciplines within colleges and universities.

Academic entomology entered the American science curriculum through three avenues. The first avenue developed without public fanfare in and around Harvard College (later Harvard University) as Peck and Harris taught small groups of students who were interested in insects. This teaching took place in connection with natural history lectures, or more often, at informal gatherings in the home.[90] The second avenue came through the influence of European-trained zoologists and comparative anatomists like Agassiz and Henry Goadby, who instituted instruction in animal morphology, physiology, and comparative anatomy at Harvard, Albany University, and Michigan Agricultural College. Agassiz, who came to Cambridge in 1847, had the greatest influence. His contributions were made through the founding of the Museum of Comparative Zoology, with its matchless insect collection, his instruction of entomological assistants, Scudder, Packard, and Uhler, and his hiring of Hagen as curator of insects and professor of entomology at Harvard.

· Hagen had studied medicine in Königsberg, Prussia (where his father and grandfather had been professors), and later at the Universities of Berlin, Vienna, and Paris before settling in Königsberg as a physician and surgeon. His doctoral thesis on the synonymy of the European dragonflies was the first of many publications on the Neuroptera, his specialty. In 1861, he contributed a *Synopsis of the Neuroptera of North America* to the Smithsonian's series of monographs on insect orders. Hagen was also a leading bibliographer in the expanding field of entomological literature. In 1862–1863, he published the *Bibliotheca Entomologica,* an attempt to list all the entomological literature up to that time.[91] When Agassiz hired Hagen for the museum in 1867, Hagan was told that his title would be "professor." It was evidently the first time that a professor of entomology carried this formal title in the United States. In 1870 he commenced giving instruction in entomology at the museum.[92]

Academic entomology at Harvard was further strengthened in 1871–1872 when Sanborn was appointed instructor of entomology and microscopy in the Bussy Institution, the agricultural and forestry department of Harvard, located outside Cambridge. His instruction there foretold the eventual fulfillment of Harris's dream, in the 1840s, of formal instruction in agricultural entomology at Harvard.[93]

The third avenue developed from the movement for agricultural and industrial education and the establishment of land grant colleges. This movement began with Solon Robinson and others who in the 1840s organized the National Agricultural Society with the goal of directing the Smithsonian bequest to agricultural research. In the 1850s, Marshall P. Wilder organized the U.S. Agricultural Society to lobby for the establishment of land grant colleges and the creation of a department of agriculture. The movement for agricultural education gained momentum in 1851–1852, when Jonathan B. Turner, Illinois educator, fruit grower, and co-editor of the *Prairie Farmer,* delivered a series of lectures on industrial education. Success finally came in 1862, when Congress passed the Morrill Land Grant Act, granting land to the states for agricultural and industrial colleges.[94]

The efforts of agriculturists and entomologists to secure a place for entomological teaching and research within the programs of the new agricultural schools achieved early successes in New York and Michigan. Fitch, who by the early 1850s had determined to find full-time support for his entomological work, hoped for appointment as state entomologist. In case that plan failed, however, he had an alternative plan to lobby for the establishment of an agricultural college in New York state with himself as instructor in agricultural entomology. In the late 1840s, a movement developed to found a great national university at Albany, New York. This effort ultimately failed for lack of funding from the New York legislature, but some classes started in 1850–1851. These included entomological instruction under Henry Goadby, a physician and former microscopical dissector in the Royal College of Surgeons in London. Fitch, who traveled to Albany to observe Goadby's instruction, concluded that his methods were not suited to the needs of the agricultural community because Goadby's identification of specimens was often wrong, and Goadby stressed laboratory dissection rather than field investigations. At various times in the 1850s, Fitch himself delivered lectures on agricultural entomology at his alma mater, Rensselaer, at Troy.[95]

Michigan Agricultural College, founded in 1855 at the urging of the state agricultural society, served as the model for the new land grant colleges established under the Morrill Act of 1862. Entomological instruction began in 1859, again under Goadby, who meanwhile had established his medical practice in Detroit. Goadby left his practice to accept an appointment at the college as professor of animal and vegetable physiology and entomology, but he served only one year. He was succeeded by Manley Miles, a physician who had served as an assistant state geologist in Michigan. Miles was placed in charge of the college farm in 1864 and named professor of agriculture in 1865, perhaps the first in the world to be so designated. His course in agricultural entomology, which began in 1863, was also apparently the first of its kind. Miles taught at Michigan until 1875, where he investigated agricultural pests such as the Colorado potato beetle, the army worm, and the cankerworm. It was his student, Albert John Cook, however, who did the most to develop entomology at the college. Cook graduated from Michigan Agricultural College in 1862. After managing a fruit farm near Sacramento, California, he returned in 1867 to teach at Michigan. In 1869 he became head of the department of zoology and physiology and professor of entomology, a position that he developed into what was in effect the Michigan state entomologist. In addition to teaching classes, he answered farmers' inquiries, spoke before the State Pomological Society of Michigan, developed a kerosene-soap emulsion insecticide, and wrote reports on injurious insects for the Michigan State Board of Agriculture.[96]

Cornell University, founded in 1867, developed a second model for agricultural entomology in the colleges. During the search for Cornell faculty, President Andrew D. White consulted with Louis Agassiz regarding science appointments. Lintner hoped for appointment to a chair of agricultural entomology, and he contacted his friend Scudder (a former Agassiz student) to secure Agassiz's support. Lintner proposed the establishment of such a chair as a precedent for other agricultural colleges to follow.[97] Though Lintner was not chosen, John Henry Comstock, a student entering Cornell two years after its founding, developed the program at Cornell into a leading center of applied entomology.

Attracted by the science program advertised by Cornell (including a proposed course in entomology), Comstock entered in 1869 with plans to pursue medical studies. He made such good progress in his studies with Burt G. Wilder, an Agassiz student who headed the department of zoology and physiology, that within two years he was chosen to be Wilder's assistant. In 1870 Comstock obtained a copy of Harris's *Treatise,* and he became so interested in entomology that he decided to make a career in that field rather than in medicine. Comstock was encouraged by Riley, who in 1872 gave a series of lectures on economic entomology at Cornell. He was also inspired by Agassiz, who lectured at Cornell that year. The following summer, perhaps on the advice of Agassiz, he went to Cambridge, Massachusetts, where he studied entomology under Hagen at the Museum of Comparative Zoology. In his junior year at Cornell, he was allowed to lecture on economic entomology in the natural history faculty, and from that time he was determined to make a profession of college teaching. In preparation, he visited Fitch to learn more about his chosen field, and following his graduation from Cornell in 1874, he studied under Addison E. Verrill at Yale University. He soon developed strong ties to the Western New York Horticultural Society and other horticultural and agricultural groups. In 1876, he traveled to Philadelphia, where he visited the insect display of the U.S. Department of Agriculture at the Centennial Exposition and the insect collection of the

American Entomological Society (which he noted had no biological material).[98]

With the support of White, who favored the expansion of economic entomology at Cornell, Comstock became assistant professor of entomology and invertebrate zoology in 1876. From that time until his retirement in 1914, Comstock and his wife, Anna Botsford Comstock, built the entomological department at Cornell into one of the outstanding departments in the United States. Comstock also played an important role in the organization of the entomological department at Stanford University and other schools. Through his teaching, his writing, and personal contact, Comstock had a decisive influence on the generation of entomologists who were trained in the United States in the late nineteenth and early twentieth centuries.[99]

In Minnesota, applied entomology entered the university via the back door of the state survey. The Minnesota State Natural History Survey (1872–1888) was designed by university administrators as a means of adding a department to the University of Minnesota, and all the appointments for the survey went to professors and students of the university. Winchell, who had just resigned from the Michigan survey, was hired as state geologist, but he actually functioned as a professor at the university. Though the original plan for the survey called for investigations in geology, botany, and zoology, in that order, the survey/department was flexible in responding to popular demands like the call to investigate outbreaks of the Rocky Mountain locust. In 1876, Alexander Whitman was appointed as entomologist with special duties to report on the locust and other injurious insects, and the position was retained in the university.[100]

By the early 1870s, course offerings in agricultural entomology had appeared in half a dozen agricultural colleges scattered from Maine to California. In 1872 and 1873 courses in entomology were offered at the University of Mississippi, the Ohio Agricultural and Mechanical College, Columbus ("pests of the farm"), Claflin University, South Carolina, the University of California, Berkeley, and the University of Kansas. A majority of this first generation of instructors in college classes were agricultural entomologists like Riley, Packard, and Thomas who had acquired their expertise to a considerable extent outside formal college classrooms. "Graduate studies" in entomology for their generation meant stints with the Hayden survey, or field investigations for state departments of agriculture, although sometimes (as in Packard's case) they included seminars and expeditions led by university-based teachers like Agassiz or Hagen. These founders of academic agricultural entomology included Fitch, who taught at Rensselaer School; Riley, who lectured at Kansas Agricultural College, Cornell University, and the University of Missouri at Columbia; and Packard, who gave entomological instruction at Massachusetts Agricultural College, Maine State Agricultural College, Bowdoin College, and Brown University.[101] In the late 1870s and 1880s, the ranks of these founders with their eclectic educational backgrounds were augmented by instructors who were trained primarily in academic settings and whose career paths within the colleges and universities were secured by the growing public support for applied entomology. This was part of a larger transition in the American professions in which the leadership was transferred from those who were trained as apprentices under individual masters to those who were trained in college and university classrooms. In some professions, like mechanical engineering, the transition involved considerable conflict between competing elites. Mechanical engineers trained in the older "shop culture" resisted the attempts of those trained in "school culture" to assume leadership of the profession.[102] In entomology, however, the transition seems to have been accomplished smoothly. By the 1880s, the discipline was finding a secure home in the university setting. The first college degree in entomology was awarded to C. M. Weed in 1891, and the first Ph.D. in entomology was granted in 1901.[103]

American academic entomology thus developed from a blending of European and American practices and traditions, but the most distinguishing characteristic of the new discipline derived from American experience in agricultural change and the resulting challenges in applied entomology. The influence of the Europeans, Agassiz, Hagen, and Goadby, was on the whole less important in the formulation of research programs and entomological instruction than the pressing problems arising from insect pests. Typical texts used in entomology courses were Packard's *Guide to the Study of Insects* and state entomologist's reports by Fitch and Riley, rather than European literature. Research in the new entomological departments most frequently addressed current problems with insect pests and various proposed remedies, such as tests of insecticides.

By the 1870s, approximately one of every three publishing American entomologists received primary support from the agricultural community. Between 1840, when Harris was the only American who had received payment for entomological services, and 1870, public support increased to approximately twenty thousand dollars per year for full- and part-time entomological research and instruction.[104] While this total was only twice what the chief engineer of the Illinois Central Railroad earned annually in the 1850s, it represented an impressive sum for entomologists, who were accustomed to paying their own way. It was also a far larger commitment to agricultural entomology than in any other country. In England, for example, where John Curtis had received a pension in 1840 for agricultural entomology, the total in 1870 was still John Curtis's pension.[105]

This level of support in the United States was a product of the needs arising from massive agricultural change. It was also the product of a large, flexible, and decentralized political system. The states, the federal government, and the agricultural colleges offered multiple institutional possibilities for those seeking such support. This support translated into substantial expansion of employment for entomologists. The careers of agricultural entomologists during the 1860s and 1870s typically included service with two, three, or more of the new institutional "homes" for the emerging discipline such as state or federal surveys, positions for state entomologists, college and university positions, and museums.

The careers of agricultural entomologists illustrate this multiple institutional support. Lintner, the second state entomologist for the state of New York, followed a career in business for thirty years while studying entomology in his spare time. In 1847, while assisting Fitch in the entomological survey of Washington County, he began collecting insects systematically. By 1860, Lintner was specializing in the Lepidoptera, and in 1869 he began to correspond with William Henry Edwards, who was at that time initiating a comprehensive work on the butterflies of North America.[106] Following his unsuccessful bid for a chair of entomology at Cornell in 1867, Lintner found a position as zoological assistant to James Hall in the New York state museum, where he pursued his entomological studies and contributed important studies of Lepidoptera and other subjects to the museum publications.[107] When Fitch's failing health forced him to resign as state entomologist in 1870, many persons hoped Lintner would be named Fitch's successor; however, the New York legislature delayed funding his replacement.[108] In 1871, Lintner commenced a series of articles on agricultural entomology for the *Cultivator and Country Gentleman,* a leading agricultural paper later known as the *Country Gentleman.* Through this column and other agricultural and horticultural papers, Lintner became well known among agriculturists. Their support prompted Lintner's special appointment, in 1874, as an entomologist in the state museum, a position that was regarded as replacing Fitch.[109] Lintner and his supporters, citing the need for a more secure legislative authorization for the state entomologist, asked for a permanent office under the regents of the state university. They succeeded in 1880 when Lintner was appointed by the university regents as New York state entomologist, a position he held until his death in 1898.[110]

Following Walsh's death in 1869, LeBaron, Thomas, and Forbes served successively as state entomologists of Illinois. LeBaron studied medicine and practiced in his hometown of North Andover, Massachusetts, meanwhile studying natural history on his own, turning his attention to birds, plants, and finally to insects. In 1844, he moved west to Geneva, Illinois, where he practiced medicine, taking some time off to return to Harvard Medical College for additional schooling. About 1850 LeBaron began to write about entomology for the agricultural press. His account of the destruction of crops by the chinch bug and the known remedies was so thorough that it was used by Fitch in his second New York report. Because of the popularity of his agricultural writings, LeBaron was named editor of the influential *Prairie Farmer* in 1865. In 1870, he was appointed to succeed Walsh as state entomologist. He produced four annual reports and was working on a fifth in 1875 when he became ill and was forced to resign. He died the following year.[111]

Thomas, the third state entomologist of Illinois, was born in Kingsport, Tennessee, in 1825. Intent on achieving fame by means of a professional career, Thomas trained himself first in the ministry, then in natural history (specializing in entomology), and finally in ethnology. Thomas moved to Illinois in the 1840s, where he farmed, then practiced law, then entered the clergy as a Lutheran minister. He began the serious study of injurious insects in 1856, at about the same time as Walsh, and he wrote for agricultural papers including the *Prairie Farmer,* the *Rural New Yorker,* and the *American Agriculturist.* Thomas saw the potential of agricultural entomology as an emerging scientific profession that he could master at little expense and without formal schooling. It offered the possibility of making original contributions from materials close at hand. In 1850, Thomas married a sister of John "Black Jack" Logan, later a famed Union general and political boss of Illinois. When Logan initiated funding for the Geological and Geographical Survey of the Territories (the Hayden survey) in 1869, Thomas joined the survey as naturalist.

Thomas's work with the survey coincided with the locust invasions that plagued settlers on the plains in the 1870s. He published extensive treatments of the order Orthoptera, which included the locusts, and he wrote the first American monograph on the Acrididae (straight-winged forms of Orthoptera).[112] In 1874, while still working for the survey, Thomas was named professor of natural history at Southern Illinois Normal University, Carbondale. In 1875 he succeeded LeBaron as Illinois state entomologist, and he resigned his teaching post the next year, but he still retained his connection with the Hayden survey. From 1877 to 1881 he served on the U.S. Entomological Commission. In 1882, Thomas changed professions a final time when he took up archaeology and ethnology as a scientist in the U.S. Bureau of Ethnology. He worked there until his death in 1910.[113]

Forbes, who succeeded Thomas as Illinois state entomologist in 1882, began his work in agricultural entomology in the late 1870s. This work involved some pioneering quantitative studies of the contents of birds' stomachs, a topic that in the 1880s became a major focus of research within the U.S. Department of Agriculture. Born and reared on a farm in Stephenson County, Illinois, Forbes served in the Union army, then studied medicine for a time in Chicago and took up school teaching. Becoming (in his word) "infatuated" with natural history, he served as curator of the State Natural History Society of Normal, Illinois, in 1872 and as professor of zoology at the Southern Illinois Normal University from 1875 to 1878, where he overlapped a year with Thomas. He was appointed director of the Illinois State Laboratory of Natural History in 1877, and when Thomas resigned as state entomologist, Forbes was given that position also. When the state agencies were reorganized in 1917, Forbes became chief of the Illinois State Natural History Survey.[114]

The career of Packard, a major figure in nineteenth-century American biology whose full biography remains to be written, was identified with each of the institutional structures in agricultural entomology.[115] Born in Brunswick, Maine, in 1839, Packard followed a career that has some suggestive parallels to that of his fellow New Englander, Harris. Like Harris, Packard developed an early interest in natural history, studied medicine, and desired a career in science, and like Harris he turned to entomology as his specialty.[116] In fact, Harris was Packard's main inspiration for taking up entomology. In 1856, upon reading Harris's *Report on the Insects of Massachusetts,* Packard turned from his earlier interest in conchology to take up entomology, noting in his diary that entomology in the United States is "greatly neglected."[117]

Packard earned a degree from Bowdoin College, where his father was professor of Latin and Greek, an M.D. from Maine Medical School, and a B.S. from the Lawrence Scientific School at Harvard, where he studied under Agassiz. Searching for a career in science, Packard became librarian of the Boston Society of Natural History in 1865, then curator of the Essex Institute (later the Peabody Academy of Science) at Salem, Massachusetts, in 1867, and finally professor of zoology and geology at Brown University, where he taught from 1878 to 1905. In 1867, while at the Essex Institute, Packard helped found the *American Naturalist,* which soon became a leading American scientific journal. He was also one of the originators of the "Neo-Lamarckian" school of evolutionary theory.

Unlike Harris a generation earlier, Packard found solid support for entomology in the scientific and agricultural communities. Packard corresponded with LeConte, Scudder, Osten-Sacken, Uhler, Thomas, Riley, and others who provided encouragement and assistance. In 1859, Packard began writing articles on entomology for the *Maine Farmer,* a series that ran for forty-five years. Packard worked for the Maine Geological Survey in 1860, and he published one report for the Maine Board of Agriculture. He also lectured frequently on economic entomology for agricultural organizations in Maine and Massachusetts. In 1869, Packard published *Guide to the Study of Insects,* a monograph that for many years remained the most complete reference to entomology in America. It was through this *Guide* that Packard became well known in America and Europe. Packard published two reports on the insects of Massachusetts (1871 and 1873) that compared favorably with Riley's Missouri reports in the same period.[118] In 1875–1876, Packard worked for the Hayden survey, traveling extensively in the West, where he studied and reported on the insects of the Great Salt Lake region, the range and migrations of the Rocky Mountain locust, and the geometrid moths of Colorado.[119] From 1877 to 1881, Packard served with Riley and Thomas on the U.S. Entomological Commission, and he wrote most of the final report of the commission, on insects injurious to trees, published in 1890.

The new career paths were only one of the visible changes that were reshaping the entomological community. The pervasive influence of agricultural change on the people and the institutions associated with the development of American entomology were apparent in every aspect of the discipline. The ranks of publishing entomologists increased from two or three in the 1840s to approximately thirty professionals in the 1870s, almost half of whom received their primary support from agriculture. Through the leadership of agricultural entomologists like Fitch, Walsh, Riley, LeBaron, Sanborn, Thomas, and Packard, the discipline changed from a largely self-supported "amateur" activity to one dominated by professionals.

By the early 1870s, American entomologists exhibited clear signs of professional consciousness. Informal discussion among entomologists reflected their desire to take their place among traditional and emerging professions, with the same status as physicians and engineers. In 1872, for example, Sanborn wrote to Edwards: "I enclose my professional card, indicating a 'new department' or rather the establishment of a new Profession, on the same basis as Medicine, Analytical chemistry or professional Geology. This I do partly in self defense, and partly in the hope to benefit my pupils who are or may be hereafter qualified to practice such profession. Tell me what you think of it. Baird, Riley, Hagen, Putnam and others highly approve the plan."[120] Sanborn's desire to elevate entomology to professional status and the approval of the "Brethren" clearly reflected the new public service consciousness associated with agricultural entomology.

Another sign of maturing professional status was the formation in the mid-1870s of the Entomological Club of the American Association for the Advancement of Science (AAAS). This movement reflected widely diverse interests among American entomologists, including the rise of economic entomology, the interest in evolutionary theories, and the need for a standard entomological nomenclature. Entomologists had actively participated in the AAAS from its inception, as indicated by the 1843 proposal of Morris to form an American society. The rapid expansion of agricultural entomology since the founding of the AAAS, however, was a primary unifying factor among those in the 1870s who called for an entomological section. The first proposal for an entomological club, in 1872, came at the AAAS meeting at Dubuque, Iowa, a frontier community whose inhabitants were familiar with insect problems associated with rapid agricultural change. Growing interest in agricultural entomology prompted a major address by LeConte before the 1873 meeting of the AAAS in Portland, Maine, in which he reviewed the rise of agricultural entomology in the nation and called for a major mobilization of the discipline to be spearheaded by a reorganized Department of Agriculture.[121] In Portland, the entomologists drew up the proposal for an entomological club that was formally adopted the next year in Hartford, Connecticut. The club included in its membership all members of the AAAS with an interest in entomology. Special invitations were issued to members of the American Entomological Society and the Entomological Society of Ontario. In the 1870s and 1880s, the club provided a forum for discussion of common concerns, such as outbreaks of destructive insects during the current season.[122]

The pervasive influence of agricultural change was present in other areas besides the dramatic levels of funding, the public service orientation, the tentative steps toward self-definition by entomologists as professionals, and the organization of the Entomological Club. Public support for agricultural entomology provided major new funding for all activities of the emerging scientific community, such as publications, curation of collections, and other desiderata of the growing scientific discipline. The generous support from agriculture transformed the network of institutions originally conceived according to European models into a distinctive set of American institutions that now served as the model for Europe and the rest of the world. The influence of agriculture also reoriented American entomology away from the primary emphasis on taxonomy toward the study of life histories and the economy of nature, or ecology. This reorientation in turn influenced the reception of evolutionary theory by American entomologists and the incorporation of evolutionary biology into the American science curriculum.

5 The Balance of Nature

The rapid institutional growth of agricultural entomology, most evident in the state, provincial, and federal entomologists' offices and in the introduction of entomology courses and departments into higher education, owed much to the conceptual arguments employed by its proponents. The idea of the "balance of nature," in particular, provided a theoretical framework that was widely accepted among entomologists and agriculturists. Before examining the concept and how it was used, it is worthwhile to take a look at the general relationship between farmers and entomologists as they exchanged ideas about nature, agriculture, and the changing American environment.

While most American farmers in the mid-nineteenth century were more intent on cultivating the abundant resources of the continent than in cultivating their minds, there was an articulate group of agricultural editors, progressive farmers, and especially horticulturists who claimed to speak for the agricultural community. Their vehicle of communication was the agricultural press, and their effective organization was the agricultural society. As agricultural change accelerated, the influence of this articulate minority increased. This trend was reflected in the reading habits of farmers. In the pre–Civil War era, the reading material most popular among farmers was the farmer's almanac, which was consulted regarding the phase of the moon that was best for that year's planting. By the Civil War era, the agricultural papers competed seriously for the readership of farmers. To be sure, these papers contained a liberal share of folk wisdom, but they also contained an increasing amount of accurate scientific information. This broadening and enlightening influence of the agricultural press played a critical role in the spread of ideas favorable to the acceptance of agricultural entomology by the public.

Entomologists, as has been noted, used their writing skills effectively to promote their discipline in the agricultural press. In the 1840s and 1850s, as various regions experienced insect outbreaks, notices of the nature and extent of the damage made the rounds in the agricultural papers, along with numerous, often contradictory suggestions from farmers and editors on how to control the outbreak. Articles by Harris, Fitch, and other knowledgeable entomologists appeared also with increasing frequency. By the late 1850s, entomologists like Walsh objected to this mixture of folk wisdom and science in the farm papers and pointed to the need for entomological experts who through official reports could furnish accurate information. Walsh called upon entomologists to evaluate the various control measures suggested in the agricultural press, no matter how controversial such practice might be. Walsh explained that farmers were generally trusting; they believed what they saw in print and were therefore unable to distinguish between the "loudmouth impudent charlatan and the unobtrusive naturalist, who usually shrinks . . . from . . . controversy."[1] Disagreeable as the task was, Walsh insisted that agricultural entomologists must expose quack remedies for insect pests and educate farmers concerning the role of the entomologist:

When a farmer wants crackers, he goes to a baker for them; when he wishes for a new coat, he goes to a tailor; and if he lacks a new pair of boots he usually calls on the shoemaker. But . . . most farmers have a lurking idea . . . that the place to go for information about the Natural History of Noxious Insects is, not to those who have made such matters the study of their lives, but to the first impudent mountebank that comes along with a precious story about a Cock and a Bull.[2]

On one occasion, Walsh wrote of a fertilizer which, it was claimed, was: "sure death . . . to the Cankerworm, the Curculio, the Apple Moth, the Potato Bug, the Cotton Worm, the Tobacco Worm, the Hop Louse, the Army Worm, [and] the Current Bug. . . . we hear nothing of the ninety and nine cases where the Universal Remedy was applied and found to do no good, while in the one case where the medicine worked . . . the happy experimenter lauds it to the skies."[3]

With the establishment of official entomologists in several states and provinces, agricultural entomologists turned out reports that were designed to correct such errors in the press. These entomological reports followed a standard format in which the author first recounted the taxonomy of the insect and compared it to similar species by means of drawings and descriptions, then described its transformations, its life history, and its habits (often including a historical summary of its destructive career in North America), and finally evaluated the various control measures available. There were four general categories of suggested control measures. Natural controls focused on the encouragement, or at least noninterference with insects, birds, and other organisms that preyed on the insect pest. Cultivation methods like crop rotation, fall plowing, fertilizing, and burning of stubble encouraged sturdy crops and/or discouraged the multiplication of insect pests. Mechanical methods included hand picking of larvae, shaking beetles loose from trees, and rolling fields to kill insect pupae. Chemical controls included the application of sulfur, camphor, kerosene and soap emulsions, and later arsenic-based compounds to kill undesirable insects.

It seems that initially the agriculturists expected the entomologists to provide them with information of two distinct but related kinds. First, agriculturists expected the entomologists to provide them with information about the identification, habits, and life histories of injurious insects, and second, they expected the entomologists to suggest effective remedies based on the information about the insect. At first, they expected the entomologist not to solve problems for them (as later in the case of eradication programs undertaken by the federal and state governments) but to tell them the most effective means by which to solve problems themselves.

The entomologists claimed they could provide the desired information but with two significant provisions: first, the gathering of accurate information would demand sustained investigations over many seasons, which required continued funding of full-time positions, and second, the success of the control measures would depend on the agriculturists' appreciation of the complexities of the natural order and their willingness to adjust their practices to conform to these phenomena. The entomologists held out little hope for a simple, quick, and complete solution to insect outbreaks.

What effect this advice had on actual farming practice is difficult to assess. It seems clear, however, that prior to the extensive use of arsenic insecticides in the 1870s, the entomologists could claim few dramatic victories over insect pests. The results of their advice were apparently modest. The question then arises, why did the agricultural community continue to place explicit faith in the entomologists and their work?

The answer lies partly in the agricultural leaders' faith in science as a method of general improvement. A more specific reason in the case of entomology was the appeal to the concept of the "balance of nature" (or the "economy of nature") for an explanation of the fluctuations in insect populations as the ultimate guide to control measures. This widespread and popular concept drew assent from many quarters. In the nineteenth century, entomologists used the concept more frequently perhaps than any others in the scientific community.

The "balance of nature" or "economy of nature" presupposed a beneficent, orderly God who had created the world of plants and animals with natural checks and balances and who continued to regulate it through these mechanisms. It further assumed a basically harmonious relationship among all living species, each of which was endowed with traits that would enable it to maintain its essential shape (morphology) and to inhabit its geographic range despite challenges from the other species or from the environment. These ideas date from antiquity, but they were given their classical statement in the eighteenth century. William Derham, an English clergyman and disciple of John Ray, apparently first used the term "balance" in an ecological context in 1713: "The Balance of the Animal World is . . . kept even, and by a curious Harmony and just Proportion between the increase of all Animals, and the length of their lives, the World is through all ages well, but not overstored."[4] The concept was employed widely by naturalists in the eighteenth century, including Linnaeus, who framed his science of ecology according to its organizing principles, and his student, Peter Kalm, who in his travels helped to introduce it in North America. The concept was adopted by leading American scientists like Benjamin Franklin, Thomas Jefferson, and John and William Bartram.[5]

The classic statement of the balance of nature in the entomological literature was formulated by Kirby and Spence in their *Introduction to Entomology* (1816): "The common good . . . requires that all things . . . should bear certain proportions to each other. . . . none act a more important part than insects [which] thus are fulfilling the great law of the Creator, that of all which he has made nothing would be lost."[6] Similarly, Kirby and Spence gave a teleological explanation for the transformation of insects:

A very important part assigned to insects in the economy of nature . . . is that of speedily removing superabundant and decaying animal and vegetable matter. For such agents an insatiable voracity is an indispensable qualification, and not less so unusual powers of multiplication. But these faculties are in a great degree incompatible. An insect occupied in the work of reproduction could not continue its voracious feeding. Its life, therefore, after leaving the egg, is divided into three

stages. . . . [the larva's] sole object is the satisfying of its insatiable hunger . . .
[while the sole object of the adult is the multiplication of its kind].[7]

One popular offshoot of the concept of the balance of nature was
phenology, which was the practice of recording the seasonal unfolding of
biological phenomena on a calendar, like the annual dates for the blooming of flowers, the appearance of insects, or the migration of birds. Such
observations hinted at practical applications: biological calendars might
help farmers detect changes in the climate or assist physicians in their
search to correlate environmental influence and disease. The most popular book on phenology was *Florula Bostoniensis* (1814) by Jacob Bigelow, a
Boston physician and student of John Bartram. Besides inspiring numerous floral calendars in other parts of the country, Bigelow's book served
as a model for Harris, who at one time considered writing a book entitled
Faunal Insectorum Bostoniensis.[8] Accounts of the phenological genre persisted in the nineteenth-century entomological literature, for example, in
Lintner's calendar of butterflies for New York, which was based on the
observations of volunteers throughout the state.[9]

It is important to note that the concept of the balance of nature as it was
generally received in the nineteenth century consisted of a set of assumptions rather than a scientific hypothesis to be tested. For this reason, the
concept survived intact well into the twentieth century without being
seriously questioned, despite the development of data that would otherwise have demanded revision. For example, the balance of nature assumed the immortality of species, but despite the fact that the majority of
naturalists after about 1800 believed in the extinction of species, this
information was not used to question the balance of nature.[10] Because the
balance of nature was based on unchallenged assumptions rather than
stated as a hypothesis to be tested, additional data, even if inconsistent
with the assumptions, could be and were incorporated into this view of
nature. This uncritical acceptance of the balance of nature helps explain
why, despite the rapid acceptance of evolutionary theory by American
entomologists following Darwin's *Origin of Species* (1859), the discussions
of injurious insects continued to be framed in terms of this concept.

Entomologists employed the idea of the balance of nature to explain
the ecological relationships of insects in the natural world, but they also
insisted on the need for practical control measures when insects upset the
balance in a field or orchard. The eighteenth-century naturalist Richard
Bradley, while subscribing to the idea that all species were necessary for
the harmonious working of nature, had no qualms about exterminating
insect pests when they threatened his garden.[11] In doing so, he implied

that the "balance" could be manipulated for man's benefit. American entomologists adopted a similar position.

American entomologists viewed the fluctuations in animal populations as the product of an intricate functioning of climate and predator-parasite-prey relationships. Temporary fluctuations were part of the natural order, and the appearance of unusual numbers of insects in any given year was to be expected. As Riley pointed out, "Every collector of insects knows that scarcely a single season elapses in which several insects, that are ordinarily quite rare, are met with in prodigious abundance . . . , but the same holds true with predators on those insects."[12]

According to American entomologists, the unusual destructiveness of insects in North America was due to the fact that the balance of nature had been upset by human intrusion. This balance had been upset in two ways. First, Europeans and their descendants had replaced the natural forest and grasslands with single crops, concentrated in specific areas devoted to monoculture. Second, they had introduced new insects from the Old World, which had no corresponding predators to keep them in check. As Walsh told the Illinois State Agricultural Society in 1859, insects in a state of nature were regulated by natural checks, but "whenever man, by his artificial arrangements, violates great natural laws . . . , he pays the penalty."[13] In a similar vein, Riley held that insects originally played their proper role in the economy of nature in North America, "But civilized man violated this primitive harmony. His agriculture . . . is essentially the . . . cultivation, in large tracts, of one species of plant [which] gave exceptional facilities for the multiplication of such insects as naturally fed on such plants."[14]

While native species temporarily increased at unusual rates because of concentrated sources of food or favorable weather conditions, imported insects increased even more rapidly, because they had few, if any, predators to keep their numbers in check. For this reason, entomologists agreed that introduced pests posed the greatest problem.[15]

The control measures proposed by entomologists followed from their discussion of the balance of nature. The agriculturists should first of all be informed as to which insects were their "friends" and which were their "foes." Only the careful observation of insect life cycles would yield the basic information necessary for control. Once the friends and foes had been identified, the agriculturist could modify his practice to assist natural controls by rotating and diversifying his crops, fertilizing to strengthen crops against insects, and similar means. Such basic knowledge would also ensure that the application of chemicals would target only the harmful species at the most vulnerable part of their life cycles

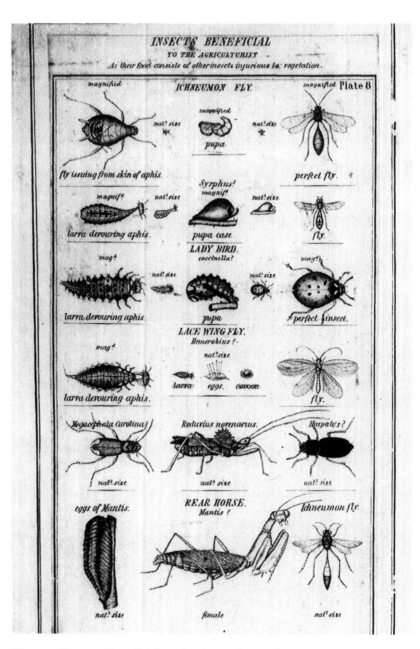

Plate 8, "Insects Beneficial to the Agriculturist," Townend Glover Notebooks, ca. 1850–1878, Cotton Insects volume, box 8, Record Unit 7126, Smithsonian Institution Archives, Washington, D.C.

and would not harm the beneficial species. Because of these considerations, entomologists sometimes advised against popular control measures, like burning stubble or "sugaring" to attract moths, because these methods destroyed the pest's natural enemies. By interfering with natural controls, the farmer would intensify, rather than alleviate, the problem. Above all, the entomologists assured the farmer that, in the long run, the economy of nature would reassert itself, bringing abnormal populations under control.

In the case of insect pests that had come from outside North America and therefore had no natural controls, American entomologists proposed the importation or breeding of imported predators or parasites. Walsh, who first made this proposal in 1859, offered his own theory to explain the unusual destructiveness of Old World insects in America. Combining assumptions about the balance of nature with speculations based on Darwinism, he postulated that the New World fauna had its origins in a more ancient geological era than the fauna of the Old World and was therefore less hardy. It followed that New World predators would be less able to control the populations of the sturdy Old World imports.[16] The solution, Walsh argued, was to import Old World predators, so that they could reassert the balance of nature in the new environment. In 1859, he advised the Illinois State Agricultural Society that "one of the most effectual means of controlling noxious insects is . . . the artificial propagation of such cannibal species as . . . prey on them."[17] He proposed public support for entomologists to search for predators of the worst pests, to import them, and to raise them for distribution. In addition to the importation of predators, or "cannibal species" that devoured unwanted pests, American entomologists also proposed the importation of parasites that weakened the host species, thus reducing its capacity to multiply. Walsh insisted, "The plain common sense remedy . . . is . . . to import the European parasites."[18] In the 1860s, Fitch, Riley, Packard, and others advocated the importation of parasites.[19] Riley was apparently the first to raise and distribute insect parasites. In 1870, he sent parasites of the plum weevil *Conotrachelus nenuphar* to various places in Missouri, and in 1873 he sent a predacious mite *Tyroglyphus phylloxera* to France to combat the phylloxera grapevine louse. Neither was successful, but Riley and others continued to experiment along these lines.[20]

Another variation was the use of insect disease as a control measure. John L. LeConte suggested this expedient at the meeting of the American Association for the Advancement of Science in 1873, and in 1879 Riley, John Henry Comstock, and J. H. Burns tested, but did not confirm, Hagen's suggestion that yeast could infect and kill insects. In the 1880s and 1890s, Stephen A. Forbes performed extensive experiments on insect disease as a control measure but with limited success.[21]

In general, the entomologists were optimistic about nature's ability to readjust to new conditions like the introduction of new crops. They noted that although severe insect depredations often occurred in the years immediately following the clearing of forests and planting of new crops, it usually required only a few seasons for the pest's natural enemies to increase and reestablish the balance. Packard noted that "certain injurious insects become suddenly abundant in newly cultivated tracts. The balance of nature seems to be disturbed, and insects multiplying rapidly in newly settled portions of the country, become terrible pests. In the course of time, however, they seem to decrease in numbers and moderate their attacks."[22] Similarly, Thomas reassured Illinois farmers that injurious insects were part of nature's way of readjusting the balance. When nature's balance was upset by clearing land and raising crops, the insects turned to consuming the introduced plants. Soon, however, the old enemies of the noxious insects would increase, and in time their enemies also.[23]

In the 1860s and 1870s, the relation of birds to insects and crops, popularly known as the "bird question," was the most widely discussed aspect of the balance of nature. Since colonial times, humans had considered birds primarily as competitors for cultivated crops and as legitimate game for anyone with the means of destroying them. By the 1850s, gardeners, farmers, and horticulturists, besieged by insect pests, were beginning to rethink the role of birds in the balance of nature. Many now viewed them primarily as insect eaters and hence as potential friends and allies. Agricultural and horticultural societies held frequent discussions on the bird question, in which they attempted to establish which bird species were "friends" and which were "foes." At an 1864 meeting of the Missouri State Horticultural Society, for example, the bird question was debated at length. Mr. Hull moved that the society recommend the destruction of the "oriole, Cat bird, woodpecker, blue jay, and cherry bird" but that the "hairy woodpecker" and the "large woodpecker, standing a foot high [Pileated Woodpecker]," be spared. This motion "excited a long and very earnest . . . discussion . . . in which . . . each bird was taken up. . . . Some contended that these birds also destroy insects, as well as fruit, but others said the insects destroyed were helpful, not harmful; [they were] the 'cannibals' among insects."[24] Similar discussions were held in other agricultural and horticultural societies. E. A. Clifford of the Illinois State Horticultural Society listed the following as horticulturists' enemies: the "butcher birds," jays, orchard orioles, catbirds, cedar birds, and the sapsucker. Among the horticulturist's friends he named the swallows, martins, wrens, woodpeckers, and nuthatches.[25] In New York, Gurdon Evans insisted that the farmers made a mistake when they killed flocks of "yellow birds" that, he contended, were feeding on the "wheat fly."[26]

The first legislation protecting birds came in 1850 when legislators in New Jersey and Connecticut established fines for killing insectivorous species or destroying their eggs. By 1864, twelve states and the District of Columbia had similar laws.[27] Despite the passage of such legislation, very little was known about the feeding habits of the various species; hence ornithologists and entomologists were called upon to provide more precise information. The demand for such expert opinion served as an additional argument for the establishment of positions for state entomologists, and the official entomologists responded to these demands by devoting considerable attention to the study of relations between birds and insects.

By the 1860s entomologists and other investigators had reached general agreement that the indiscriminate slaughter of birds should be stopped. This annual slaughter was apparently as much for sport as from economic motives. For example, Philo R. Hoy of Racine, Wisconsin, advocated the slaughter of sapsuckers, which arrived in Racine about April 15 each year. "Then is prime 'boy time,'" he said, when the boys "armed with bows and arrows, stones, clubs, cross guns, and the like wage . . . war on the sapsuckers."[28] It seems clear that many men and boys did not discriminate between species when the annual campaign began. Packard noted the "barbarous custom" according to which young men organized in the fall for the indiscriminate slaughter of every species of bird.[29] Thomas also asserted that farmers often waged war on those very birds that were destroying their worst enemies.[30] A Kansas settler reported that in 1874 his neighbors had slaughtered prairie chickens for food and for sale on eastern markets. He concluded that the killing of prairie chickens that fed on insects and insect larvae had contributed to the destruction caused by young locusts the following year.[31]

The entomologists who were called upon to distinguish between avian friends of the farmers and avian foes at first had scarcely more accurate information than the farmers themselves, and their advice was not always the best. Riley, for example, proposed that state and local governments offer rewards for hawks' heads, as was being done in Colorado, on the assumption that they preyed on insectivorous birds. He also proposed the introduction of the English grackle and English rook to North America because of their presumed utility in devouring destructive insects.[32] In light of the unfortunate results that followed the introduction of the

English sparrow during the same period, Riley's suggestions seem in retrospect to be particularly ill advised.[33] One also wonders why Riley did not consider the long-term implications of upsetting the balance of nature when he advocated the slaughter of hawks. Subsequent experience and research have demonstrated their useful role in checking rodent populations that destroy cultivated fields and pasture.

In contrast to Riley's flamboyant advocacy of introducing new avian species to North America, Walsh approached the proposition cautiously. He first asked ornithologist Joel A. Allen to draw up a list of bird species in Illinois with a description of what they ate, but he later expressed some doubt about the conclusiveness of Allen's report. Before any action was taken, Walsh insisted, it was necessary to "draw up a careful Debtor and Creditor account, and ascertain on which side the balance lies."[34] Walsh tried to persuade the Illinois legislature to appoint a special commission to study the insect-bird relationship, but the legislators ridiculed the suggestion.[35]

Meanwhile, a growing number of American investigators in various quarters had begun the systematic examination of the contents of birds' stomachs as a means of establishing more precisely their role in the balance of nature. One of the first was J. W. P. Jenks, a naturalist in Massachusetts, who in 1858 published the results of his examination of robins' stomachs.[36] In 1870, LeBaron reported the investigations of a farmer in Illinois who had observed prairie larks eating grubs and, upon an examination of their stomachs, determined that they were eating harmful insects.[37] Packard cited an ornithologist in Ipswich, New York, who had opened the stomachs of three thousand birds and estimated the number of cankerworms eaten by them.[38]

Public concern over the massive destruction of crops by the Rocky Mountain locust in the 1870s served to transform these individual efforts into a distinct new scientific discipline of economic ornithology. The U.S. Entomological Commission, appointed primarily to deal with the locust emergency, devoted considerable attention to economic ornithology, including a thirteen-year study of bird-insect relations by Samuel Aughey of Nebraska.[39] In Illinois, Forbes received support from the Entomological Commission for the Illinois State Laboratory of Natural History to undertake detailed studies of the contents of birds' stomachs. Forbes's investigations brought for the first time a consistent, detailed, quantitative approach to the study of birds' eating habits.[40] In 1878–1880, Forbes published the first results of these studies. His examination of the stomachs of some fifteen hundred birds, including all the species of economic importance, supported the growing conviction among farmers, entomologists, and ornithologists that predation by birds played a significant role in limiting insect populations.[41]

These studies by Forbes and others fit into a pattern of growing interest in the statistical analysis of quantitative data that characterized American investigators from the Jacksonian period onward. Quantitative investigations took a prominent place in the work of new biological agencies appearing in the Department of Agriculture and elsewhere beginning in the 1870s. These included separate divisions or bureaus for entomology, economic ornithology, and mammalogy, many of which were eventually incorporated into the U.S. Biological Survey. These agencies based their scientific and regulatory activities largely on an ongoing program of observing and measuring the eating habits of predators.[42]

The confidence that agriculturists placed in the early investigations in economic ornithology reflects a general acceptance of the balance of nature as a reliable frame of reference. The assumption of the agriculturists is suggested indirectly by the fact that entomologists who knew their audience employed this concept in their writings. Further evidence comes from writers of farm journals, who at times referred directly to the balance of nature. The editor of the *Prairie Farmer* in 1853, for example, summarized the balance of nature in a way that suggests his readers' familiarity with the concept:

Birds, reptiles, and predacious insects were doubtless provided to keep the most injurious species in check. . . . But man interfered, and reversed the natural order. . . . The forests which fed the destructive insect and sheltered the devouring bird and reptile, have been swept away, and the countless billions of the one and the frightened remnants of the other, seek food where it can be found, or safety in less "civilized" regions; and the insect has now settled in our gardens and fields, and his enemies, and our friends are fast disappearing before the tide of "improvement," and . . . ignorant prejudice or wanton barbarity.[43]

Similar statements in the agricultural press of the period indicate a common acceptance of the balance of nature as an underlying principle. In 1846, the editor of the *Prairie Farmer* quoted Fitch's report on the wheat fly, noting that the fly's natural enemies, like the "yellow bird," were "intended by Providence as the . . . corrective" that would reassert the balance if given a chance by farmers.[44] A writer in the *Prairie Farmer* in 1852, as a preface to his discussion of the "philosophic" and practical importance of birds in the control of insect populations, quoted the *North East Cultivator*, which stated: "The great laws of nature . . . are carried on through insects."[45] Gurdon Evans, who published an agricultural survey of Madison County, New York, wrote in 1851, "It is . . . an all pervading principle of want and supply in nature [that] when one kind of insect gets the ascendance [sic] . . . its enemy soon increases in the same ratio, and succeeds in gaining the victory."[46]

Though most writers in the agricultural press accepted the balance of nature theory, there were some critics. Some agriculturists and horticulturists pointed out that it took nature too long to reassert the balance, and therefore the concept was of no practical value. One Missouri horticulturist calculated the rapid multiplication of curculio (weevils) in his fruit orchard and concluded that within two years this population could destroy more fruit than could be grown on one thousand trees. He sarcastically admonished horticulturists to "cut down your fruit trees . . . and the much admired balance of nature . . . will be restored."[47] The corrective in such cases, he concluded, would be to establish an artificial balance by developing and using more effective insecticides. Henry Shimer of Illinois criticized entomologists like Walsh and Thomas for overemphasizing the effectiveness of predator insects as the central feature of the balance of nature and underestimating the direct influence of weather as a check on insect populations. On the basis of his observations of the Colorado potato beetle, Shimer concluded that changes in the weather produced massive death from epidemic disease among insects and that weather was a more effective agent in the control of insect populations than "cannibal insects."[48] Shimer's criticism that entomologists laid too much stress on predator-prey relationships in defining the balance of nature was well founded. As has been noted, however, the set of assumptions that underlay the balance-of-nature concept was flexible enough to incorporate divergent views like the relative efficacy of climate as a regulating mechanism or the empirical results of new insecticides, without altering the overall concept. For example, Riley later cited the warmer, more humid climate of the Mississippi Valley as a more effective barrier than any predator to the spread of the Rocky Mountain locust east of the Mississippi River.

Despite the criticisms and modifications, most farmers and agriculturists who expressed opinions on the subject subscribed to the balance-of-nature concept as the basis for the entomologists' work. The entomologists were expected to suggest strategies the farmer could use to manipulate the natural balance in a way more beneficial to him. The entomologist's knowledge of the insect's life cycle thus became immediately practical, for as LeBaron wrote, "there is a period in the lives of most . . . noxious insects . . . when some one or other of the common remedies, such as soap, tobacco, lime, or ashes, is effective."[49] The agricultural entomologist assumed the task of identifying the vulnerable points in the life of harmful insects. Once these were known, he could recommend the appropriate remedy.[50]

The importance of the theory of the balance of nature in agriculturists' support for economic entomology may be suggested by comparing agricultural chemistry and agricultural entomology. In the 1840s, agricultural editors promoted soil analysis as a quick and easy way to improve and expand farm production. When soil analysis proved to be much more complex than advertised by its promoters, however, and when the payoff proved to be much less immediate and dramatic than first promised, its promoters in the agricultural press quickly abandoned the campaign.[51] Their disenchantment came about in part because of the growing distance between the scientific experts on the one hand and the promoters and users on the other. Agricultural chemists like John Pitkin Norton typically received their graduate training in Germany and the British Isles. Norton warned against encouraging young men to become chemists unless they were prepared to undertake specialized studies for seven years or more.[52] Despite their familiarity with farmers' problems and their ability to communicate in the language of the agricultural press, agricultural chemists found that their specialty required them to use scientific terms and chemical formulas beyond the grasp of farmers. The gulf in learning between agricultural chemists and agricultural promoters also meant that, by the 1850s, agricultural chemistry had developed to a stage where farmers' empiricism was of little use in the advancement or practice of the scientific specialty. Indeed, agricultural chemists insisted that their science could be useful only by maintaining a sharp distinction between trained scientists on the one hand and their scientifically untutored clients on the other. Norton warned, "when you meet with a man who makes [soil chemistry] easy . . . , distrust that man, for he is either intentionally imposing upon you, or he thinks he knows what he does not."[53]

In contrast to the growing theoretical and empirical gulf dividing agriculturists and scientists in the field of soil science, agricultural entomologists continued to explain their work in the familiar principles of the balance of nature, and they continued to value the empirical information gathered by agriculturists. Forbes, lamenting his limited opportunities for field observations of birds, appealed to the members of the Illinois Horticultural Society to record their observations each day in a diary and to submit these to him or other investigators for study.[54] Entomologists continued to solicit information from agriculturists and other observers, and they credited these observers for their valuable contributions to science, including them in effect as auxiliary members of the entomological fraternity.[55] Their use of observations and descriptions that were familiar to the educated elite of gentlemen farmers, ministers, and physicians helped ensure the continued support of this elite.

Given this general agreement on principles and methods, the introduction of more effective chemical insecticides like Paris green, which became the standard remedy against the Colorado potato beetle in the 1860s, was not seen as being inconsistent with the general reliance on predator-prey relations up to that time. Though chemical insecticides eventually became the most important component in the entomologists' remedies, and biological control was relegated to the background until recent times, the basis of all control measures, whether "natural, mechanical, or chemical," was still assumed to be the balance of nature.

A Weevil, a Fly, a Bug, and a Beetle

6

Armed with the scientific knowledge and institutional framework of their discipline, and guided by generally accepted principles enshrined in the balance of nature, American entomologists responded with confidence to the insect outbreaks that spread with increasing rapidity and force on the American agricultural scene. Among these insect outbreaks, certain species posed especially persistent threats to growers and their entomological consultants. Five such species will be treated in some detail. These are the curculio, the Hessian fly, the chinch bug, the Colorado potato beetle, and the Rocky Mountain locust. Outbreaks of each of these species prompted significant advances in knowledge or professional organization within the American entomological community. The cumulative result of these advances was the emergence, by the 1870s, of the American entomologists as the world's leading practitioners of applied entomology.

The curculio, a destroyer of plums, peaches, pears, nectarines, apricots, and cherries, threatened the development of horticulture, while the Hessian fly and the chinch bug at various times laid waste to major field crops like wheat and corn. The Colorado potato beetle, though its host plant was more often a home garden vegetable than a commercial crop, merits attention because its control by arsenic poison initiated the first era of chemical insecticides. Finally, the crisis caused by the Rocky Mountain locust, which in the 1870s devastated large areas of the newly settled plains region, called forth the support of entomology by professional organizations and government on an unprecedented scale. In particular, the locust outbreak led to the formation of the U.S. Entomological Commission, the prototype for government scientific bureaus.

In the years between the revolutionary war and the Civil War, fruit growing engaged the attention and energies of many of the nation's leaders including George Washington, Thomas Jefferson, and Henry Clay. In the 1830s and 1840s, with the development of improved transportation, this gentlemanly pursuit expanded into a major commercial sector that centered in New York state and around the Great Lakes. Commercial canning increased sixfold during the Civil War.[1] Apples, the largest single fruit crop, were comparatively hardy and easy to raise, while the soft fruits, such as plums, pears, and peaches, caused the horticulturist a host of problems, one of the most persistent of which was the curculio, or "Little Turk."[2]

The curculio is a member of the weevil family, Curculionidae, from which it derives its name. The species overwinters as an adult, hibernating in ground cover in the vicinity of the orchard. In the spring, as the fruit trees blossom, the insects emerge from hibernation and congregate in the host trees. The females deposit their eggs in the developing fruits and cut a crescent-shaped scar at the point of oviposition. This characteristic scar, which gives rise to the name "Little Turk," presumably prevents the rapidly growing cells of the fruit from crushing the eggs.

After the eggs hatch, the larvae feed by tunneling through the fruit. Fruits attacked early in their development drop to the ground, while those attacked at later stages of development cling to the tree, so that the fruit is wormy at harvest. The late injury provides entry for rot organisms which destroy the fruit as a secondary effect of insect injury. In this way, the entire crop of fruit may be lost. The mature larvae abandon the fruit and enter the soil, where they pupate and emerge again as adults to repeat the cycle. So detrimental was the curculio to the interests of horticulturists that in some regions the raising of soft fruits was partially abandoned for a time.[3] Prior to the 1850s, the search for measures that could protect the fruit growers from the ravages of the curculio elicited the most extensive discussion of any insect pest in America.[4]

For many years there was practically no reliable information about the curculio's history and control. Andrew J. Downing, one of the most knowledgeable horticultural writers, advocated as a control the building of a nine-foot-high fence around the orchard.[5] In 1848, the president of the St. Louis Horticultural Society reported the failure of frequently proposed remedies for the curculio such as applying horse manure to the tree trunks, salting the ground, smoking the trees with sulphur, and

"Shaking Trees to Release the Plum Curculio," from J. P. Trimble, *A Treatise on the Insect Enemies of Fruit and Fruit Trees* (1865), plate 7

washing the trees with tobacco, whale oil, or soap. The only sure method he had found was to spread sheets around the trees and tap the trees with a mallet, then collect and kill the weevils which fell on the sheets. Remedies reported to be partially successful were paving the orchard floor and running poultry and swine in the orchard so that they would eat the larvae.[6] Other proposed controls for the curculio were reported regularly in the agricultural press, but the only sure method appears to have been the practice of jarring the trees and killing the adult weevils.[7]

The horticulturists' problems with the curculio directly influenced the appointment of at least two state entomologists, Fitch in New York and Walsh in Illinois, both of whom were directed by their agricultural society sponsors to devote their first annual reports to the insects affecting orchards and gardens.[8] Fitch, Walsh, and other entomologists were able to unravel the life history of the curculio in great detail. In the years up to 1880, however, the only control methods entomologists could recommend with confidence consisted of more elaborate technological versions of the jarring and collecting method already in common practice by the 1850s.

Despite the lack of new control measures for the curculio, the horticulturists' support of government-sponsored entomology was soon vindicated in other areas. In 1869, William Saunders experimented with sulphur dust to control rot and mildew on grape orchards. His success marked the beginning of an era in the use of poison dust and sprays in the control of important orchard diseases. In the early 1870s, LeBaron, prompted by the growing use of the arsenic-based insecticide Paris green in the control of the Colorado potato beetle, recommended the use of Paris green on various orchard pests. By 1872, C. M. Hooker of Rochester, New York, had reported success in the use of Paris green in the control of canker worms and current worms. A few years later J. S. Woodward, a horticulturist in Lockport, New York, introduced the use of arsenical sprays in orchards. Based on these and similar successes, horticulturists continued to support government-sponsored applied entomology.[9]

The Hessian fly has the distinction of being the first insect pest to plague American farm crops on a massive scale. Unlike the curculio, which attacked fruit, to some extent a luxury crop, the Hessian fly laid waste the wheat crop, the very staff of life. For a time in the 1870s, the Rocky Mountain locust surpassed all other insects in destructiveness, but considering the nineteenth century as a whole, only the chinch bug was counted a worse farm pest than the Hessian fly.[10]

The adult Hessian fly is a "small gnat-like two winged creature, about half as large as a common mosquito, which it resembles."[11] It is the larval stage that destroys grain crops, especially wheat but occasionally also rye and barley. The female lays the eggs on the leaves of the young wheat plants, where they hatch into larvae and travel down to the sheath and attach themselves to the stalk. There they remain, sucking the juices and weakening the wheat stem at the joint. With the first heavy wind or rain, the weakened stems break, transforming fields of ripening grain within hours or minutes into worthless jumbles of straw. The larvae, still in the straw, change into brown prepupae, called "flax seeds," and remain for a variable time before pupating and emerging as adult flies. There are two main generations of the Hessian fly, one generation that matures in the spring and lays eggs on spring wheat, followed by a fall generation that lays eggs on winter wheat.[12]

The first appearance of the Hessian fly in America occurred in 1778 on Staten Island, New York, where it destroyed a portion of the wheat crop. The name "Hessian fly" was applied at that time because of the suspicion that it had been introduced inadvertently by Hessian mercenaries in the straw of horses quartered in that area during the revolutionary war. Fitch later confirmed this suspicion, calculating that the "flax seed" stage of the larvae was transported in the straw of the Hessian troops in August 1776.[13] In the last two decades of the eighteenth century, the Hessian fly spread throughout the wheat district of the eastern United States. Agricultural societies and legislatures on both sides of the Atlantic sought means of controlling the Hessian fly. In 1788, the British Privy Council, fearing the spread of the fly to Great Britain, prohibited the importation of American wheat. In response, the Executive Council of Pennsylvania asked the Society for the Promotion of Agriculture to investigate the matter. The society's report, which stated that the insect could not be transferred in grain, persuaded the Privy Council to lift the imposition.[14]

The most comprehensive early report was written in 1792 by Judge Jonathan N. Havens for the New York Society for the Promotion of Agriculture, Arts, and Manufacturing. He proposed the burning or plowing under of wheat stubble as a control measure. He also encouraged the sowing of bearded varieties of wheat, especially one called "Underhill" wheat, which in the 1780s had demonstrated resistance to the Hessian fly.[15] During the first two decades of the nineteenth century, when the Hessian fly did little damage, the fly attracted the interest of taxonomists rather than agriculturists. In 1817, Say described it and gave it the scientific name *Cecidomyia destructor* (Say).[16] The fly resumed its widespread destruction of wheat crops in New York in the 1820s, then abated somewhat in the 1830s. In the 1840s a major outbreak occurred in

New York and Pennsylvania and spread rapidly throughout the new wheat district around the Great Lakes and in the Mississippi Valley, eventually reaching the plains and the Pacific coast states.[17] In the years from 1844 to 1847, the Hessian fly destroyed much of the wheat crop in Indiana, Illinois, Michigan, and Wisconsin.[18]

With the renewed destruction caused by the Hessian fly in the 1840s, new studies were published by Edward Claudius Herrick, librarian of Yale College, and by Fitch.[19] Though Herrick and Fitch could add little to what was already known about the life history of the Hessian fly, they did evaluate the various control methods and recommend those they considered to be most effective. Some methods they rejected as useless, like soaking or sun drying the wheat seed or buying seed from unaffected areas. These methods were given a boost through the publication, in 1840, of a study of the Hessian fly by Margaretta Morris, an entomologist in Germantown, Pennsylvania. Morris proposed the theory that the fly laid its eggs on the wheat berry rather than on the stalk as had been generally accepted. Herrick and Fitch rejected Morris's theory, and their position was subsequently sustained. It seems likely that Morris confused the larval stages of the Hessian fly and the wheat midge, another species she investigated. The adults of the two species are easily confused, and the larva of the wheat midge are associated with the wheat berry.[20] Other methods rejected by Fitch included the practice of drawing alder branches over the young wheat and the application of salt, ashes, or lime.

Control methods recommended as partially effective were planting wheat only in rich, well-fertilized soils, deep planting, grazing livestock on spring and fall wheat, using a roller on the young wheat plants to crush the eggs and larvae, and using selected patches of oats or wheat as a decoy crop and then destroying them. Fitch had at first recommended burning the wheat stubble, but he later changed his mind, arguing that it destroyed too many of the fly's natural enemies. Fitch strongly recommended the late sowing of wheat and the selection of resistant varieties, though he noted that no variety was completely flyproof. Fitch concluded that no complete "remedy" for the Hessian fly was known, or probably ever would be known. He maintained, however, that certain of the recommended measures, if used in the right combination, would reduce the damage to a minimum and ensure good crops.[21]

The Hessian fly continued its destructive career at various times and places, with particularly heavy damage in the West. From the 1850s on, the fly was investigated extensively by the new generation of entomologists, like Thomas, Riley, and Packard, from their posts at state

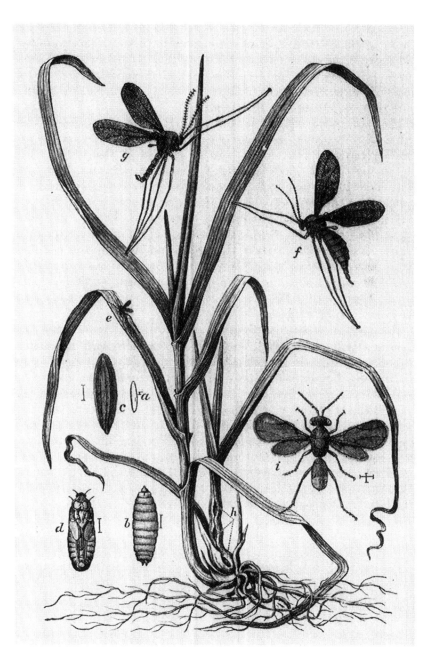

"The Hessian Fly and its Transformations," from A. S. Packard, Jr.,
"The Hessian Fly," in *Third Report of the United States Entomological
Commission* (Washington, D.C.: U.S. Government Printing Office,
1883), pp. 198–248, plate 4: *b* by Charles V. Riley, *d* and *f* by Edward
Burgess, and *a, g, e,* and *i* by A. S. Packard, Jr.

entomological offices, the Hayden survey, and the U.S. Entomological Commission. In 1880, Packard summarized the information available in a bulletin on the Hessian fly published by the U.S. Entomological Commission. A comparison of his conclusions with those of Fitch in the 1840s and with modern authorities shows a gradual refinement of methods, with some important new directions.

The first new direction came from investigations by Thomas and Riley that indicated strong links between outbreaks of the Hessian fly and weather patterns. They found that the Hessian fly flourished in wet seasons but that it ceased to be a problem in dry seasons. They also concluded that natural predators increased rapidly enough to render it unlikely that outbreaks would occur in the same area for more than one year in a row. On the basis of these observations, Thomas and Riley began predicting outbreaks for the coming season in specific localities with a fair degree of accuracy.[22] These predictions were the first instances of what eventually developed into annual insect pest surveys. By the twentieth century, such surveys constituted a primary service that entomologists provided to farmers in specific areas of the Midwest as a basis for planting more rationally.[23]

The second new direction was the establishment of "fly-free dates" for planting, the times in spring and fall before and after which the Hessian fly appeared in specific localities. According to Packard, planting during fly-free periods promised to be a very effective control method.[24] The establishment of fly-free dates for planting, however, represented an enormous task, because the number of generations and the life span of each stage of the Hessian fly vary from place to place. Fly-free dates vary considerably even within individual states.[25] Nineteenth-century entomologists lacked the personnel and other means to carry out these investigations, so they could only recommend early and late planting and hope that the farmers would thereby avoid massive damage from the Hessian fly. By 1880, the goal of ascertaining fly-free dates for planting and the basic methodology for doing so had been established. In the twentieth century the charting of fly-free dates for the various states developed into one of the most important control measures for the Hessian fly.[26]

The third new direction deriving from nineteenth-century Hessian fly investigations was the search for resistant varieties of wheat and the improvement of these varieties through selective growing. Eventually the development of resistant varieties came to be the single most effective measure in the control of the Hessian fly. Even more than the charting of fly-free dates, the full development of resistant varieties lay beyond the reach of nineteenth-century entomologists because it required the development of sciences like agronomy and plant pathology, which came only later in the century. This search for resistant varieties involved the development of specific varieties for different regions; one variety, for example, was found to be more effective in Kansas and another in North Dakota.[27]

Packard's 1880 report on the Hessian fly indicates that the practical measures available to farmers at that time represented mainly refinements of those recommended earlier by Fitch in the 1840s and 1850s. Packard advocated "farm culture" methods, such as late sowing, the pasturing of sheep, the planting of decoy crops, and increased fertilization and cultivation of the seedbed to produce strong plants.[28] Packard, like Fitch, advised against the burning of stubble because fire destroyed insect enemies of the Hessian fly. Later entomologists recommended burning, however, provided that it was delayed to allow the parasites to run their course but before the adult Hessian fly emerged.[29]

In summary, the investigations of entomologists in the years from 1840 to 1880 indicated the potential of some important control measures against the Hessian fly, such as the annual insect surveys, the establishment of fly-free dates, and the development of resistant varieties. These investigations also provided valuable advice on the effectiveness of various control methods available and on the conditions under which each of the controls would be most effective.[30]

To what extent farmers heeded the advice of entomologists, or what success they had in controlling the Hessian fly in the years prior to 1880, is difficult to ascertain. The use of resistant seed was apparently widespread, especially the Mediterranean variety that was introduced in 1819 and remained the standard variety throughout the period under study.[31] The burning of stubble was also widely practiced, despite the warnings of Fitch and Packard that burning destroyed the ichneumon wasps that preyed on the Hessian fly.[32] In New York in the 1850s, farmers generally switched from winter to spring wheat, a move that succeeded in controlling the Hessian fly and other insect pests of wheat.[33] Farmers also began plowing wheat stubble under, though to what extent is unclear. It is likewise unclear whether farmers fertilized specifically to control the Hessian fly or for other purposes, whether they rotated crops in the hopes of cutting losses from the Hessian fly, or whether they were forced to drop wheat from their planting after suffering repeated crop failures. Some farmers reported that rolling the fields helped control the Hessian fly, but this method was probably not widely practiced because it required very level fields and specialized machinery.[34]

Some control methods recommended by entomologists were never widely practiced. These included planting of decoy strips of another crop and then destroying them, pasturing sheep on the young wheat, and the introduction of natural predators in the hopes of establishing biological control of the Hessian fly.[35] Fitch, Packard, and later Herbert Osborn all speculated on the possibility of introducing and distributing parasites from abroad, but the experiment was apparently not made.[36]

Even after the introduction of highly resistant varieties and the determination of fly-free dates for planting in the twentieth century, the cultural methods developed by nineteenth-century entomologists remained basically the primary control methods for the Hessian fly practiced by American farmers.[37] In the control of insect pests of field crops like the Hessian fly and the chinch bug, farm culture methods were still the only ones that were economically feasible, given the low cost of application and the low per acre value of the crops.[38]

The chinch bug was once an inconspicuous member of the eastern woodland insect fauna, but by the middle of the nineteenth century it had become the single most destructive species to American agriculture. In contrast to the Hessian fly, which has spread to widely separate wheat-growing regions of the world, outbreaks of the chinch bug have been confined primarily to the midwestern states of Illinois, Indiana, Iowa, Kansas, Missouri, Ohio, Nebraska, Oklahoma, and Texas, with occasional outbreaks in Michigan, Minnesota, Wisconsin, South Dakota, North Carolina, and South Carolina.[39] In outbreak years the chinch bug attacks fields of wheat, corn, timothy hay, oats, millet, barley, and rye, all of which are crops of the grass family.[40]

Adult chinch bugs are about one eighth of an inch long, with white wings folded on their backs in an X. The adults hibernate in masses within clumps of grasses, under logs or boards, and in other protected places. In the spring when temperatures rise above seventy degrees Fahrenheit, they swarm to a nearby wheat field, where they suck the sap from the stems and exposed roots. They then mate, and each female lays up to 500 eggs on the plant stem, just above or below the ground. The eggs hatch in seven to forty-five days, depending on the weather. The tiny bugs that emerge from the eggs attach themselves to the stalks and begin sucking sap. All stages of the insect are found together, and they gradually move up the stalks. As the stalks wither, the bugs move to fresh stalks, progressively destroying the field. Where corn and wheat are grown in adjacent fields, the chinch bug remains on the wheat until the end of June. Then, when the cornstalks shoot up above the withered wheat, the bugs begin a mass migration, mostly on the ground and occasionally in the air, to the new source of food. There they suck the sap from the corn and breed a second generation. In the fall, as the corn withers and drys, the bugs move to clumps of grass in nearby pastures, where they overwinter and begin the cycle again.[41]

In the first recorded outbreak in North Carolina in 1783, the chinch bugs attacked the wheat fields with such severity that after several years farmers ceased to plant wheat. Sporadic outbreaks were reported over the next fifty years. In the 1840s, the chinch bug first appeared in outbreak numbers in the West, in Indiana, Illinois, and Wisconsin. The outbreaks were typically confined to one or several counties in any single year rather than spreading to an entire state or region. At such times the bugs could be seen moving across the fields of wheat, corn, and hay, inexorably bringing ruin to whole farming communities.[42] In the first outbreak year in Illinois, a resident of Knox County reported the bugs covering the ground "in myriads, defying extermination."[43] LeBaron, who witnessed the outbreak in Kane County, Illinois, in 1850, wrote, "so sudden is the invasion and so rapid the progress of these insects, that it is scarcely probable that any preventative . . . will ever be discovered."[44] During 1850 the losses in Illinois were estimated at four dollars per person, and in 1864, when the chinch bug reached its peak of destruction, the richest farming sections in the western states lost half of the corn and three-quarters of the wheat crops. Widespread damage came again in 1871, 1874, and 1881. In the latter year a chinch bug convention was convened in Windsor, Kansas, where delegates passed a resolution urging farmers to discontinue raising wheat until the emergency was over.[45]

In contrast to the Hessian fly, there were no early studies of the chinch bug to provide farmers with information about its life history and possible control. In fact, during the first outbreaks in North Carolina, the chinch bug was frequently confused with the Hessian fly. In 1831, Say named the species *Lygaeus leucopterus,* from a single specimen taken in Virginia, but it remained for the western entomologists to investigate its life history and to develop effective controls. When the first outbreaks occurred in Illinois, the editor of the *Prairie Farmer* noted that the pest was well known to farmers in the East, but it had not yet been treated fully by any naturalist and its habits were imperfectly known.[46] In December 1845, the *Prairie Farmer* printed a woodcut that was advertised as the first accurate illustration of the chinch bug.[47]

The first extensive scientific discussion of the chinch bug appeared in an article by LeBaron in the *Prairie Farmer* in 1850. LeBaron described accurately its presence in all stages following its appearance in June and its migration to fields of wheat, corn, and finally to pastures of wild grass. He also speculated, correctly, that the eggs were laid in the ground.

Unaware of Say's description (from a little-known pamphlet published at New Harmony), LeBaron proposed the name *Rhyparochromus devastator.* LeBaron also noted that the great increases in the chinch bug population came in dry seasons.[48]

The early stages in the life history of the chinch bug eluded investigators for many years. The presence of all stages of the insect at the same time led investigators to the erroneous conclusion that there were up to six generations per year.[49] In 1866, Riley showed that there were two generations in northern Illinois, and sometimes three in southern Illinois, and in 1869, Riley and Walsh first established conclusively that the chinch bug overwintered in the adult phase in clumps of grass. In 1875, Riley was finally able to give a full description of the early stages in the life cycle of the chinch bug.[50]

The control methods proposed by entomologists fell into four categories: (1) natural controls such as predators, parasites, and diseases; (2) farm culture; (3) barriers to stop migrations; and (4) annual surveys of insect populations and weather conditions conducted to predict and prepare for outbreaks the following year. At the time of the first outbreaks in the West, LeBaron held out the hope that the balance of nature would solve the problem. He predicted that natural enemies of the chinch bug would soon increase and provide relief to the farming community.[51] During subsequent years, however, entomologists were disappointed in the apparent lack of effective predators on the chinch bug populations. In 1861, Walsh reported on four species of ladybird beetles that preyed on the chinch bug, but no internal parasites were identified until the twentieth century, and none of these proved very helpful in controlling outbreaks of chinch bugs. Though some vertebrate predators like quail, prairie chickens, and red-winged blackbirds were reported to eat chinch bugs, entomologists reluctantly concluded that natural enemies were not as effective in the control of the chinch bug as for other destructive species.[52]

The idea that disease would prove to be an effective control of the chinch bug was advanced by Henry Shimer, the champion of weather influences and other environmental factors (as opposed to predators) as the main controlling agents of noxious insects. Shimer reported that the record chinch bug outbreak in 1864 was halted by the spread of insect disease. The role of disease in the limitation of chinch bug populations caught the imagination of entomologists like Walsh, Riley, and Packard. In 1882 Forbes initiated an intensive study of diseases of the chinch bug as a practical control measure. By the early twentieth century this line of investigation had led to limited success in chinch bug control through the artificial introduction of a chinch bug fungus.[53]

Having failed to identify effective predators and other natural controls, entomologists encouraged adjustments in farm culture for the control of damage from chinch bugs. These adjustments included cleaning up areas adjacent to fields, burning stubble and other substances to remove sheltered places where chinch bugs overwintered, manuring, and diversification and crop rotation.[54] The two most innovative methods were the erection of barriers to stop the migration of the adult insects and the adjustment of planting to avoid chinch bug infestations.

Many of those who observed the mass migrations of chinch bugs across open land from wheat to corn fields concluded that this migration presented the most vulnerable point in its cycle and the most susceptible to control. In 1866, Walsh reported the success of farmers who had erected board fences coated with coal tar to stop the migrations. Barriers of different kinds were tried, the most effective being ditches filled with water, creosote-saturated fences, kerosene-soaked soil, and (later) contact poisons like calcium cyanide. Such barriers, if erected prior to the migrations and properly maintained, proved effective in saving crops of corn or hay threatened by the first brood of chinch bugs.[55]

As in the case of the Hessian fly, the most important new direction in the control of the chinch bug was the correlation of weather data with outbreaks in order to predict outbreak seasons. In the 1840s both Thomas and LeBaron noted that one or more dry years allowed the chinch bug population to build up to outbreak proportions.[56] Fitch, Riley, Packard, and others noted further that dry seasons allowed population buildups but that wet seasons effectively limited their numbers.[57] Thomas investigated the relationship between weather conditions and insect outbreaks extensively, and in 1879–1880 he presented his views on chinch bug outbreaks and their control.[58] He based his findings on weather observations in Illinois and the adjacent areas of Iowa and Missouri covering the forty-year period from 1840 to 1880 and upon records of insect outbreaks recorded by naturalists in the area during the same period.[59] He found that chinch bugs became destructive during seasons of low rainfall and high temperatures. Two consecutive seasons when these conditions prevailed led to massive outbreaks. According to his analysis, seasons of low rainfall occurred regularly in a seven-year cycle in this region, but seasons of high temperatures occurred irregularly. Given the irregular occurrence of warm seasons, predictions of outbreaks were not yet possible, but Thomas held out the hope that advances in meteorological techniques would eventually make predictions for the coming season possible.[60]

Even in the present state of knowledge, Thomas's findings could be used to good advantage by growers. For example, if the previous summer had been wet, chinch bugs would most likely cause little damage to crops, and wheat and corn (the most susceptible to chinch bug depredations) could be planted. If the previous season had been dry and warm, growers were advised to plant oats and other crops that were less susceptible to chinch bug damage.[61] The season of 1880 was warm and dry with considerable chinch bug damage, and Thomas warned that 1881 would most likely bring an outbreak of chinch bugs. His prediction proved only too accurate. Riley chided farmers in the affected region for not taking heed of Thomas's warning and changing their planting accordingly.[62]

Thomas's studies and predictions were the first tentative steps toward the now well established practice of annual insect pest surveys for the chinch bug and several other injurious species prepared by government entomologists and distributed to farming communities, especially in Midwest states where the chinch bug remains a problem. Such surveys require considerable manpower and time to conduct. They now help entomologists to prepare emergency control measures and farmers to plan their season's planting more rationally.[63]

The demand for bulletins issued by government entomologists on the chinch bug indicates that farmers practiced the methods recommended, or at least sought information about chinch bug control. The chinch bug bulletin issued by Thomas for the U.S. Entomological Commission in 1879 was quickly exhausted. In 1888, Howard prepared a new bulletin on the chinch bug, which he said was necessary because the old reports of Riley and LeBaron were out of print.[64] Despite this interest, during the nineteenth century, the efforts of farmers and entomologists to combat the chinch bug were to little avail. The tentative steps toward annual surveys, which eventually proved effective, required resources and organization beyond the means available in the 1870s.

The appearance of the Colorado potato beetle as a pest in the 1860s and its control through the use of insecticides dramatized two aspects of applied entomology. First, the beetle was a striking example of an insect pest adapting to introduced crops, and second, the chemical control method devised in response to it initiated the pervasive trend toward the use of chemicals in the control of insect pests that for the next century remained the main practice in applied entomology.

Americans were slow to accept the potato in their agriculture and diet. The raising and marketing of potatoes was very labor intensive in comparison with the labor required by other crops, and potatoes therefore did not attract American farmers until the appearance of large urban markets. Furthermore, potatoes did not occupy an important part in the diet of the American colonists, partly because they did not experience famines like those in central Europe in the eighteenth century, which forced populations there to turn to potatoes for food. Despite this resistance, the raising of potatoes for home consumption became well established by the 1830s because the yield per acre for potatoes was the highest of any known crop.[65] In the years 1843 to 1845, the potato blight swept over the northern United States and Canada, reducing yields and discouraging farmers from cultivating potatoes. In 1845, the blight was carried across the Atlantic, where it destroyed the Irish potato crop and sent up to one and a half million desperate Irish immigrants to America. By then the worst effects were over in America. Prices for potatoes rose and production climbed. Thereafter, potato production rose rapidly, especially in the Northwest, stimulated by the growth of markets and by the immigration of Irish and Germans, who relied on the potato as a major staple in their diet.[66]

Since 1824, entomologists had known of a small, bright yellow beetle with black wing stripes, which Say had collected on the Upper Missouri and named *Leptinotarsa decemlineata* Say. This beetle was of interest only to collectors of Coleoptera until about 1861, when it suddenly appeared in numbers sufficient to devour the potato plants of settlers in Nebraska and Iowa. By 1864, this new pest was destroying potato crops in Missouri and Illinois, and by that time it had come under investigation by Walsh, Riley, and other entomologists.[67]

Entomologists had little difficulty in discovering the history and habits of the beetle. Walsh attributed its appearance in farming regions where it had not been present before to an extension of its range made possible by the settlement of the plains. The beetle, he explained, was originally confined to semidesert areas in the West, where it fed on wild plants of the family Solanaceae, the family to which the domestic potato belonged. When domestic potatoes were introduced in the mining settlements of Colorado in the 1850s, the beetle added this food to its diet. Being a poor flyer, however, the beetle was incapable of extending its range until the settlements in Colorado, Nebraska, and Kansas were close enough to allow the beetle to move from one potato patch to another. Once it reached the settled farming region of the Mississippi Valley, the potato beetle could be expected to extend its range steadily eastward at a rate Walsh first calculated at fifty miles per year.[68]

The question of the beetle's original range was of practical importance,

Walsh and Riley insisted, because if the beetle were native to the East, as Fitch and Thomas at first stated, it would likely pose no new danger to the eastern potato-growing regions. If it were a distinctly western species, however, as they believed, it would proliferate as it extended its range eastward because of the lack of natural predators. They predicted that the "new" potato bug would soon endanger potato growers in the eastern United States and Canada.[69] Other entomologists soon conceded that Walsh and Riley were correct, and the potato beetle came to be called the Colorado potato beetle because of its origin in the West. Later calculations indicated that the beetle extended its range at about eighty miles per year, even faster than Walsh first predicted. It crossed from Michigan into Ontario, Canada, in 1870, and it reached the Atlantic in 1874. Walsh pointed out that this was the first known instance of a noxious insect extending its range from west to east. He predicted that American farmers would have to contend with other western species as settlement continued westward, a prediction that was soon fulfilled in the locust plagues of the 1870s.[70]

The Colorado potato beetle caused such consternation among potato growers that by 1866, according to Townend Glover, this insect caused the most frequent complaints about insect pests that reached the commissioner of agriculture. In 1872, Riley reported that the beetle was the primary pest species in Missouri.[71] The spread of the Colorado potato beetle, like that of the Hessian fly, had international repercussions. In 1871, the Canadian commissioner of agriculture and public works appointed William Saunders to head a commission to report on the damage caused by the beetle and to recommend controls. By that time the beetle was already causing extensive damage in Ontario. It was already too late to implement Bethune and Riley's proposal to establish a ten-mile potato-free buffer zone along the American-Canadian border.[72] By 1875, several European nations, including Belgium, France, Germany, and Switzerland, had prohibited the importation of American potatoes because of the fear that the Colorado potato beetle would be spread to those countries.[73]

The life cycle of the beetle as worked out by Walsh and Riley in 1862 and 1863 proved to be relatively simple. The adult beetles were found to hibernate in the soil, becoming active in May. The females then laid several hundred eggs in batches on the undersides of the potato leaves, and from these the larvae hatched in four to nine days. The larvae were orange-red with black spots; they fed voraciously, passing through four instars until they were fully grown larvae. They then burrowed into the ground to pupate. After ten to nineteen days the adult beetles emerged

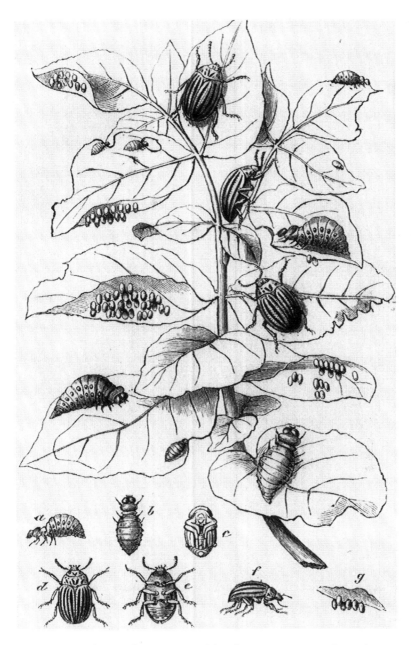

"The Colorado Beetle: Larvae and Perfect Insects at work on the Potato Plant . . . slightly enlarged," from Charles V. Riley, *The Colorado Beetle, with Suggestions for its Repression and Methods of Destruction* (London: G. Routledge, 1877)

from the soil and commenced feeding. Within a few days the females laid eggs for a second generation, the two generations usually overlapping in a single field. If left uncontrolled, the larvae and adults fed on the leaves until the entire potato patch or field was stripped, leaving only stems covered with insect droppings.[74]

Though the insect's extension of its range and its life cycle were easily explained by the entomologists, its control proved more difficult. As in the case of other insect pests, the entomologists expected that natural checks would assert themselves. Natural predators such as toads, snakes, crows, and quail, however, did not prove very effective, and tame poultry such as turkeys, ducks, and chickens did not relish the beetles.[75] It was found that the beetles flourished in cool wet seasons and that their numbers were reduced in hot dry seasons, but this knowledge offered little in the way of practical controls. Early and late planting or refraining from planting potatoes for a season or longer were also suggested, but these were likewise unsatisfactory.[76] For many years the only effective control measure was hand picking or shaking the beetles from the plants into a container of hot water. Some growers in Iowa experimented with horse-drawn machinery that was intended to knock down and crush the beetles, but apparently the experiment was unsuccessful.[77]

All of the standard chemical insecticides in use by midcentury, such as ashes, lime, hellebore, quassia, and tobacco, were tried on the Colorado potato beetle but to no avail. Walsh and other entomologists pronounced these useless.[78] Then, in 1867, some farmers in Illinois and Indiana experimented with a dust containing Paris green, an arsenic-of-copper compound that was used to color green paint. They reported that the poison was effective against the beetle. Responding to their reports, Walsh, Riley, Glover, and LeBaron conducted experiments and gathered testimony on the new remedy and they confirmed its effectiveness, though they recommended that it be used with caution.[79] At first Paris green was mixed with various dusts and powders in dry applications, but by 1872 a more convenient and efficient liquid spray had come into use.[80]

Despite the reported effectiveness of the new method, many growers were hesitant to use a known poison on their crops, fearing that the poison would be absorbed by the tubers and transferred to those who ate them. For many years after the spread of the beetle to the eastern United States, hand picking remained the most common control measure. Men, women, and children would walk through the fields, knocking the beetles into pans of hot water. By 1880, however, most growers had adopted the use of Paris green in preference to the "stick-and-pan" method.[81]

Entomologists generally were less cautious. LeBaron, Glover, and

Riley expressed some initial concern about the widespread use of arsenic poison, but their experiments and their correspondence with farmers convinced them that these fears were groundless. Within a few years Riley, who had originally been skeptical, was the strongest promoter of Paris green, insisting that the plants transformed the poison as it was taken in.[82] When LeConte, in an address before the National Academy of Science, protested the use of poison, Riley refuted the claim, calling upon chemists at the Michigan Agricultural College and the U.S. Department of Agriculture for supporting testimony.[83] As James Whorton, historian of early pesticides, has pointed out, the entomologists' endorsement of arsenic insecticides as harmless to human health was unwarranted, because they did not consult the medical literature, which had much to say about chronic arsenic buildup. At the same time, medical doctors, who might have insisted on examining the public health aspects of pesticides, found themselves divided and insecure within their own profession, and they avoided the issue.[84]

The successful use of Paris green in the control of the Colorado potato beetle was the first widespread use of chemical insecticides in the world, and its application was viewed as a triumph for American applied entomology. American entomologists accepted the credit and promoted chemical insecticides almost as actively as the chemical suppliers themselves. The success with Paris green inspired the search for other chemical insecticides. In the 1880s, London purple, another arsenic compound, and Bordeaux mixture, a fungicide, came into widespread use. These and other chemical pesticides initiated the major trend in applied entomology for the next century.

The American example spread quickly to other countries. In France, where arsenic insecticides had been in use from the late 1700s but had been banned in 1846 because of fear of danger to public health, the American success with Paris green and the testimony of entomologists about the safety of the insecticide persuaded the French to resume the use of arsenic-based pesticides in the 1880s. The control of the Colorado potato beetle thus initiated fundamental changes in American entomological science and set an example, albeit with dubious long-range results, for the worldwide use of chemical pesticides.[85]

The four insect species reviewed thus far, though by no means the only major insect pests, were among the most significant in terms of their overall destructiveness and their importance in the organization and development of applied entomology in North America.

First, all the species reviewed demonstrate in various ways the potential destruction from both native and imported species when growers began to specialize in commercial crops and to introduce large acreages of monoculture into new geographic environments. A primary result was the new demand for professional help to meet the insect emergency. The damage caused by the curculio, in particular, led the horticulturists to push for expert help from entomologists in the control of destructive insects.

Second, the entomologists made significant advances in their knowledge of the life history, habits, times of appearance, and number of broods of pest species in specific localities. This knowledge enabled them to advise farmers on more effective farm culture practices and specific control measures. Entomologists demonstrated the importance of weather conditions, natural predators, and insect diseases as control measures. These pointed the way toward conducting annual insect surveys for the chinch bug and the establishment of fly-free dates for the Hessian fly and toward the eventual use of natural predators and insect disease as biological controls.

Third, the function of the applied entomologist in the discovery and the confirmation of effective control methods was firmly established. The authoritative information supplied by professional entomologists was useful to growers, but in the case of the entomologists' enthusiastic promotion of arsenic insecticides, it had some negative effects in the area of public health.

Fourth, a number of effective control methods, like the adjustment in planting schedules and the use of resistant varieties for the Hessian fly, the use of barriers for the chinch bug, and Paris green for the Colorado potato beetle, came into general practice. Other control measures, like annual insect surveys for chinch bugs, were introduced.

Finally, the cumulative effect of these and other insect pests convinced both growers and entomologists that the insect menace was increasing and that existing entomological institutions were inadequate. The appearance of yet another and greater insect pest, the Rocky Mountain locust, confirmed this opinion and dramatized the need for increased funding and effective organization at the federal level to bring the benefits of applied entomology to American agriculture.

The Rocky Mountain Locust Plague

7

The locust plague of the 1870s is the most dramatic event in the history of American entomology. Accounts have appeared in novels, children's books, and feature motion pictures, as well as in the writings of professional historians and scientists. Professional entomologists have described the locust plague in their reminiscences, presidential addresses, in histories of entomology, and in scientific writings about locusts. Locust specialists have based scientific discussions of the theory and control of locusts on the American locust plague of the 1870s. Agricultural historians have examined the impact of locusts on agricultural settlement in the tradition of Turnerian frontier influences. And historians of science have identified the locust plague as the impetus for a new "problem-centered" approach in federal science and the U.S. Entomological Commission as the organization model for post–Civil War federal science.[1] The present account relates the locust plague to organizational and theoretical developments in the American entomological community that by 1880 placed American applied entomology at the forefront of world science.

Entomologists had been urging fundamental changes in America's agricultural institutions prior to the locust problem. They used the locust emergency to persuade public officials to enact the proposed reforms. These proposals called for a federal Agricultural Department organized along scientific lines. The implementation of proposals by entomologists and others by the 1870s transformed the Department of Agriculture (as it was eventually called) into the leading federal scientific agency. Moreover, the locust investigations themselves represented important advances in scientific organization, theory, and control. These theoretical and organizational achievements helped generate sustained public support for applied entomology in the federal government.

It was appropriate that the advances in organization, theory, and control should come about through the agency of locusts, one of mankind's most ancient and formidable insect adversaries. According to modern definitions, locusts are polymorphic shorthorn grasshoppers that under certain environmental conditions take on different forms in structure and behavior.[2] In 1921, Sir Boris Uvarov, working with the migratory locust of Asia, showed that certain grasshoppers exist in different forms that he called the solitary and gregarious phases. These phases are extremes that occur at the far ends of a continuous series of forms, but they are so different that they were often thought to be distinct species.

The most significant difference between the phases is in their behavior. As the name implies, members of the solitary phase behave relatively independently from each other, responding primarily to environmental stimuli such as food, vegetation cover, and weather conditions. When favorable conditions like an optimum breeding environment, mild seasons, or freedom from natural enemies cause population densities to build up, however, gregarious behavior replaces solitary behavior and hoppers respond more to the massed population than to outside stimuli. The hoppers then congregate in bands that move as single units as the individuals imitate each other's movements. Bands soon join together and move in a single direction, propelled by their own momentum. Hoppers that grow and molt under these conditions of crowding undergo changes in their structural development. The most important morphological change is the development of longer wings that enable them to fly long distances.[3] As the hoppers molt and become winged adults, marching is replaced by swarming. Both marching and swarming commence when daytime temperatures rise sufficiently, but whereas marching ceases at night, swarming continues by day and night, often transporting the locusts for hundreds of miles. Eventually either external conditions or the depletion of internal energy causes the swarm to land, where the locusts commence feeding and depositing eggs. Most remarkable of all, the structural and behavioral characteristics of the gregarious phase are passed on to the succeeding generations as long as the crowded conditions prevail.

The practical effect of the locusts' behavior and transformations was well known to growers of crops. If the swarm landed in an agricultural sector, the locusts would devour the crops and other vegetation. If the swarms found favorable reproductive conditions in the new habitat, the destruction continued for two or more seasons.

Settlers in North America had experienced periodic destructive outbreaks of grasshoppers; however, it was not until the settlement of the plains that they first experienced the full fury of true migratory locusts.[4] Canada experienced the first extensive plague in 1818–1821. In 1856–1858, locusts destroyed crops on the plains settlements in Manitoba, Saskatchewan, Alberta, and Minnesota and to a lesser degree in the central and southern plains in the United States. In 1866–1867, locusts laid waste crops in Kansas, Nebraska, and Missouri. In 1873, the first of a prolonged series of locust invasions destroyed crops over a wide area in Minnesota, the Dakotas, and Iowa. The locusts struck with full fury in 1874, along the plains from Canada to Texas, resulting in damage that was described by Riley as a "national calamity." The swarms of adult locusts abated somewhat in 1875, though the Missouri River settlements from Omaha to western Missouri suffered from newly hatched hoppers. In 1876, the swarms returned with a destructiveness similar to that in 1874. The year 1877 saw a respite, and though it could not be known at the time, it was also the last serious locust outbreak for almost a decade.[5]

The destructiveness of locusts was described by many eyewitnesses. One description by Riley conveys the awesome power of the swarms:

The voracity of these insects can hardly be imagined by those who have not witnessed them, in solid phalanx, falling upon a cornfield and converting, in a few hours, the green and promising acres into a desolate stretch of bare, spindling stalks and stubs. . . . On the horizon they often appear as a dust tornado, riding upon the wind like an ominous hail storm. . . . In alighting, they circle in myriads about you, beating against everything animate or inanimate; driving into open doors and windows; leaping about your feet and around your buildings, their jaws constantly at work biting and testing all things in seeking what they can devour.[6]

Other observers reported that trains were unable to move because of crushed locusts on the tracks, that tree limbs had broken from the weight of the insects, that clothes had been eaten from the line, and that barnyard fowl had gorged themselves on locusts until they could eat no more.[7]

Frontier settlers, who were poorly financed and ill prepared to sustain losses of crops and livestock, suffered extreme hardships. In some localities, such as north central Kansas, westward settlement was temporarily reversed as hundreds of settlers abandoned their homes and returned east. So severe was the suffering that the U.S. Army commander at Omaha distributed emergency rations to destitute settlers in Minnesota, Dakota, Nebraska, Iowa, and Colorado. In Kansas, a relief committee distributed money, food, and clothing. Several state legislatures convened in special session to vote funds for emergency relief.[8]

The extent of the disaster caused the states to look to the federal government for aid. In 1873, the governors of Minnesota and Iowa requested federal assistance, and in 1874 the governors in Nebraska and Kansas joined them. In 1875, the U.S. Congress appropriated $150,000 to prevent the starvation of those suffering from the effects of locusts.[9] As the plague continued with no end in sight, Governor John S. Pillsbury of Minnesota issued a call for the governors of the states and territories affected to meet in Omaha in October 1876. They sought to identify the most effective measures against the locusts, and they petitioned the federal government to provide relief from the scourge.[10] By that year, Riley estimated that the damage caused by the locust invasion totaled more than $200 million.[11]

The locust plague lent a sense of urgency to the proposals of prominent entomologists and others to restructure the U.S. Department of Agriculture. Barely a decade old and lacking either specific legislative mandate or creative executive leadership, the department still operated as a clerical agency, much as it had under the patent office. In the 1860s and 1870s, as the general awareness of insect damage to agriculture grew, and as entomologists demonstrated new applications of their science, the department came under increasing criticism for its failure to lead, or even to keep pace, in the field of agricultural entomology.

LeConte made a public appeal for improvement of entomology at the department in a speech delivered to the American Association for the Advancement of Science in Portland, Maine, in 1873.[12] He called for the department to be reorganized on a "scientific basis," so that it would promote the farmers' interests in the same way that the U.S. Coast Survey promoted the interests of the shippers. LeConte's remarks reflected a widespread conviction among scientific critics that correcting the department's deficiencies would have to begin with the replacement of the commissioner of agriculture, Frederick Watts.

Criticism of commissioners of agriculture was not new, but with President Ulysses S. Grant's appointment of Watts, a Pennsylvania judge and railroad executive, as commissioner in 1871, the criticism from scientists and some sectors of the agricultural community had changed to open hostility. The scientists resented Watts's arbitrary decisions in matters that they considered outside his expertise, such as his abrupt firing of C. C. Parry at the national herbarium. The national herbarium had been established upon the urging of Asa Gray and John Torrey prior to Watts's arrival at the department. Parry had begun organizing the department's plant specimens, but Watts decided the herbarium lay outside the proper jurisdiction of the department and removed him, thereby arousing the ire of American botanists. Riley and other scientists also resented what they considered pretentious observations of Watts on technical or scientific subjects.[13] After one such encounter with the commissioner, Riley wrote to LeConte that Watts was "perfectly incompetent, with no appreciation of science whatever, and no ability to discriminate between true and bogus science."[14] Agricultural leaders in the West, miffed because Watts was an easterner and because he was chosen on the basis of state political patronage rather than with their consultation, made common cause with the scientists against Watts.[15]

In the view of scientists like LeConte, Riley, Samuel W. Johnson, and Eugene W. Hilgard, the commissioner of agriculture should be a technically trained expert, appointed upon the recommendation of the National Academy of Sciences. This procedure would avoid the spoils system that placed political appointees like Watts in a position better filled by a scientist. In an obvious reference to Watts, LeConte said the commissioner should be the "highest scientific ability that can be procured [and not a] semi-sinecure for persons of political influence."[16] A scientist as commissioner, LeConte maintained, could better coordinate the work of the state entomologists and ensure the scientific accuracy and usefulness of the department's reports.[17]

Riley and LeConte at first hoped that vigorous opposition spearheaded by agricultural leaders would force the removal of Watts. "We must first strike at the head," Riley wrote to LeConte; "we must work to have Watts replaced by someone who will institute the needed reform."[18] Riley warned that President Grant and the Congress were too deeply enmeshed in the spoils system to be moved by appeals to science, but he was convinced that sustained agitation on the part of agricultural leaders would force Watts's removal. There was already a great opposition to Watts, Riley reported, centering in New York and in the West. Riley assured LeConte that he was "personally acquainted with . . . all the leading agriculturists . . . of the country" and that with their help Watts could be replaced. Riley himself gave a series of lectures in Washington, D.C., and wrote articles for the *New York Tribune* toward that end.[19]

Dislodging Watts proved more difficult than anticipated, however. A year after the campaign had begun, Riley wrote, "I have about come to the conclusion that no change can well be brought about until Gen'l. Grant is succeeded."[20] Grant's stubborn loyalty to faithful subordinates prevailed against the opposition to Watts, and the commissioner remained in office until the end of Grant's term.

Reform of the department also posed the problem of what to do about Townend Glover, the federal entomologist. Glover had served in the department since 1854 (when agriculture was in the patent office), and he had a credible record as an agricultural entomologist. By the 1870s his health was failing, however, and he was unable to carry out the rapid expansion of fieldwork called for by Riley and others. When Glover's long-delayed publications of insect etchings began appearing in 1873, they were criticized rather severely by LeConte, Osten-Sacken, and other entomologists. LeConte noted that the government entomologist had prepared "a great mass" of notes that he said should be revised under "proper scientific supervision" and distributed to farmers in order for them to be useful.[21]

In 1874, as the full extent of the locust emergency became apparent, the lack of response from Watts and Glover supplied added ammunition to those calling for reform of the department. To the exasperation of Riley, the department's main response was to increase the distribution of seeds, a stopgap measure that did nothing to address the locust problem directly. "We have a Department of Agriculture . . . supported by large appropriations," he wrote, "and I have yet to learn that the Commissioner . . . has made any effort to subdue the locust plague."[22] Thomas, who was then with the Hayden survey, suggested that the department begin field investigations, but Glover was unreceptive. In 1876, as the locust plague reached its climax, Glover busied himself preparing a departmental exhibit for the Centennial Exposition in Philadelphia.[23] Thomas wrote to Hayden, "You might as well try to get a prairie dog out of his hole as to get Glover out of his nest."[24] To Scudder, he wrote, "Ought not something be done to get a *live* Entomologist at the Agricultural Department?"[25] As a federal employee Thomas hesitated to say anything openly against Glover, but he expressed his hope privately that entomologists outside the government would do something. Thomas suggested that Glover might be retained as curator of the museum, while "an *active, working* entomologist" could be appointed to conduct field operations.[26] Riley charged that Glover not only failed to respond adequately to the locust threat but even confused the issue by lumping reports of destructive grasshoppers from all over the country instead of distinguishing between migratory and nonmigratory species. In doing so, he said, Glover did more harm than good.[27] The ailing Glover, industriously compiling annual reports for the commissioner, could not match the public relations campaign spearheaded by Riley and LeConte.

The massive locust invasion in 1874 linked the locust problem inextricably to the reform of the department. It was by now clear, however, that neither Glover nor Watts would be replaced. Having conceded failure in this approach, Riley and LeConte proposed instead the formation of an independent entomological commission that would carry out the program they had proposed for the department. They called for a commission of entomologists and other scientists to be appointed upon recommendation of the National Academy of Sciences. LeConte made a public appeal for the formation of an entomological commission when the American Association for the Advancement of Science met in August 1875.[28] The commission would investigate the injurious insects of national importance, like the locust, the chinch bug, the Hessian fly, the army worm, and the cotton worm. Anticipating objections that the commission would duplicate the Department of Agriculture, Riley suggested the counter argument that the department in its present form was "about useless . . . and that . . . a commission . . . during a specific term of years, would do more for . . . agriculture than . . . the Department, and therefore be more economical."[29] Riley outlined this position in his annual report for 1875:

In cases, as with the Locust, the Chinch Bug, the Cotton Worm, etc., where the evils are of a national character, a National Commission, appointed for the express purpose of their investigation, and consisting of competent entomologists, botanists, and chemists, is necessary, and . . . steps have been taken . . . to memorialize Congress to create such a Commission, the members to be chosen by the . . . National Academy of Science.[30]

Riley and LeConte drafted a memorial to Congress that was supported by scientists and agricultural leaders and organizations.[31] Substantial support came from the membership of the American Association for the Advancement of Science, where the entomologists had organized as the Entomological Club of the AAAS. The club provided a forum where entomologists could discuss matters of common concern, such as the means of combating injurious insects and publicizing the need for applied entomology among the membership and the public. Entomologists played a prominent part at the AAAS meeting in Detroit in August 1875. LeConte was retiring president of the AAAS, and both LeConte and Riley delivered papers in which they urged the creation of an entomological commission.[32] LeConte and Riley were gratified by the number of scientists who signed the petition.

By the time the memorial made the rounds at the AAAS meeting, however, most agricultural organizations such as the National Grange of the Patrons of Husbandry and the state agricultural societies had already held their annual meetings. It was too late to gain their endorsement in time for the 1875 legislative session.[33] Nevertheless, the next year the agricultural organizations were contacted in time to support the proposal, and at Riley's request, Kansas Senator John J. Ingalls introduced the bill embodying the petition in Congress. Riley and LeConte appeared before the agricultural committees in both houses to urge passage of the bill.[34]

The debate over the Ingalls bill was the first time the problem of injurious insects had been discussed extensively in Congress. Entomology was now a matter of national importance. The Ingalls bill did not fare well for two reasons. First, Riley and Thomas disagreed as to the necessity of a commission, and, second, fiscal conservatives in Congress opposed any expansion of federal agencies and especially a commission whose jurisdiction overlapped or replaced that of the Department of Agriculture.

Supporters of the measure argued that a special commission was necessary because the locust invasion was a national emergency and because Glover, the federal entomologist, was already too busy to do field investigations and to distribute locust control information rapidly to farmers. Furthermore, the investigations involved the protection of people and property in the territories where the federal government had a direct responsibility. The Hayden survey, where extensive locust investigations were already under way, was authorized to do only scientific studies and had no authority to implement control measures.[35]

The bill's most vociferous opponent, Illinois senator John A. Logan, was advised by Thomas, Logan's brother-in-law. Logan charged that the bill was intended for the "benefit of someone" (presumably LeConte or Riley). He maintained that the existing agencies, in particular the Hayden survey, were adequate to meet the problems of locusts and other injurious insects.[36] Thomas clearly viewed the LeConte-Riley proposal as competition to his work on the Hayden survey. He argued against the proposal in the pages of the *Prairie Farmer* and in private correspondence with Hayden and others. Insect problems were essentially local, Thomas insisted, and they were therefore the responsibility of the states. Migratory locusts were the one exception. In this case, where travel across several states and territories was necessary, the job could be handled by a single investigator attached to one of the surveys. The proper theory, Thomas told Hayden, was that each state should have its own entomologist, and the federal entomologist "should be a central point for . . . gathering and distribution of information."[37] Thomas considered preposterous "the idea of sending out a commission of three or four entomologists . . . at the expense of the Govt. with positions secured for five years."[38] A commission, he said, would spend half of its time identifying new species and making learned reports, when all that was needed was the life history of a single species, and this could be done by one worker attached to a survey. If the current federal entomologist was incompetent, as Thomas charged, he should be replaced, but there should not be a separate, competing agency.

Thomas apparently influenced others besides Logan, because when the bill reached the Senate Committee on Agriculture, its scope was significantly reduced. Instead of three commissioners appointed for a period of five years at a salary of $5,000 per year, there would now be only one commissioner appointed for one year at $4,000, and this person was to investigate only the locust rather than the list of insects originally proposed.[39] When the amended bill was reported out of committee, Senator Samuel B. Maxey moved to authorize the commissioner to study the cotton worm also. The Texas senator complained that residents of other sections did not appreciate the seriousness of the problem in his state, where they suffered from both locusts and cotton worms.[40] Next, the senator from Virginia, Robert E. Withers, moved to insert "tobacco worm" after cotton worm. Apparently some senators did not share their colleagues' estimation of the seriousness of the issue, as reflected in the following exchange:

MR. AARON A. SARGENT (California): If it be in order to move to insert another worm—

PRESIDENT: The Chair will entertain any question that is relevant.

MR. SARGENT: There is one more destructive than another: I move to insert "worm of the still" (laughter).[41]

Sectional rivalries over which insects to include were resolved by an amendment that renamed those in the original bill, but this opened the amended bill to objection on the grounds that one entomologist could not possibly study all the named insects in one year. Senator Frederick T. Frelinghuysen complained of the illogicalness of those who first claimed that the insect problem was not significant enough to warrant a commission but then insisted that the problem was too great to be handled by one commissioner.[42] When another amendment placed the entomological commissioner under the commissioner of agriculture, Senator Justin Morrill objected to the "anomaly" of paying the subordinate twice as much as his superior. When the Ingalls Bill passed the Senate, the editors of the *Nation* ridiculed the measure, writing that it represented an unnecessary expansion of the federal government:

The Republicans in the Senate . . . have passed a bill to investigate insects injurious to vegetation. . . . The act, should it pass the House . . . will be a new application of the great principle of the division of labor, for in the future the Agricultural [*sic*] Commission will scatter the seed broadcast over the land, while the entomologist will follow closely on his trail and exterminate the various bugs that may attack the ripening grain. We only want now another Commission to harvest the crops, and another to see that they get to deep water, and the husbandman will be entirely relieved from grinding toil.[43]

Riley responded to the *Nation's* comments by objecting to what he considered unwarranted satire of a serious agricultural problem. He estimated that destructive insects cost the nation a minimum of $100 million each year. Riley agreed, however, that the bill as amended was badly flawed, and he indicated that he would not support it.[44] The amended bill passed the Senate, but Morrill asked for a reconsideration, and the bill apparently died without another vote.[45]

By the fall of 1876, the locust plague in the western states and territories had reached such crisis proportions that local and state authorities could not wait for congressional action.

In October, as the western governors prepared to meet in Omaha to devise ways to meet the locust emergency, Thomas and Riley met to resolve their differences. They drafted a proposal for an entomological commission, which they presented to the governors. Because the subject of the conference was the locust, and not other insect pests, they agreed to limit the scope of the proposed commission to the locust problem, at least for the present. Thomas prevailed in having the proposed commission attached to the Hayden survey, but Riley convinced Thomas that the locust problem required three scientists, not one. The tentative proposal they agreed upon was for an additional $15,000 in the sundry civil bill (by which Hayden's survey was funded) "for the purpose of paying the salaries and expenses of two competent entomologists and one ornithologist to investigate . . . the history . . . of the destructive grasshoppers . . . and to recommend . . . remedial measures."[46] The inclusion of an ornithologist in the draft proposal reflected Thomas's opinion that a permanent solution to the locust problem would come only with the aid of insect-eating birds. "The ornithologist would be needed," Thomas told Hayden, "to investigate the food habits of [birds] so as to find what species eat grasshoppers. Hence it will not require [an Elliott] Coues or

[Spencer F] Baird but a good dissector of crops, who has some knowledge of entomology and a sufficient knowledge of ornithology to determine species accurately."[47]

The plan suggested by Riley and Thomas was adopted by the governors with slight modifications. The final resolution called for the appointment of three entomologists (rather than two entomologists and one ornithologist) and two western men. The inclusion of two western men, Thomas explained privately, was a "sop" to the governors of Minnesota and Nebraska, each of whom had a favorite he wanted appointed.[48] The appropriation request was increased to $25,000 to cover the additional two members. On December 21, 1876, a memorial with this request was sent to the Forty-fourth Congress by the governors and by Riley and Thomas.[49]

The strategy of linking the entomological commission to the survey proved effective. Additions to an existing agency did not raise the same congressional objection as creating a completely independent commission.[50] Also, the inclusion of the appropriation in the survey budget meant that the request would go immediately to the subcommittee that recommended Hayden's annual appropriation, thereby avoiding the agricultural committee and floor debate that had frustrated Riley and LeConte earlier.[51] Hayden, Thomas, and Riley worked to organize support for the inclusion of the entomological commission in the survey, and the measure passed on March 3, 1877.[52] In its final form, the bill authorized the secretary of the interior to appoint "three skilled entomologists . . . to report upon the depredations of the Rocky Mountain Locusts [and] the best . . . methods of preventing . . . or guarding against their invasions." An appropriation of $18,000 was added for the commissioners' salaries and expenses.[53]

The $18,000 appropriation for the entomological commission almost doubled the total amount annually budgeted for entomology by all the states and the federal government combined in the 1870s, an amount already the largest of any country in the world. Although modest by comparison with salaries in some established professions—in the 1850s for example, the chief engineer of the Illinois Central Railroad alone earned $10,000 per year—the appropriation for the Entomological Commission represented major new funding for entomology. In the 1870s, state entomologists earned about $1,200 per year in salary, about the amount earned by a second-rank engineer. Salaries for the three commissioners amounted to approximately twice that amount.[54] Understandably, there was considerable rivalry among the entomologists over the staffing of the commission.

Until the final version of the bill, Hayden was to have chosen the commission members, and Thomas expected Hayden to make him head of the commission. In the final version, however, the secretary of the interior was named as the one to appoint the commissioners. When Carl Schurz, Riley's fellow Missourian, was chosen as secretary of the interior by Rutherford B. Hayes, the incoming president, Riley was assured of becoming the head of the commission, with Thomas as the second member.[55] Thomas and Riley both favored Packard as the third member of the commission, but LeConte expected to receive the appointment because he had been one of the initiators of the original proposal. To LeConte's claim, Riley replied that the locust commission was "special, and quite different in scope" from what they had proposed together in 1875.[56] Thomas was more outspoken. Voicing the practical entomologist's pique at being snubbed by eastern systematists, Thomas charged that "LeConte has never paid attention to the Orthoptera in his life. The fact is there is not one of those noted Entomologists of the East except Packard who is really practical. . . . It is not book knowledge that is needed but practical information and experience."[57] LeBaron was gentler in his criticism of LeConte's recommendations for practical applications. Upon receipt of LeConte's 1873 address ("Hints for the Promotion of Economic Entomology"), LeBaron expressed his satisfaction that LeConte was "giving the weight of [his] authority to . . . making our science available for practical purposes," but he advised LeConte that some of the control methods he advocated were "much less valuable in practice than . . . in theory." He warned that the attraction and destruction of insects by fire was self-defeating, because many of the harmful insects, such as the potato beetle, were diurnal, whereas many useful insects were nocturnal, "thus . . . we destroy myriads of harmless and indirectly useful insects where we capture one which is injurious." The same objection held for "sugaring" with poison.[58]

The change in administrations in 1877 also meant a probable change in the commissioner of agriculture, so LeConte, having been passed over as a member of the Entomological Commission, redirected his efforts toward being appointed commissioner of agriculture. On the day the Entomological Commission was established, LeConte wrote to Julius Erasmus Hilgard, head of the U.S. Coast Survey, "now would be a good time to modify the organization of the agricultural department by putting a scientific man as the head of it who could obtain the support of other men of the country for the benefit of the agricultural interests."[59] LeConte's bid for the commissionership reflected a certain proprietary attitude the entomologists felt toward the department. LeConte's desire to place a scientist (preferably an entomologist) such as himself at the head of the department was reinforced by his need for a regular salary to bolster his dwindling income. LeConte relied mainly on rent from real estate he had inherited. This property was declining in value as early as 1860, and by the 1870s LeConte was concerned about his personal finances.[60]

Many influential scientists supported LeConte's bid for the commissionership, but LeConte lacked significant backing from farmers' groups. The position taken by Joseph Henry, secretary of the Smithsonian Institution, demonstrated the limited effectiveness of scientists' endorsements in the absence of similar backing from other groups. Henry agreed with LeConte that the spoils system had been responsible for the appointment of an incompetent like Watts, but he refused to speak directly to the incoming president, Hayes, to point out Watts's shortcomings.[61] LeConte's lack of support among the agriculturists was not surprising in light of his attitude toward agrarian reform. To Eugene W. Hilgard he wrote, "If [the Grangers,] and the 'practical farmers,' . . . combine, in a few years there will be little to eat, and that of bad quality."[62] Riley, who might have helped LeConte's cause with the agricultural organizations, preferred Willard C. Flagg, an agricultural and horticultural leader from Illinois, as commissioner.[63]

LeConte presented the incoming president with a long list of scientists' endorsements, and at the same time he published (without his signature) a critical "abstract" of the expenditures of the department for the years 1862 to 1876 in leading newspapers.[64] This abstract purported to show, among other things, that half of the departmental funds went for printing and for the distribution of seeds. The department was characterized as a "very well-devised scheme of mutual benefit." In its functioning it was unfavorably compared with the Smithsonian Institution, the Army Signal Service, and the U.S. Commission of Fish and Fisheries. Entomology in the department came under special censure. Since 1865, twenty-four thousand dollars had been spent for an entomologist, he charged, yet "when it was desired to have scientific information concerning the locust . . . , it was found necessary to make a *special* appropriation . . . and appoint a *special commission* . . . to investigate."[65]

LeConte quickly lost interest in the contest. A believer in leadership by an educated elite, he was indignant at the rude treatment his wife received from Washington reporters when she delivered letters recommending LeConte to the president.[66] Distraught with a political process that ignored scientific merit, yet driven by a sense of service, LeConte confided to his friend Scudder, "Now that I have gotten into the mess . . . I feel it my duty to go on and do all that I can to get the place (which I will make only of hard work, and not a sinecure)."[67]

LeConte's chances of being appointed were never very good. Within two months, Hayes had chosen General William Gates Le Duc, a native of Ohio who had made good on the Minnesota frontier as a salesman, lawyer, and land promoter, as his commissioner of agriculture. Le Duc, a graduate of the president's alma mater, had campaigned actively for the position and recruited the critical support of William K. Rogers, the president's private secretary. His appointment was made in the well-established post–Civil War pattern of political patronage.[68] LeConte continued to seek an appointment with the Hayes administration, and in 1878 he received a position at the U.S. Mint in Philadelphia.[69]

Riley's relations with the Department of Agriculture indicate that he too considered this department the entomologists' special charge. In 1878, Glover's ill health forced him to resign, and Riley, now the head of the Entomological Commission, was appointed chief of the Division of Entomology as well. Riley was a resourceful lobbyist, and his efforts soon showed results in expanded appropriations. In 1878 he sought and gained congressional funding to expand the work of the Entomological Commission to include the study of other insect pests, thus elevating the commission to the role that he and LeConte had originally envisioned. Riley also appealed directly to Congress for funds to expand the ento-mological work in the department. Riley's direct dealings with Congress so infuriated Le Duc that the commissioner fired him in 1879. For two years, Riley continued to operate the Entomological Commission out of his home in Washington, D.C., in competition with John H. Comstock, his replacement as chief of the Division of Entomology. When the administration changed again in 1881, Riley's friends made sufficient clamor to have him reappointed as chief of the Division of Entomology.[70]

The locust menace and the creation of the Entomological Commission marked a turning point in scientific organization within the federal government. In his study *Science in the Federal Government,* A. Hunter Dupree credits the entomologists with initiating a "problem-centered" approach to agricultural research and the creation of a prototype for scientific government bureaus. According to Dupree, scientists and administrators began, for the first time, to organize scientific activities with respect to problem solving rather than with respect to fields claimed by traditional scientific disciplines (like entomology, botany, and chemistry). A bureau organized in response to a problem had the independence and flexibility to shift personnel and funding as the problem changed. Other characteristics that marked the numerous bureaus organized along the pattern set by the Entomological Commission were the presence of a stable corps of personnel loyal to the bureau, the development of harmonious relationships with other bureaus and groups, the support of an organized professional group outside government, the exercise of routine regulatory functions, and a working relationship with one or more universities.[71]

The further development of entomology as a model government bureau lies beyond the scope of this study, but the innovations introduced by the entomologists are noteworthy. They included a rapid increase in appropriations for entomology in the Department of Agriculture, which rose from $7,000 in 1881 to $42,000 in 1885, and the creation of other divisions modeled after the Entomological Commission and the Division of Entomology. Dupree cites the Bureau of Animal Industry, the Division of Vegetable Physiology, and the Bureau of Plant Industry, among others, that used the problem-centered approach.[72] The clearest example is the work in economic ornithology and mammalogy that grew out of studies of the vertebrate enemies of insects sponsored by the Entomological Commission. In 1885, a separate branch in economic ornithology and mammalogy was established within the Division of Entomology, and in 1886–1887 this became a separate division. In 1896, it was renamed the Division (later Bureau) of the Biological Survey. Under the direction of Clinton Hart Merriam, a pioneer animal ecologist, the U.S. Biological Survey developed into a major scientific research and regulatory agency within the Department of Agriculture.[73]

While the significance of the Entomological Commission in the organization of science has been noted, the scientific findings of the commission and their importance in the history of locust research has received less attention. The work of the commission during the five years from 1877 to 1882 demonstrated the possibilities for applied entomology on an unprecedented scale. Some of the results, such as the studies of the Hessian fly and the chinch bug, were noted in the preceding chapter. My discussion here will focus on the locust investigations.

By 1877, entomologists already knew much about the Rocky Mountain locust. Walsh, who had personally investigated the locust invasion of Kansas, Nebraska, and Missouri in 1866, provided the first extended account of its range and natural history and gave the species a scientific name, *Caloptenus spretus*.[74] After Walsh's death in 1869, the locust investigations were carried on by state entomologists in Missouri and Illinois (Riley, Thomas, and LeBaron) and by entomologists attached to the Hayden survey (Thomas, Scudder, Packard, and Philip R. Uhler). One of

Hayden's primary concerns was opening up the central plains and Rocky Mountains for settlement. The locust that constituted a significant barrier to settlement therefore became a central concern in the survey's investigations and reports.[75]

The results of the investigations up to 1877 may be summarized under two headings. First, the known control measures were effective only against the unfledged hoppers but not against mature locusts, and second, a feasible plan had been developed to investigate the origin and movement of the swarms of adult locusts and was expected to lead to eventual control of locust plagues.

Direct control methods that the entomologists suggested for the unfledged hoppers were actually little improved over those used in antiquity. These methods all had the disadvantages of requiring intensive manual labor and the general cooperation of people over an entire region. These methods consisted of (1) the destruction of eggs by plowing or the payment of bounties for locust eggs, and (2) the destruction or control of unfledged young by crushing them with mechanical rollers or by burning, or by halting their march with ditches. To these primitive control methods, the entomologists added recommendations for the encouragement of natural enemies, especially birds like quail and prairie chickens (which the entomologists said should be protected by law), and by feeding the hoppers to turkeys and other barnyard fowl. The only significant improvement introduced in the 1870s was the "hopperdozer," a large pan that was mounted on runners and pulled by horses and filled with tar or water and coal oil and fitted with a screen to catch the hopping insects. The hopperdozer caught astounding numbers of hoppers, but it was really only a mechanical extension of the hand net. All the known insecticides, including Paris green, were ineffective against hoppers.[76]

The entomologists had been more successful in suggesting indirect ways to minimize the damage from locusts. Riley, Thomas, and Packard, for example, studied the relationship between local weather conditions and the rates of successful hatching and maturing of the hoppers. On the basis of these studies, Riley was able to make fairly accurate predictions about the expected severity of damage in certain regions. This enabled farmers to know whether to plant early or late and, if the first planting was destroyed, whether to replant that season. These studies drew upon questionnaires Riley distributed to persons in each of the counties in the locust district.[77]

The primary problem of swarms of winged locusts remained unsolved. Once these swarms had formed and commenced their movement, entomologists could offer no real hope of averting widespread destruction. By 1875, the entomologists had concluded that the real solution was to prevent swarms from forming. This would require a thorough study of the locust's life history, including an exact delineation of its breeding grounds, the conditions that produced the swarms, and the factors that directed the flights of the swarms.

American entomologists were keenly aware of the historic challenge of investigating locust outbreaks. They rejected the fatalism of previous generations as unbecoming a scientifically enlightened age. In 1875, for example, the editor of the *St. Louis Republican* voiced his opinion that the appearance of young hoppers that year had come in divine retribution for the people's wickedness and their lack of repentance following the plague of the previous year. Governor Charles H. Hardin proclaimed June 3, 1875, as a public day of humiliation, fasting, and prayer for deliverance. Riley challenged them publicly, refuting the judgment theory and arguing that the appearance of the hoppers could be explained on the basis of the insect's natural history. He insisted that practical measures for destroying the young locusts before they could fly would be more useful than prayer. Riley predicted that the hoppers would be weakened by disease and that they would be gone in time for a second planting in June. His prediction was confirmed by the events of the season. The exchange between Riley and Hardin was frequently cited by entomologists and others who wished to demonstrate the advance of science and its triumph over superstition.[78]

At the American Association for the Advancement of Science meeting in 1875, Riley called attention to the opportunity for American science. In the Old World, where locust plagues had occurred for thousands of years, there was still no precise knowledge of how the devastating swarms originated. But now, he announced, American entomologists could answer the questions about the locusts of the American West and set an example for European nations to follow.[79]

The achievement of Riley, Thomas, and Packard consisted in their documenting for the first time the origin and nature of a major locust outbreak. Specifically, the commissioners were able to set forth a working theory of locusts that included the precise location and role of the breeding grounds in locust outbreaks and the relation of weather to the subsequent movement of locust swarms. The principles they developed in these two areas furnished the foundation for locust theory from that day to the present.[80]

By 1877, when the Entomological Commission began its work, entomologists had developed a working theory of locust swarms based on the twin concepts of a permanent breeding ground where swarms originated, and their dispersal by flight to adjacent or distant areas where they typically inflicted the most damage. The notion of a permanent breeding ground was suggested by the American naturalist Alexander S. Taylor in 1858 and was further developed by Walsh, Thomas, and Packard and by the Russian scientists Fedora Petrovich Köppen and Victor Ivanovich Motschulsky. By 1875, Thomas and Packard had reported their findings from the Hayden survey that the Rocky Mountain locust required dry soil for ovipositing and therefore its permanent breeding ground was restricted to the high, dry plateau of the Rocky Mountains. They postulated that swarming began with a series of wet seasons that provided food for the survival of abnormally large numbers of hoppers, followed by a dry season, when the hoppers crowded together on the remaining food and were prompted to migrate.[81] Prior to 1877, however, there had been no systematic reconnaissance undertaken that would enable one to map the breeding area and describe the breeding conditions of any locust species accurately. In Russia, the government had mobilized troops in an attempt to repel locust invasions, but as yet no government had supported a scientific investigation into the source of the swarms. Such an investigation required a combination of trained experts, the demand for relief from locust outbreaks, and sufficient financing from government or other sources to support such an investigation. All three of these conditions were met in the Entomological Commission.

Within three years, the members of the Entomological Commission made good these promises. Specifically, the commissioners were able to learn (1) the location and extent of the breeding grounds where swarms of Rocky Mountain locusts originated and (2) how weather patterns affected the buildup and movement of locust swarms.[82]

Within one year after its establishment, the commission reported that the breeding grounds of the Rocky Mountain locust consisted of an area of some 300,000 square miles in the high plateau country of the eastern Rocky Mountains in Colorado, Wyoming, Idaho, Utah, western North Dakota, Montana, and Canada.[83] The entomological commission then delineated a "subpermanent" breeding region extending east to include portions of Missouri and Illinois, which, during seasons' peak breeding conditions, the locusts would occasionally invade and in which they would produce one generation of young. Entomologists of the commission stated that moist conditions in the Mississippi Valley would prevent the Rocky Mountain locust from establishing a permanent range outside its biological limits. The entomologists thus assured farmers that the Rocky Mountain locust would never migrate eastward as the Colorado potato beetle had recently migrated from its home in the Rocky Mountains to the Eastern Seaboard to wreak havoc among American potato growers. The question of where the locust swarms originated was a mystery no more.[84]

The definite mapping of the breeding ground helped resolve the question whether the Rocky Mountain locust was indeed a distinct species. Scudder had maintained that the Rocky Mountain locust was not a distinct species at all but only one of three forms of one species of grasshopper, the other forms being the red-legged locust of the Mississippi Valley and the Atlantic migratory locust of the Atlantic seaboard. Riley, on the other hand, maintained that the Rocky Mountain locust was a separate species. This was an extremely difficult point to settle, because (as noted earlier) the morphology of grasshoppers changes as they move from the solitary to the gregarious phase. The only way to prove the continuity or discontinuity between the populations was to follow their movements in the field. The question had important practical implications, for if Scudder was right, destructive swarms of the same magnitude as those in the West could be expected to develop anywhere within the range of the red-legged or Atlantic grasshoppers. The data collected by the commission confirmed Riley's position that the Rocky Mountain locust was a separate species. This migratory species could breed successfully only in specific locations in the West. The others were not so confined. The Rocky Mountain locust was migratory, the others were not.[85]

Twentieth-century entomologists repeated the species debate regarding the Rocky Mountain locust. Until after midcentury, most experts considered the Rocky Mountain locust to be a gregarious form of an extant species, *Melanoplus Mexicanus* (Sussure) that, owing to altered environmental conditions, had not appeared in outbreak numbers since the nineteenth century. More recently, taxonomists have concluded that the Rocky Mountain locust was indeed a distinct species that is now probably extinct.[86] The twin concepts of the breeding ground and the distinction between migrating and nonmigrating species are still fundamental in the investigation and control of locusts.[87]

The second question, as to what guided the movement and direction of the swarms, intrigued Riley, Thomas, and Packard because they correctly saw the possibility of combining the investigation of locust migrations with the investigation of weather and climate currently being carried out by pioneer American meteorologists. It was in this area that the commission produced its most original contributions to locust theory.

The scientific study of insect migration began in Europe in the 1840s, and it addressed a range of complex phenomena for which there was, and to some extent still is, no satisfactory explanation. The term "migration" has been used to describe any mass population movement over geographical distance. It thus includes those regular seasonal movements of insects and birds. Migration has also been used to describe the gradual extension of the range of a species, such as the Colorado potato beetle. Finally, migration is used to describe the semirandom movements of locusts under varying internal and external influences as treated here.[88]

The science of meteorology was a field in which, as in entomology, Americans demonstrated an early competence.[89] Beginning with the "storm controversy" of the 1830s and 1840s, Americans had constructed increasingly sophisticated "meteorological systems" involving thousands of observers that were designed to produce scientifically accurate explanations of American weather phenomena. The appointment of James P. Espy to head the army medical department's meteorological work in 1842, making him in effect the first national meteorologist, and the Smithsonian meteorological project in the 1850s and 1860s marked significant advances in the growth of American meteorological systems. The development of the telegraph as a powerful new tool in meteorological science led to telegraphic weather messages by the Smithsonian Institution in the 1850s and Cleveland Abbe's local weather forecasts for the Cleveland, Ohio, area in the late 1860s. In 1870, Congress authorized a national weather service, which came to be directed by Colonel Albert J. Myer, the head of the Army Signal Service, and Cleveland Abbe, the civilian scientist who guided the scientific development of the weather service during its first two decades. By the mid-1870s, the new weather service constituted the largest such institution in the world and was taking the lead in international cooperation in meteorology.[90]

The systematic weather data now available from the new weather service offered unprecedented possibilities for prediction and control of injurious insects. As early as 1875, Packard, Thomas, and Riley had each contacted Myer and Abbe, asking for assistance in the correlation of weather data with the appearance and movement of locust swarms. Shortly before the establishment of the Entomological Commission, Packard called attention to the fact that the appearance and flights of locusts had never before been studied by using extensive meteorological data but that scientists in the United States now had the opportunity to make these investigations.[91]

The three entomological commissioners agreed that weather influenced the buildup of swarms and that weather data could be used to help predict the timing and severity of locust years. They disagreed, however, about the relation of weather to the movement of locust swarms. Riley maintained that the flights outward from the breeding grounds were initiated by "internal" factors, primarily an instinct to search for new sources of food, and that the return flight was guided by a similar instinct to return to the breeding grounds. He cleverly combined "internal" and "external" factors by claiming that the locusts' homing instinct prompted them to fly only at times when the wind blew in the direction of the breeding grounds.[92] Packard, on the other hand, held that wind alone directed the swarms once they were airborne.[93]

Packard and Thomas assumed primary responsibility for the commission's investigation into the relationship of climate and weather to locust swarms. They worked closely with Abbe, making extensive correlations of meteorological data and locust behavior.[94] They concluded that daily temperatures induced the swarms to take flight but that, once aloft, the swarm's direction and duration were controlled by the prevailing winds.[95] Soon Riley was convinced by their arguments, and he abandoned the theory that instinct, such as a "search" for food or habitat, had anything to do with the direction of swarm movements. These findings have been confirmed by subsequent investigations, and the use of climatological data such as prevailing wind currents remains one of the key tools in locust control.[96]

Though it could not be known at the time, the year 1877 was the last year that the Rocky Mountain locust would cause widespread damage in North America. That year the swarms reached Montana, Dakota, and Iowa but no farther. The last reported invasions in the United States occurred in Montana and northwest Dakota in 1885 and in one county in Minnesota in 1888. In Canada, the last major invasion occurred in 1875, though a limited invasion occurred in 1898–1899. The last specimen of a living Rocky Mountain locust was collected in Canada in 1904, and the species is now considered extinct.[97]

The termination of the locust years in the 1870s was not considered unusual, because locust outbreaks by their very nature are of limited duration. There is some evidence that locust outbreaks in America have been controlled by a fungus.[98] The reasons for the insect's permanent demise, however, are probably man-made changes in the environment. Ironically, the extinction of the Rocky Mountain locust seems to have been related to the near extermination of the buffalo in the 1870s. In the area identified as the locust's permanent breeding ground, the buffalo had periodically overgrazed the range and pulverized the soil, thus reducing the available food while at the same time providing a suitable bed for ovipositing eggs, both of which effects favored a buildup of populations. When the buffalo herds disappeared, the conditions favoring locust buildups also disappeared.[99] Other environmental changes such as the introduction of large-scale agriculture in the high plateaus of the Rocky Mountains may also have played a part.[100] Similar environmental changes have diminished locust populations elsewhere in the world. The Asiatic locust, which once invaded Western Europe from the Black Sea area, no longer poses a threat to that region.[101]

Though the Rocky Mountain locust has disappeared, other species of grasshoppers have replaced it as serious agricultural pests. Major outbreaks have occurred in Canada in the 1890s and in both Canada and the United States in the 1920s and 1930s. The last sustained outbreak occurred on the Great Plains in the 1930s.[102]

Though the control measures developed by the Entomological Commission played no significant role in the suppression of the Rocky Mountain locust, these methods were used effectively against other species of locust and grasshopper, and they marked the beginning of effective locust research and control. Specifically, the commission proposed that the U.S. Signal Service and the North West Mounted Police assist the entomologists by conducting annual surveys of conditions in the breeding grounds. The Signal Service and Mounted Police could then warn of the danger of locust outbreaks and supervise the destruction of eggs and young in the breeding ground before there was a buildup of swarming insects.[103] These measures emphasized the prevention of population buildups by means of regular observation of conditions in permanent breeding areas and the control of such populations when they increased to dangerous levels. The annual surveys of locust and grasshopper conditions begun by the commission became a standard program of the Departments of Agriculture in the United States and Canada and, beginning in the 1880s, the techniques and personnel of the U.S. Department of Agriculture were used in locust outbreaks in other countries, such as Cyprus, Algeria, and Argentina.[104]

The grasshopper investigations of the commission and the Department of Agriculture were popular among farmers, especially in the Plains region, and though scientists eventually concluded that the commission had nothing to do with the disappearance of the Rocky Mountain locust, this pronouncement did not detract from the general impression at the time and later that the Entomological Commission had been effective in the control of these pests.[105] Toward the end of the commission's work in 1880, there appeared a note in the *American Naturalist,* probably by Packard, that stated: "It is believed that this locust will never be so destructive as in the past, and due credit has been given by disinterested persons in Kansas, Nebraska, Colorado, and Utah to the practical value of the U.S. Entomological Commission in obtaining and diffusing such a knowledge of its breeding habits, migrations, and distribution as to abundantly justify Congress in ordering the investigation."[106]

In responding to the emergency presented by the Rocky Mountain locust plague of the 1870s, the American entomological community demonstrated its world leadership in the organization and research methods of applied entomology. Most significant in the organization of science were the creation of the U.S. Entomological Commission and the introduction of the problem-centered approach and the scientific bureau as central features of the Department of Agriculture. In terms of research methods and findings, the American entomologists were the first fully to document the phenomenon of locust outbreaks, an achievement that laid the basis for subsequent locust research and control. The American entomologists were the first fully to document the importance of a permanent breeding ground in the cycle of periodic locust outbreaks, the first to correlate climatological data with outbreak conditions, and the first to establish the role of wind patterns in the migration routes of locust swarms. These successes impressed the agricultural community and ensured the continued support for and increased funding of applied entomology. The locust plague of the 1870s and the American entomologists' response confirmed American world leadership in applied entomology.

8 Profile of the American Entomological Community About 1870

By the 1860s, the American entomological community had developed its own characteristic profile. The community as described here is drawn from a study of entomological authors cited in the *Record of American Entomology*.[1] During the six years that the *Record* was published (1868–1873), a total of 141 authors were cited, of whom 33 lived outside North America. The 108 authors from the United States and Canada thus comprised the leadership of the American entomological community in about 1870.[2] The data obtained from this group were analyzed and compared with those from contemporary American scientists for whom quantitative information was available. Finally, the select group of entomological authors cited in the *Record* was compared with the full community of Americans with entomological interests and capabilities, that is, the Brethren of the Net.

Biographical information for the 108 entomological authors was assembled with the aid of standard biographical reference works and specialized references to entomological biographies. Pamela Gilbert, *A Compendium of the Biographical Literature on Deceased Entomologists*, yielded bibliographical citations for 69 of the 108 American entomological authors from the *Record*. Additional information for 26 individuals was obtained from other sources.[3] For 13 individuals, the only information available is the fact that they wrote at least one piece that is cited in the *Record of American Entomology* (1868–1873).

It should be noted that the "community" of 108 authors cited in the years 1868 to 1873 included 4 who died prior to 1868. During the years the *Record* was kept, 5 more of the 108 died, and 6 more died before 1878, the year the second edition of the *Naturalists' Directory* appeared. These mortalities are handled statistically by computing percentages on the basis of those "for whom we have information." If a particular author died between 1866 and 1878, for example, he is included in the analysis only in those years when he lived. In those categories, such as educational background, where physical presence was irrelevant, the community is treated as a whole despite the fact that not all were living in 1870.

The treatment of this group of living and deceased authors as a "community" is justified on the basis that authors need not be living in order to contribute to the intellectual enterprise of advancing knowledge about American insects. It is their intellectual contribution, published during their lifetime or soon thereafter, that is significant to the community. It is also important to recognize that an intellectual community is dynamic, not static. Younger members, who were on the threshold of prolific publishing careers, are not represented in the *Record*. These "living" but not yet published members balance the "published" but recently deceased members. To leave out the latter would detract from our analysis and from our understanding of the dynamics of the community during that specific time.

The core group of 108 entomologists was analyzed according to the following categories: geographic location at the time of birth, youth, and adult residence; the occupation of parents; educational background (informal and formal); and occupation and institutional affiliation. In addition, the individuals have been ranked according to priority within the group of 108. Inasmuch as this priority ranking is, to my knowledge, the first time this technique has been employed in analyzing American scientists, a fuller explanation is in order.

In order to rank the authors according to their "priority," or overall importance, for American entomology, I entered and ranked them on a "priority grid," a tool frequently used in applied psychology to help an individual choose a career path. All the individual's dreams, desires, plans, and so forth are listed at random on a chart designed in such a way that each item can be compared to each of the others. At the end of the process, the "choices," or decisions, are tallied, and the results of the tally give the ranking, or priorities.

In applying the priority grid to the entomologists, I compared each individual with every other individual according to the general question "which individual of the two was more important to the American entomological community?" These choices were "subjective" insofar as I made them all, and they were necessarily limited by my own knowledge and influenced by my biases as to what constitutes "importance" or "significance." It is best to state the "biases" as succinctly as possible. In many cases, the choices were clear, as for example, when comparing a figure like Riley with an unknown or relatively unknown person. Other choices were more difficult. How does one compare, or prioritize, the contributions of an individual who has published profusely with one who has published relatively little but who has played key roles in scientific organization, say as a secretary of an entomological society? Or how does one decide between an individual who has described hundreds of species within a distinct area of specialization with another who has written primarily in agricultural journals and has described no new species? In such cases, I made a conscious effort to give as much weight to activities in scientific organizations as to sheer numbers of publications. In cases where all other contributions seem about equal, and one individual has contributed to agricultural entomology and the other not, I gave priority to the one with agricultural contributions on the grounds that agriculture, and not systematics, was the most innovative area of American entomology for this period. The priority ranking of entomologists appears in Appendix 2.

Extensive analyses of three contemporaneous groups of American scientists provide useful comparisons with the entomologists. Robert V. Bruce analyzed 477 American scientists who were active during the period 1846 to 1876 and who were included in the *Dictionary of American Biography (DAB)*.[4] Sally Gregory Kohlstedt has analyzed the group associated with the American Association for the Advancement of Science in the 1850s, including separate analyses for 377 leaders and 2,068 members of the AAAS.[5] Clark A. Elliott has analyzed a group of 503 American scientific authors, all of whom wrote three or more articles listed in the *Royal Society Catalogue of Scientific Papers* (1800–1863).[6]

As these three groups provide the primary basis for comparison, the similarities and differences to the entomologists should be noted. Bruce's *DAB* scientists are closest in time to the entomologists. His period (1846–1876) is very close to the entomologists (who published about 1870), and the median year of birth for the *DAB* scientists (1831) approximates the median year of birth for the entomologists (1834).[7] Both Bruce's *DAB* scientists and the entomologists represent the Civil War generation, born in the 1830s, coming of age in the 1850s, and reaching their greatest productivity in the 1860s and later.

Kohlstedt's AAAS leaders and members as a group fall about one decade earlier than the entomologists. These savants were born in the Era of Good Feeling, came of age in the 1840s, and reached their peak output and leadership positions in the formative years of the AAAS in the 1850s.[8]

Elliott's scientific authors comprise a less compact group chronologically because of the way they were selected. Elliott's group consists of a steadily increasing pool of scientists, beginning with 55 scientists who were over age twenty in the first decade of the century, and increasing to a peak number of 364 over age twenty in the 1850s.[9] Elliott's authors thus cluster, as a group, a full generation to a generation and a half earlier than the entomologists. Most of them were born around 1800, came of age around 1820, and reached a peak of activity in the 1830s and 1840s. By the 1850s, Elliott's selection of scientific authors, those who were over age twenty and who published before 1863, was beginning to decline, because those whose publications appeared after 1863 are excluded from the group.

The authors of all three of the comparative studies provide analyses (to varying extents) of their groups according to decades when subgroups were born. Such analysis has been undertaken for the entomologists on a very minimal basis, as it was concluded that the "intellectual community" under study was compact enough chronologically to study as a group without a breakdown according to decade of birth.

How the four groups compared in terms of their relative selectivity, or exclusiveness, is more difficult to determine. Bruce's *DAB* scientists would seem to be the most exclusive: no other group had a *DAB* entry, or the equivalent, as an entrance requirement. Kohlstedt's AAAS leaders and members appear to be roughly equivalent in degree of selectivity to the entomological authors and the larger group represented in the *Naturalists' Directory*, respectively. There is reason to believe that Kohlstedt's AAAS leaders and the entomological authors represented the same leadership class within their respective areas. Kohlstedt's 337 leaders comprise 16.3 percent of the total of 2,068 AAAS members. By comparison, the 108 entomologists listed in the *Record* represent 17.2 percent of the total number of entomological entries listed in the index to the *Naturalists' Directory* for 1878. As one would expect, there was some overlap between the entomologists and the AAAS scientists. Five of Kohlstedt's AAAS leaders and 9 of her AAAS members are represented in the list of 108 entomological authors.

Elliott's scientific authors would likewise seem to be roughly equivalent to the entomological authors in terms of their selectivity. Elliott's criterion, the publication of three or more articles listed in the *Royal Society Catalogue of Scientific Papers* (1800–1863), compares roughly to the criterion for the entomological leaders of one or more pieces listed in the *Record of Entomology.* Scholars who have compiled statistics on American scientific authors have found that a minority of authors have been responsible for the majority of publications; therefore, the leadership among both groups is about the same.[10] In other words, any individual who published one article had a very high probability of publishing many more. A quick glance at the authors from any year of the *Record* reveals a preponderance of the same individuals, as would be true for any representative year of publication for Elliott's authors or similar groups.

In summary, Bruce's *DAB* group is contemporaneous with but more selective than the entomologists listed in the *Record,* Kohlstedt's AAAS leaders fall a decade earlier in time but are roughly equal in selectivity; and Elliott's scientific authors are a generation or more earlier but also roughly equivalent in terms of their exclusiveness.

In table 1 the geographical locations of entomologists are plotted over time, from birthplace through youth to adult residence in about 1870. Looking first at the birthplace, we find an unusually large proportion

Table 1. Geographic Location of Entomologists at Birth, in Youth, and at Adult Residence About 1870

Region	Birth no.	Birth %	Youth no.	Youth %	Res66[a] no.	Res66[a] %	Res78[b] no.	Res78[b] %
Europe	20	26.7	18	24.0	—	—	—	—
Canada	5	6.7	7	9.3	9	11.3	7	9.3
N. England	24	32.0	23	30.7	24	30.0	23	30.7
Mid-Atlantic	21	28.0	22	29.3	34	42.5	31	41.3
Midwest	3	4.0	4	5.3	9	11.3	7	9.3
South	2	2.7	1	1.3	1	1.3	2	2.7
West	—	—	—	—	3	3.8	5	6.7
Total	75	100.0	75	100.0	80	100.0	75	100.0

Note: New England: Maine, New Hampshire, Massachusetts, Rhode Island, Connecticut, Vermont. Mid-Atlantic: New York, New Jersey, Pennsylvania, Delaware, Maryland, Washington, D.C., West Virginia. Midwest: Ohio, Indiana, Illinois, Missouri, Kentucky, Tennessee, Minnesota, Wisconsin, Iowa, Michigan. South: Virginia, North Carolina, South Carolina, Georgia, Alabama, Mississippi, Louisiana, Arkansas, Florida. West: California, Oregon, Washington, Texas, Kansas, Nebraska, South Dakota, North Dakota, Wyoming, Colorado, New Mexico, Montana, Idaho, Utah, Arizona, Nevada.
[a]Res66 = residence in 1866. More addresses are known for 1866 than for other years (see discussion on pp. 184–87).   [b]Res78 = residence in 1878.
Sources: Entomologists' biographies were compiled from sources listed in note 3 above. Residence in 1866 and 1878 is based on Cresson, *American Entomological Society;* the *Naturalists' Directory, Part 2, North America and the West Indies,* edited by F. W. Putnam (Salem, Mass.: Essex Institute, 1866); and *The Naturalists' Directory for 1878, containing the Names of the Naturalists of America North of Mexico,* edited by Samuel E. Cassino (Salem, Mass.: Naturalists' Agency, 1878).

born outside the United States. One-third of the entomologists were born in Europe or in Canada (column 2, birth percentage for Europe and Canada).[11] This proportion of foreign born is considerably higher than in other contemporary groups of American scientists. The comparable figures are 12.8 percent for *DAB* scientists, 14.4 percent for AAAS leaders, and 14.4 percent for scientific authors, 1800–1863.[12]

In table 2, entomologists of foreign origin are compared with the foreign born among the scientific authors, 1800–1863, and with the proportion of foreign born among the general population in 1850. Whereas the foreign born among scientific authors were about five percentage points above the national average of the foreign-born nonslave population in 1850, the foreign-born entomologists were almost twenty-three percentage points greater than, or three times as high as, the foreign-born portion in the nonslave population (table 2, subtotals for foreign born). Both Europe and Canada rank disproportionately high as birthplaces when compared with birthplaces for other groups of scientists. Approximately 10 percent more entomologists were born in Europe than in the comparable group of scientific authors. Canada stands out just as prominently as a birthplace of entomologists. Whereas 6.7 percent of the entomologists were born in Canada, in other groups of scientists only marginal numbers were born there. Neither Bruce nor Kohlstedt considered Canada important enough to single out as a birth region separate from "foreign," and Elliott found only that 0.4 percent of the scientific authors, 1800–1863, were Canadian born.[13]

Why the entomologists included such a high proportion of European- and Canadian-born members in comparison with other groups of American scientists is an interesting question. The conspicuous proportion of those with British roots and upbringing among both the European and Canadian born provides one indication of a common source. Fully 18.6 percent of the entomologists were born in the British Isles (the majority of these in England), and as many spent at least part of their youth in the British Isles (tables 1 and 2). All the Canadians (except for one in French-speaking Quebec) had a British background and settled in English-speaking communities in Canada. Taken together, the immigrants from the British Isles and the Canadians with British roots make up 25.3 percent, or one-quarter, of the entomologists publishing in the 1870s. The prominent British representation reflects the keen interest in zoology in the early nineteenth century among a wide spectrum of the British populace, both as a popular recreation and as an emerging scientific and professional specialty. Individuals from all social classes shared a keen interest in Britain's expanding colonial empire. Exotic fauna being

Table 2. Foreign-Born Entomologists Compared with American
Scientific Authors, 1800–1863, and U.S. Population in 1850

Birthplace	Ent.		Sci. auth.		US1850[a]	
	no.	%	no.	%	no.[b]	%
British Isles	14	18.6	32	6.5	1.3	5.6
Continental Europe	4	5.3	32	6.5	0.7	3.0
Canada	5	6.7	2	0.4	0.1	0.5
Other	1	1.3	6	1.2	0.1	0.5
Subtotal foreign born	24	32.0	72	14.7	2.2	9.6
U.S. born	51	68.0	417	85.2	20.9	90.4
Total	75	100.0	489	100.0	23.2	100.0

[a]US1850 = Foreign born in United States in 1850.
[b]Millions.
Sources: Table 1 above; *The Statistical History of the United States from Colonial Times to the Present* (Stamford, Conn.: Fairfield Publishers, 1962), series A 17–21, "Area and Population of Continental United States: 1790–1950," p. 8, and series C 218–283, "Foreign-Born Population, by Country of Birth: 1850–1950," p. 66; Elliott, "American Scientist" (1970), table 8, "Birthplaces of Authors," pp. 59–60, and table 10, "Foreign Born Scientists Compared to U.S. Foreign Born Population in 1850," p. 68.

reported from far-flung places intensified the British infatuation with zoology. Britain's flourishing middle classes, which developed claims on political and social leadership earlier than those on the Continent, cultivated intellectual pursuits appropriate to hardworking, purposeful, and inquisitive citizens. Invigorated by evangelical religion and ingrained with values of industry and thrift, these classes in the early Victorian era produced a new breed of naturalists for whom collecting became a compulsion. The publication of an *Introduction to Entomology* by Kirby and Spence (1816–1826), for example, triggered such an enthusiasm for entomology in Britain that by the 1820s it "threatened to swamp natural history's other branches."[14]

Whatever the sources in Britain, the high proportion of entomologists with British origins is certainly in line with findings of Elliott that antebellum American scientists in general came from families with British backgrounds.[15] One should note, however, that the families with British origins cited by Elliott came mostly from long-established families in the United States, whereas the entomologists with British origins were all first-generation immigrants. Despite this difference in longevity in North America, the two groups are similar with respect to the relative ease and rapidity with which these families of British origin settled into secure financial and social positions on the Eastern Seaboard of North

America and thus laid the foundation for their descendants' entry into scientific pursuits. Another factor linking entomologists and other natural historians to British institutions is the strong presence of British interests in trade and commerce. These were also fields favored by entomologists and their families, as I will note presently.

Except for the unusually high foreign, especially British/Canadian, contingent, the entomologists as a group compare in most geographical respects to contemporaneous groups of American scientists. Table 3 shows that most entomologists (like scientists in general) were born and raised in the New England and mid-Atlantic states. New England produced more entomologists in proportion to its population than other regions, but the mid-Atlantic region gained proportionately over time as a residence of entomologists. The growing importance of Washington D.C., as a scientific center during the Civil War and post–Civil War era is the most conspicuous development in the shift toward the mid-Atlantic states (table 3, entomologists' youth and residence, and U.S./Canadian population, 1850, 1860, and 1870). The growing importance of the Midwest and West as a residence of scientists is to some extent a reflection of the westward shift of population, but note that by the 1870s the West (California) had a higher proportion of entomologists than it did of the combined U.S./Canadian population.

The South, often cited as the region least active in science, figured even less prominently as a source of entomologists than for scientists in general. Whereas 20 percent of the combined U.S./Canadian nonslave population resided in the South, only 2.7 percent of the entomologists were born in the South, and only 1.3 percent spent their formative years there. The same holds true when one looks at the South as the place of residence of entomologists following the Civil War (table 3, "South"). Other groups of scientists had a somewhat higher representation in the South. Bruce found that approximately 6 percent of the *DAB* scientists were born in the south Atlantic and Gulf states; Kohlstedt found that 8.3 percent of the AAAS leaders were born in the South; and Elliott found that 6.9 percent of American scientific authors, 1800–1863, were born in the South.[16]

In an examination of the South's support of science compared with that of other regions, Ronald L. Numbers and Janet S. Numbers found that the north central states (Midwest in my terminology) fell behind the South in its percentage of scientists relative to population.[17] As the figures in table 3 demonstrate, the situation for the entomologists was exactly the reverse. The Midwest figured more prominently as a resi-

Table 3. Geographic Location of Entomologists at Birth, in Youth, and at Adult Residence in Comparison with U.S. and Canadian Population

Region	Ent.,[a] birth no.	%	US/C,[b] 1830 mill.	%	Ent., youth no.	%	US/C, 1850 mill.	%	Ent., res66[c] no.	%	US/C, 1860 mill.	%	Ent., res78[d] no.	%	US/C, 1870 mill.	%
Europe	20	26.7	—	—	18	24.0	—	—	—	—	—	—	—	—	—	—
Canada	5	6.7	2.0	13.7	7	9.3	2.4	9.5	9	11.3	3.2	9.3	7	9.3	3.6	8.7
New England	24	32.0	1.6	11.0	23	30.7	2.7	10.6	24	30.0	3.1	9.0	23	30.7	3.4	8.3
Mid-Atlantic	21	28.0	4.3	29.7	22	29.3	6.9	27.0	34	42.5	8.7	25.2	31	41.3	10.3	24.4
Midwest	3	4.0	3.0	20.6	4	5.3	7.3	28.8	9	11.3	11.2	32.4	7	9.3	15.0	35.5
South	2	2.7	3.6	24.8	1	1.3	5.7	22.5	1	1.3	7.0	20.2	2	2.7	7.4	17.5
West	—	—	—	—	—	—	0.4	1.5	3	3.8	1.4	3.9	5	6.7	2.3	5.4
Total	75	100.0	14.5	100.0	75	100.0	25.6	100.0	80	100.0	34.6	100.0	75	100.0	42.2	100.0

[a]Entomologists. [b]United States and Canada. [c]Residence in 1866. [d]Residence in 1878.

Sources: *The Statistical History of the United States from Colonial Times to the Present* (Stamford, Conn.: Fairfield Publishers, 1962), series A 17–21, p. 8, and series A 123–180, p. 13; *The Canadian Encyclopedia*, 2d ed. vol. 3, (Edmonton: Hurtig Publishers, 1988, "Population," pp. 1721–22.

dence of entomologists than for scientists in general. During the Civil War, the Midwest forged far ahead of the South as a residence of entomologists. This development related to the accelerated specialization, mechanization, and commercialization of agriculture in the Midwest and the attendant problems of insect pests. By 1866, 11.3 percent of the entomologists resided in the Midwest, compared with 32.4 percent of the overall population, or a "science index" of 0.34 (scientists as a percentage of population). When one looks at scientists in general, as measured by the residences of AAAS leaders in 1860, the South leads the Midwest with a science index of 0.57 compared with 0.30 for the Midwest.[18]

The most plausible reason for the prominence of entomologists in the Midwest and their absence in the South is the general interest in applied entomology in relation to agricultural change noted above and the activities of specialized lobby groups, in particular the horticulturists, in the Midwest. The South had its share of insect problems, for example the cotton worm, but these were problems of an extensive, noncapital- and nonmachinery-intensive cropping system that was not undergoing rapid change. The clientele was simply too dispersed and disorganized to mount an effective lobby to address insect problems. After 1860, the destruction of the Civil War and the dislocations of the postwar period obviously played a part in the lack of southern support for entomology, but the pattern of extensive, nonspecialized cropping and the absence of a visible entomological community had been set in the antebellum period.

Turning from the birthplaces and geographical environs of the entomologists to an examination of their parental occupations, we find a second striking characteristic of the entomologists as a group: they come in disproportionately high numbers from families engaged in commerce and trade (table 4). Approximately 42 percent of the entomologists' fami-

Table 4. Parental Occupation of Entomologists Compared with Scientific Authors, 1800–1863, and *DAB* Scientists, 1846–1876 (percent)

Occupation	Entomologists	Sci. authors, 1800–63	*DAB* scientists
Agriculture	4.0	18.49	17.5
Commerce[a]	41.9	31.37	19.0
Professions	43.2	36.09	45.6
Other	10.8	14.05	17.8

[a]Includes trade and manufacturing.
Sources: Elliott, "American Scientists" (1970), table 12, "Occupations of Fathers," pp. 79–81, and Bruce, "Statistical Profile" (1972), table 11, "Occupation of *DAB* Scientists' Fathers by Periods of Scientists' Activity," p. 82.

lies were engaged in commerce, trade, and manufacturing (in which manufacturing accounted for only 6 percent). This proportion in trade and commerce was over twice as high as among the *DAB* scientists, the group closest to the entomologists in time, if not in degree of selectivity, and one-third higher than among Elliott's scientific authors (table 4). The contrast between the groups with respect to those whose parents were in the professions is less striking. All three groups have high representations of parents in the professions, a feature common in the background of all scientific groups for that period. For families in the professions, there is a difference of less than ten percentage points among the three groups: a high of 45.6 percent for the *DAB* scientists and a low of 36.09 percent for the scientific authors, with the entomologists falling between the two groups with 43.2 percent.

The second striking figure in table 4 is the very low representation of entomologists' families engaged in agriculture: 4 percent compared with 18.49 percent for scientific authors, 1800–1863, and 17.5 percent for *DAB* scientists. These figures support the general thesis of this study. American entomological science originated with those who shared systematic interests common to the European tradition, and these individuals, at first, developed their speciality almost entirely separate from agricultural concerns. Except for Peck and Harris, American entomologists were initially not very interested in agricultural applications of their science. Only in the course of the agricultural revolution beginning in the 1840s, did the bookish, nonagricultural entomologists begin to grasp the full importance of their specialty for agricultural interests. They then used agricultural problems very effectively as a vehicle for promoting their science.

The figures in table 4 highlight another facet in the familial backgrounds of American entomologists. Despite the fact that the entomologists, as a group, lived more than a generation later in time than the American scientific authors studied by Elliott, their familial backgrounds reveal a striking similarity in the emphasis on commerce and trade. This is the first indication of a general similarity between the entomologists and scientific authors of an earlier generation, a similarity that also appears in educational and occupational patterns.

If one compares the status of entomologists' parents (as distinct from occupation) with parents of other groups, the strongly entrepreneurial disposition of the entomologists' families is reinforced (table 5). When viewed in terms of status, between 50 percent and 60 percent of both the entomologists' families and families of *DAB* scientists could be considered professional. The entomologists, however, came from families that

Table 5. Parental Status Compared for Entomologists and *DAB* Scientists, 1846–1876 (percent)

Parental status	Entomologists	All *DAB* scientists	*DAB* life scientists
Professional	57.1	58.1	54.5
Entrepreneur	31.4	17.4	19.3
Skilled worker	11.4	5.1	6.8
Other	0.0	18.5[a]	19.3[b]

[a]39 small farmers, 5 executives, 2 clerks, 1 unskilled laborer.
[b]5 executives, 2 clerks, 13 small farmers, 1 unskilled laborer.
Source: Bruce, "Statistical Profile" (1972), table 10, "Status of *DAB* Scientists' Fathers, by Periods of Scientists' Activity," p. 81, and table 12, "Status of *DAB* Scientists' Fathers by Major Fields of Science," p. 85.

were much more likely to make their living in entrepreneurial pursuits, or as skilled workers, and were much less likely to come from agricultural backgrounds. Almost one-third of the entomologists came from families with the status of "entrepreneur," compared with about one-sixth of all *DAB* scientists and about one-fifth of the *DAB* group in the life sciences. There were about twice as many skilled workers among the entomologists' families as among the *DAB* scientists' families, and virtually no entomologist was the offspring of a family with "small farmer" status, whereas about one out of six of the *DAB* scientists came from families with this status.

Another interesting aspect of the familial background is the distinctly different emphasis between the kinds of professional backgrounds of entomologists' families and those of *DAB* scientists. Entomologists tended to come from families in theology, whereas *DAB* scientists tended to come from families in medicine. Entomologists' families that were engaged in theology account for 21.4 percent of the total, whereas families of *DAB* scientists in the clergy account for only 16.7 percent. Entomologists' families engaged in medicine account for 8.5 percent of their group, whereas *DAB* scientists engaged in medicine comprise 12.3 percent of their group.[19] As will be seen, the family emphasis on theology over medicine carries over into the education and occupations of the entomologists themselves.

The figures in tables 6, 7, and 8 quantify the extent of formal and nonformal education among the entomologists in comparison with educational levels among contemporary groups of American scientists. In table 6, we see that the entomologists as a group had significantly less formal education than their scientific contemporaries. Only slightly

Table 6. College Graduates Versus Nongraduates Among Antebellum and Civil-War-Era American Scientific Communities (percent)

Educational level	Ent. ca. 1870	Sci. auth., 1800–63	*DAB* sci., 1846–76	AAAS leaders
Nongraduate	46.6	46.6	16.9	15
College graduate	53.4	53.4	83.1	85

Sources: Bruce, "Statistical Profile" (1972), table 15, "Extent of Formal Education Among *DAB* Scientists Active 1846–76," p. 89; Kohlstedt, *American* Scientific Community (1976), p. 210; and Elliott, "American Scientist" (1970), table 18, "Highest Level of Education for All Authors and for Each Decade According to When Authors Reached Age Twenty," pp. 100–103.

more than half of the entomologists graduated from college, compared with 83 percent of those listed in the *DAB* and 85 percent of the AAAS scientific leaders. This difference may be explained in part by the fact that the two comparative groups are more selective, hence more elite and more likely to have formal education. The statistics indicate that the educational level of the entomologists is practically the same as that for scientific authors of approximately a generation and a half prior to the entomologists. This supports the hypothesis stated above that in their social and educational patterns the entomologists as a group resemble their scientific predecessors of an earlier era. The figures in tables 7 and 8 further support this hypothesis. In table 7, where the educational levels are separated into more distinct categories, we find that the entomologists' educational profile resembles that of their scientific predecessors to a striking degree. Even the difference in the percentages with a Ph.D. is not as great as might appear from these figures. All of the entomologists

Table 7. Extent of Formal and Informal Education Among Entomological Authors About 1870 Compared with Other Groups of American Scientists (percent)

Highest level education	Ent. ca. 1870	Sci. auth., 1800–63	*DAB* Sci., 1846–76	*DAB* Life Sci.
Nongraduate	42.6	46.6	16.9	21.6
College Graduate	17.3	18.6	51.5	43.0
College and/or MD	33.3	32.6	20.1	28.8
Ph.D.	6.7	2.3	11.4	6.4

Sources: Bruce, "Statistical Profile" (1972), table 15, "Extent of Formal Education Among *DAB* Scientists Active 1846–76," p. 89; Kohlstedt, *American Scientific Community* (1976), p. 210; and Elliott, "American Scientist" (1970), table 18, "Highest Level of Education for All Authors and for Each Decade According to When Authors Reached Age Twenty," pp. 100–103.

with a Ph.D. held doctorates in fields other than entomology (the first American Ph.D. in entomology was granted in 1901). The entomologists tabulated here with a Ph.D. were in areas such as anatomy and physiology. They were scientists who contributed to entomology apart from their first specialty.

The figures in table 8, comparing levels of education through secondary school, modify in some areas the exact correspondence between the educational levels among entomologists and their scientific predecessors, without altering the overall similarity. The entomologists were more likely than scientific authors, 1800–1863, to have completed primary school and secondary school. The number attaining secondary-school-level education was approximately equal to that for the more contemporaneous group of AAAS scientists.

The comparison between the educational levels of all *DAB* scientists and *DAB* life scientists (table 7, columns 3 and 4) indicates that life scientists, as a group, were less likely than other scientists to have formal education and higher degrees. The entomologists, as a group, appear to occupy the far extreme of this spectrum. The life scientists apparently maintained the tradition of nonformal education longer than their fellow scientists in other specialties. The figures indicate that the entomologists may well have been the last group among the life scientists to adapt their discipline to college curricula and degree programs.

Table 8. Extent of Schooling Through Secondary Level Among Entomologists Compared with Other Groups of Scientists[a]

Level of schooling	Entomologists no.	Entomologists %	AAAS leaders (%)	Sci. auth., 1800–63 (%)
No school	N.A.	—	1.1	?
Self-taught	7	—	—	1.1
Apprenticeship	2	—	—	3.4
Common school	3	—	—	3.2
Total primary school	12	16.0	N.A.	7.7
High school or academy	6	—	—	6.5
Private school	1	—	—	0.4
Tutored	1	—	—	1.3
Total secondary school	8	10.6	11.0	8.2

[a]N = entomologists for whom we have educational information.
Sources: Bruce, "Statistical Profile" (1972), table 15, "Extent of Formal Education Among *DAB* Scientists Active 1846–76," p. 89; Kohlstedt, *American Scientific Community* (1976), p. 210; and Elliott, "American Scientist" (1970), table 18, "Highest Level of Education for All Authors, and for Each Decade According to When Authors Reached Age Twenty," pp. 100–103.

The data given in tables 6 through 8 indicate that formal schooling provided, at most, only one part of the entomologists' training in their specialty. The question arises, what then were the specific sources of training and education that characterized the preparation of this generation of American entomologists for leadership in their discipline? The answer appears to be a combination of three elements: (1) self-training, or apprenticeship under the guidance of an entomological expert or experts; (2) formal training in various disciplines in universities and museums, especially Harvard University and the Museum of Comparative Zoology; and (3) examination of European collections, study in European schools, and informal learning, such as collecting and exchanging information as a youth in Europe.

Tables 9, 10, and 11 present quantitative evidence for the multifaceted educational background of the American entomologists who were publishing about 1870. The figures in table 9 are an attempt to identify and quantify the specific contexts within which entomologists received their preparation and training in their specialty. These figures differ from a general survey of the educational background of these individuals insofar as I included only those contexts or sources that seem to have contributed directly to an individual's training as an entomologist. I then compared the individual categories of training with those for American scientific authors, 1800–1863. From these figures, I drew four tentative conclusions. First, the broad area of "self-taught" is much more important for the entomologists than for the American scientific authors, despite the fact that the two groups are fairly comparable in the extent of formal education (line 1). I will explain shortly what "self-taught" means. Second, medical studies were less important in the training of entomologists than for scientific authors, 1800–1863 (line 3). This is true in comparison with the scientists listed in the *DAB* and with the AAAS leaders as well. Third, scientific societies, academies, and museums provided significant training contexts for entomologists but were of lesser importance for American scientific authors (line 5).[20] Finally, entomological training within the formal college and university setting does not reveal a clear pattern. Undergraduate study in science may have been more important for the entomologists than for the scientific authors, but science training in preparatory school and in graduate school may have been less important for the entomologists than for the scientific authors (lines 2, 4, 8, and 12).

The most important component of entomological training might be described as a self-taught apprenticeship under an entomological expert or experts. It is manifestly clear that entomologists who were publishing in about 1870 learned their specialty as much outside as within an aca-

Table 9. Sources of Training in Entomology Compared with Sources of Training of American Scientific Authors, 1800–1863

Source of scientific training	Entomologists		Scientific authors	
	no.	%	no.	%
Self-taught in science	34	48.6	20	10.6
College graduate including science	10	14.3	9	4.8
Studied medicine	8	11.4	57	30.3
College graduate + graduate studies in science	7	10.0	30	16.0
Science studies at science society or science academy[a]	4	5.7	0	0.0
Tutored or private study	3	4.3	4	2.1
Studied medicine + additional study in science	1	1.4	16	8.5
Studied at scientific or military school	1	1.4	23	12.2
Studied science in high school or academy	1	1.4	4	2.1
Apprenticeship or work experience	1	1.4	3	1.6
Studied engineering	0	0.0	3	1.6
Studied science in college but did not graduate	0	0.0	11	5.9
Studied in private laboratory, technical school, etc.[b]	0	0.0	8	4.3
Total	70[c]	100.0	188	100.0

[a]Boston Society of Natural History, Essex Institute, etc.
[b]Elliott's category: Studied in private chem. lab.; technical or military high school/academy; or college of pharmacy.
[c]Total for entomologists differs from other tables because individuals were selected to match Elliott's categories.
Source: Elliott, "American Scientist" (1970), table 20, "Sources of Authors' Scientific Training," pp. 107–108.

demic context. What should be stressed is that the apprentice-master tradition operated in many forums: in the home, in the scientific society, in the college or university, or in the government agency. Interest in natural history ranked high among antebellum Americans. This interest translated into a sizable number who pursued careers related to their scientific interests. Advancement within the entomologists' guild, however, depended on more than an interest in insects. It depended on learning from an expert who could guide the novice in the identification of

Table 10. Fields of Study of Entomologists

Field of study	Number	Percent
Medicine	19	45.2
Theology[a]	5	11.9
Language	4	9.5
Law	2	4.8
Art	2	4.8
Zoology	2	4.8
Classics	2	4.8
Math	2	4.8
Anatomy	1	2.4
Botany	1	2.4
Geology	1	2.4
Engineering	1	2.4
Total	42	100.0

[a]Includes one philosophy as a second field.

specimens, assist in gaining access to the entomological literature, and open the door to extensive collections for study. Such "masters" in most cases received their own expertise under the guidance of a previous master, in a scientific lineage extending back to the Enlightenment and the Renaissance.

Edward L. Graef, a boyhood friend and entomological compatriot of Augustus R. Grote, one of the entomological authors listed in the *Record of American Entomology,* recalled how the two of them were schooled by entomological masters in New York in the 1850s:

About 1854 we became interested in making collections of insects. Our inspiration undoubtedly came from a small book, "Handbuch für Schmetterlingsliebhaber" von J. W. Meigen, published in . . . 1827. . . . We attempted to arrange [our insects] according to some system, but were greatly handicapped by [our] inability to get identifications of our material. . . . One may readily appreciate our delight . . . in *discovering* a man . . . from whom we learned much about rearing specimens, etc., besides being able to buy from him good insect pins, nets, setting boards, and other entomological supplies. This . . . was . . . John Ackhurst. . . . He took a fatherly interest in us and gave us much valuable information about collecting. . . . Shortly after this I made the acquaintance of an old Englishman who had a large collection of insects, and, what was much more important, an extensive, valuable entomological library. . . . this was Stephen H. Calverley. . . . [Mr. Calverley's exotics] were obtained . . . from sailors employed on the vessels controlled by Moses Taylor & Co., where Mr. Calverley was employed as a weightmaster.[21]

This account of entomological apprenticeships includes many typical elements in the training of American entomologists. The apprenticeship took place outside the context of school or college. The "masters" were of British origin, were "nonprofessionals," and had close ties to international commerce. One or more of these elements were part of the apprentice learning of most American entomological authors listed in the *Record*.

The figures in table 10 fill out the picture of entomologists' formal college training. These data represent straightforward statistics on formal education, regardless of any "end goal" of study, such as preparation for a career in entomology. They are therefore not directly comparable with the data in table 9, where distinctly entomological training is represented. For those individuals who studied in more than one field, the data have been weighted, and all figures are also expressed as percentages.

The data in table 10 indicate that 45.2 percent of those entomologists who studied in college or at a university studied medicine. Theology and language were the next most frequent fields of study, with 11.9 percent and 9.5 percent respectively. If one adds art (4.8 percent) and classics (4.8 percent) to theology and language to round out the "liberal arts," the total comes to 31 percent, approximately fourteen percentage points behind medicine. Table 11, which compares entomologists' graduate

Table 11. Fields of Graduate Study of Entomologists Compared with American Scientific Authors, 1800–1863

Level/field	Entomologists no.	Entomologists %	Sci. authors, 1800–63 %
Nongraduate	35	46.7	46.6
College graduate	12	16.0	18.6
College graduate + studied law	1	1.3	4.0
College graduate + studied theology	7	9.3	4.9
College graduate + studied science	6	8.0	7.0
College graduate + studied medicine	10	13.3	16.7
Ph.D.	4	5.3	2.3
Total	75	100.0	100.0

Source: Elliott, "American Scientist" (1970), table 18, "Highest Level of Education for All Authors . . . ," pp. 100–103.

study to the graduate study of scientific authors, 1800–1863, indicates the entomologists' continuing emphasis on theology and lack of emphasis on medicine relative to the scientific authors.

While medical study was commonplace for scientists in the making in antebellum America, the study of theology, language, classics, and art was apparently a special characteristic of entomologists. In connection with tables 9 and 11, it will be recalled that entomologists were less likely to study medicine than Elliott's scientific authors and other groups. It may also be recalled that the entomologists' fathers tended to have theological training. Though in the twentieth century, studies in theology, classics, and other liberal arts may seem an unlikely entrance into a scientific discipline, the connection with entomology is not as distant as it may seem. Descriptive entomology, to be done effectively and with authority, required a solid grounding in classical Latin and Greek, as scientific names were composed in both these ancient languages. The entomological literature of the period contains frequent discussion and debate about how to form the correct Latin or Greek name, in singular and plural, of this or that new or revised species or higher group.[22] Students of theology and classics were well qualified to take part in such debates. In addition, it should be noted that this generation came of age when the science of natural theology was pursued with as much vigor as natural history, and in fact many entomologists, like William Kirby, considered the two parts of a whole. As Kirby put it, even the insects "Preach the Gospel."[23] Although American entomologists quickly shed any vestiges of early exposure to natural theology, it is not surprising that many who prepared for the ministry in an era when natural theology flourished followed their interests in entomology as new opportunities arose.[24] Here, as in other areas, the entomologists used whatever opportunities presented themselves in preparation for their chosen specialty.

Tables 12 and 13 provide further evidence of the heterogeneous sources of entomological training. A listing of colleges and universities attended by those who received formal college training shows only two significant clusters—notably Harvard and to a lesser degree the University of Pennsylvania. Otherwise, the figures are remarkable for their lack of concentration at any single institution or group of institutions of learning. The unusually high number of European universities listed attests again to the pervasive European influence among American entomologists.

Teachers instrumental in the training of entomologists (table 13) were similarly broad in range. Harris, at Harvard, is the only teacher who stands out as a central pedagogical figure for the entomologists. Otherwise, as in the universities, the list is remarkable for the wide scattering

Table 12. Colleges, Universities, and Other Institutions Attended by Entomologists

College or University	Associated with institution	
	no.	Weighted no.[a]
Harvard College/University	10	8.33
University of Pennsylvania	6	4.50
Yale University	4	4.00
Cambridge University, England	2	2.00
Williams College	2	1.50
Art Academy, England	1	.33
Art Academy, France	1	.33
Art Academy, Germany	1	.33
Berea College	1	.50
Bowdoin College	1	.50
Brown University	1	1.00
City College of New York	1	.50
Cornell University	1	.50
Dickinson College	1	.50
Lafayette College	1	1.00
MIT	1	.33
Mt. St. Marys College	1	.50
Munich Art Academy	1	1.00
N.Y. Coll. of Phys. & Surgeons	1	.50
Oberlin College	1	.50
Rensselaer Academy	1	.33
Rutgers University	1	.33
Trinity College	1	1.00
University of Berlin, Germany	1	1.00
University of Chicago	1	1.00
University of Halle, Germany	1	1.00
University of Heidelberg, Germany	1	.50
Univ. Königsberg, Germany	1	1.00
Univ. Leipzig, Germany	1	.33
Vermont Academy of Medicine	1	.33
Virginia Military Institute	1	.50
Total associated with college/ university	50	
Not associated with college/ university	25	
Total number for whom we have educational information	75	

[a]Weights were assigned by dividing by the total number of schools attended by each individual. An individual who attended one school counted as 1 for that school, whereas an individual who attended three schools counted as 0.33 for each school.

Table 13. Entomologists' Association with Teachers

Teacher	Number associated with teachers (unweighted)	Individuals associated with 1 or more teacher(s) (%)[a]	Individuals for whom we have educational information (%)[b]
Harris	4	14.8	5.3
Agassiz	2	7.4	2.6
LeConte	2	7.4	2.6
Gray	2	7.4	2.6
Akhurst	1	3.7	1.3
Baird	1	3.7	1.3
Downing	1	3.7	1.3
Eaton	1	3.7	1.3
Edwards	1	3.7	1.3
Emmons	1	3.7	1.3
Fitch	1	3.7	1.3
Haldeman	1	3.7	1.3
Hopkins	1	3.7	1.3
Humbolt	1	3.7	1.3
LeConte, J. E.	1	3.7	1.3
Leidy	1	3.7	1.3
Markoe	1	3.7	1.3
Melsheimer, F. V.	1	3.7	1.3
Morris	1	3.7	1.3
Osten-Sacken	1	3.7	1.3
Peck, W. D.	1	3.7	1.3
Rathke	1	3.7	1.3
Ridings	1	3.7	1.3
Verrill	1	3.7	1.3
Walsh	1	3.7	1.3
Wild	1	3.7	1.3
Wyman	1	3.7	1.3

[a]$N = 27$.   [b]$N = 75$.

of teachers and influential figures who were instrumental in the training of entomologists. Both of these lists underscore the overall broad range of sources of entomologists' training in terms of institutions and individuals. Harris and Harvard University are the only significant exceptions, and Harris accounted for only 15 percent of the quantifiable sources of educational preparation of the entomologists. It was eminently possible for individuals coming from a variety of backgrounds to achieve standing in the entomological community around 1870.

The third major component in the entomologists' training, in addition to apprenticeship and selective study at college and medical school, was knowledge of European collections and other entomological knowledge gained in Europe. Table 14 indicates the extent and nature of European sources of entomological knowledge among the American entomologists, compared with American scientific authors, 1800–1863. As Elliott dealt only with American born scientists, the entomologists are divided here into those born in North America and those born in Europe in order to make meaningful comparisons.

These figures reflect once again the prominent European influence on American entomology. More than 20 percent of those for whom we have educational information were not only born in Europe but had some direct exposure to European entomological science before immigrating to America (column 2, European born subtotal). Of the American-born entomologists, 15.7 percent had direct contact with European entomologists and European entomological science. Altogether, those with European entomological experience accounted for 37.1 percent, as compared with 21.3 percent of Elliott's American scientific authors, 1800–1863, who studied in Europe.

More directly relevant to the entomologists' situation was the number of those who studied European insect collections. Such knowledge was an important prerequisite in the advancement of American entomology during the formative years of the discipline. Combining the American born and the European born, we find that 15.8 percent of the American entomologists had direct access to European museums and collections at some point in their careers (column 2). This is roughly the same proportion of Elliott's scientists of an earlier era (16.8 percent) who studied medicine and/or science in Europe. For American entomologists, knowledge of specific information in European collections, rather than European academic study or European degrees, was vital to their discipline. The American entomologists' concentration on European collections, independent of the academic context, was yet another facet of their self-taught, apprentice tradition.

Turning to the occupations by which the entomologists supported themselves, we find that they leaned toward business and entrepreneurship, with less emphasis on the professions by comparison with other groups of scientists. These patterns are consistent with the familial background and training surveyed above. In table 15, the primary occupations of entomologists are compared with the occupational pattern of the AAAS leadership in the 1850s. The entomologists show a strong emphasis on business and related occupations when compared with AAAS leaders. Among the entomologists, 22.3 percent were in business,

Table 14. European Sources of Entomological Training Compared with European Training of American Scientific Authors, 1800–1863

European experience	Ent. (no.)	Ent. (%)[a]	Sci. auth. (no.)	Sci. auth. (%)
Am. born, graduated in Europe in medicine or science	1	1.4	4	1.8
Am. born, academic training in Europe	0	0.0	2	0.9
Am. born, studied medicine or science in Europe	1	1.4	37	16.8
Am. born, visited European entomological collections	9	12.9	4	1.8
Subtotal, Am. born with European experience	11	15.7	47	21.3
Eur. born, graduated in Europe	4	5.7	—	—
Eur. born, academic training in Europe	0	0.0	—	—
Eur. born, studied medicine or science in Europe	3	4.3	—	·
Eur. born, visited European entomological collections	2	2.9	—	—
Eur. born, collected insects as a youth in Europe	6	8.6	—	—
Subtotal, Eur. born with European experience	15	21.4	—	—
Total, Am. entomologists with European experience	26	37.1	47	21.3
Total, Am. entomologists, no European experience	82	62.9	220	78.7

[a]N = 70. Total differs from other tables because individuals were selected to match Elliott's categories.
Source: Elliott, "American Scientist" (1970), Table 19, "Native Born Americans Who Studied in Europe," p. 106.

compared with only 10.4 percent among the AAAS leaders. The contrast with scientific leaders in natural history is more striking. Among AAAS leaders in natural history, only 4.8 percent were engaged in business. The entomologists show a similar emphasis on government employment by comparison with the AAAS leaders, though the difference is not as great as in business. Conversely, a relatively smaller number of entomologists made their living as teachers. Among the entomologists, equal proportions (22.3 percent) made their living in education (preparatory school through college) and in business. Among the AAAS leaders almost 39 percent found their occupational home in some branch of education. Only 18.4 percent of the entomologists made their living in professions outside education, a figure that compares with the one for the AAAS leadership as a whole. Compared with the natural history group within the AAAS, however, the entomologists had a much lower representation in the professions outside education. The explanation may lie in the fact that there were fewer physicians among the entomologists than among the AAAS leadership. The majority of AAAS professionals among the natural history specialists were physicians. The entomologists' lack of emphasis on medicine as compared with the contemporary science groups, especially AAAS natural historians, is consistent with the relative lack of emphasis on medicine in their familial and educational backgrounds.

If the entomologists' occupational structure is compared with that of scientific authors of a generation or more earlier, the emphasis on occupations outside science is more striking. Table 16 lists the entomologists according to "job description," with an indication of whether the occupation was primarily science, nonscience, or a combination of both. Table 17 compares entomologists with scientific authors, 1800–1863, in the occupational categories of science and science-related, partial science, and nonscience occupations. The entomologists' occupational profile reveals a surprising 38 percent in nonscience occupations, compared with 15.9 percent among scientific authors for the period 1800–1863 and only 7.5 percent for those authors born after 1830, those closest to the entomologists in time. About half of the entomologists made their living in some science-related occupation, the same proportion as the scientific authors. Far fewer entomologists straddled the fence between science and nonscience occupations: 11.4 percent versus 31.1 percent for scientific authors as a whole or 37.5 percent for those born after 1830.

In all the figures given so far, the physicians in both groups are considered to be in "science" occupations in order to make direct comparisons to Elliott's scientific authors. If the physicians among the entomologists are separated according to their actual work in science, as opposed to

Table 15. Entomologists Working in Various Occupational Categories Compared with AAAS Leaders

Occupation group	Entomologists		AAAS		AAAS	
	no.	%	all (no.)	all (%)	nat. hist. (no.)	nat. hist. (%)
Business	17	22.3	33	10.4	3	4.8
Government	15	19.7	47	14.9	4	6.5
Education	17	22.3	122	38.7	24	38.7
Professions[a]	14	18.4	57	18.0	21	33.8
Other	13	17.1	56	17.7	10	16.1
Total	76	100.0	315	100.0	62	100.0

Note: AAAS = American Association for the Advancement of Science.
[a]Excluding education.
Source: Kohlstedt, American Scientific Community (1976), table 18, "Leadership Scientific Interests Compared to Occupations," p. 217.

medical practice or other "nonscience" occupations, the emphasis on nonscience occupations is even more pronounced (Group II in table 17). In this scheme, even more entomologists are shown to be occupied in nonscience than in science occupations (45.6 percent versus 41.8 percent). From all these figures, the entomologists seem to represent one of the final bastions of "amateur" scientists, in the sense that roughly half made a living in nonscience or partial science occupations.

Though the entomologists lagged behind other scientists, and apparently behind other natural history specialties, in terms of the "professionalization" of their occupational structure, they were nonetheless entering into science occupations and seeking out sources of support for their entomological capabilities. By 1870, the transition from "amateur" to "professional" was well advanced, as shown in various ways in tables 18, 19, and 20. In table 18, entomologists' sources of scientific income are compared with those of the contemporary group of scientists who were cited in the *DAB*. It should be noted that half of the entomologists for whom we have occupational information had some source of scientific

Table 16. Entomologists Listed According to Job Description

Number[a]	Job description
1	ns: Catholic priest
1	ns: independently wealthy
1	ns: railroad company president
1	ns: railroad company secretary
1	ns: railroad company employee
1	ns: architect
1	ns: attorney
1	ns: bank clerk
1	ns: banker
1	ns: book dealer
1	ns: business executive
1	ns: college treasurer
1	ns: cattle breeder
1	ns: diplomat
2	ns: housewife?
2	ns: journalist
3	ns: librarian
4	ns: minister
1	ns: school administrator
1	ns: sculptor

(*continued*)

(Table 16, *continued*)

Number[a]	Job description
1	ns: secondary school administrator
1	ns: teacher
1	ns: watchmaker
1	ns: wholesale drug company owner
1	ns: wholesale glass and tin company owner
2	s: consulting entomologist
1	s: army surgeon
1	s: botanical and archaeological explorer
1	s: botanical explorer
3	s: federal government entomologist
1	s: government malacologist
1	s: insect collector
1	s: manufacturer of surgical instruments
1	s: museum assistant
3	s: museum curator
7	s: physician[b]
10	s: professor
1	s: professor and museum curator
3	s: scientific society administrator
1	s: science society secretary
4	s: state entomologist
1	s: state geologist
1	s: state mineralogist
1	s: student
2	s: taxidermist
1	s: zoological artist

Note: s = science. ns = nonscience.
[a]Total = 79.   [b]Five physicians were partial or nonscience.
Source: Based on Elliott, "American Scientist," (1970), table 23, "Authors Working at Science and Non-Science Occupations," pp. 137–38.

income, including income in areas not primarily entomological, such as a college professorship of anatomy, or a government scientist with primary duties in botany or geology. In comparison with the *DAB* scientists, the entomologists rate relatively higher in scientific entrepreneurship and wages and salary but lower in educational sources of scientific income. Government, taken as a whole, figured about equally in importance for the two groups; however, state government figured more prominently as support for entomologists and the federal government for the *DAB* scientists.

When compared with other groups of scientists, the entomologists

Table 17. Entomologists Working at Science, Partial Science, and Nonscience Occupations Compared with American Scientific Authors, 1800–1863

Science occupation	Group I entomologists		All authors		Authors, 1830–40		Group II entomologists	
	no.	%	no.	%	no.	%	no.	%
Nonscience	30	38.0	77	15.9	3	7.5	36	45.6
Science/Nonscience	9	11.4	151	31.1	15	37.5	10	12.7
Science	40	50.6	256	52.8	22	55.0	33	41.8
Total	79	100.0	484	100.0	40	100.0	79	100.0

Note: In Group I entomologists, all physicians are classified as working in science for purposes of comparison with Elliott's figures. In Group II entomologists, the physicians are divided according to whether they worked primarily in science or not in science (many "physicians" did not practice medicine at all).

Source: Elliott, "American Scientist" (1970), table 23, "Authors Working at Science and Non-Science Occupations," pp. 137–138.

Table 18. Entomologists with Scientific Income from Various Sources Compared with DAB Scientists, 1846–1876

Sources of income	Entomologists (no.)	Entomologists (%)	DAB Scientists, 1846–76 (%)
Education	11	26.8	44.5
U.S. government	6	14.6	21.0
State government	9	22.0	10.5
Wages/salary	7	17.0	6.3
Entrepreneur	4	9.8	4.4
Fees	3	7.3	4.2
Royalties	0	0.0	3.8
Other	0	0.0	0.3
Independent wealth	1	2.4	4.2
Total	41	100.0	100.0

Note: The two sets of figures are not precisely comparable: Bruce computes percentages based on weighted averages of income (scientific vs. nonscientific) for the *DAB* scientists, whereas I simply compute the percentage within each category who have some scientific income. *DAB* = *Dictionary of American Biography.*
Source: Bruce, "Statistical Profile" (1972), table 17, "Scientific Sources of *DAB* Scientists Income," p. 94.

appear to have earned scientific income in settings removed from the central government and the larger universities. This pattern is consistent with the relative lack of higher education of entomologists and with their penchant for entrepreneurship.

In table 19, the emergent professionalism of the entomological community is indicated in terms of professional and occupational affiliations with scientific institutions. Here the emphasis is first on affiliation and only secondarily on remuneration for entomological services. The position of secretary of the American Entomological Society, for example, carried at best a token monetary compensation, but the existence of this position attests to the maturing of American entomology as a scientific discipline and profession. The percentages are computed, first, in terms of the number of individuals affiliated with any one institution (regardless of whether the individuals had multiple affiliations) and, second, in terms of a weighted number adjusted for multiple affiliations. In two categories, the college and university affiliations and the state government, the weighted average is approximately equal to the unweighted. In the scientific societies and the federal government, the total, unweighted number is higher. (In private enterprise, the numbers are too small to reflect meaningful trends.) These figures seem to indicate a movement from relatively transitory professional affiliations with scientific societies to relatively stable, "full-time" positions in academia and in state government. State employment for entomologists was relatively high in this period, whereas the expansion in federal employment was just beginning. These findings would be consistent with what we know about the movement of American science in general from the scientific societies to the universities. One should recognize, however, that many of those in academic positions did not have entomology as their main specialty. For example, the paleontologist Joseph Leidy and the anatomist Charles S. Minot are both included as members of the American entomological community in about 1870, even though their entomological contributions were secondary to those in their main fields. With this caution in mind, we see that academia accounts for about 37.5 percent of the affiliations. Three other kinds of affiliations (scientific societies, state government, and the federal government) share the next level of importance, between 15 percent and 21 percent. Entrepreneurship and the private sector account for the remaining 5.5 percent to 7.8 percent of entomological affiliations.

What stands out in these figures is the importance of both scientific societies and government as early facilitators of professionalization in entomology. The scientific societies—in particular the Boston Society of Natural History—played pivotal roles not only as facilitators of scientific

Table 19. Entomologists' Scientific Affiliations

Institution[a]	Individual affiliation			Weighted percent
	(no.)[b]	(%)	(weighted no.)[c]	
Scientific society/academy				
Boston Soc. Nat. Hist.	6	8.2	2.16	5.1
Buffalo Soc. Nat. Hist.	1	1.4	1.00	2.4
Peabody Institute	1	1.4	0.33	0.8
Worcester Nat. Hist. Soc.	2	2.7	1.25	3.0
Salem Summer School Biology	1	1.4	0.33	0.8
Peabody Academy	1	1.4	0.33	0.8
American Entomological Society	1	1.4	1.00	2.4
Subtotal	13	17.8	6.40	15.3
College/university				
Harvard University	5	6.9	3.08	7.3
Museum of Comparative Zoology	4	5.5	1.91	4.5
Yale University	3	4.1	1.58	3.8
University of Pennsylvania	3	4.1	3.00	7.1
Cornell University	1	1.4	1.00	2.4
Univ. Western Ontario	1	1.4	1.00	2.4
Michigan Agr. College	1	1.4	0.50	1.2
Massachusetts Agr. College	1	1.4	1.00	2.4
Haverford College	1	1.4	0.33	0.8
Bowdoin College	1	1.4	0.50	1.2
Maryland Agricultural College	1	1.4	1.00	2.4
Johns Hopkins University	1	1.4	0.33	0.8
University of Vermont	1	1.4	0.33	0.8
Brown University	1	1.4	0.25	0.6
Southern Illinois University	1	1.4	0.25	0.6
Young Ladies Collegiate Institute	1	1.4	0.50	1.2
Subtotal	27	37.0	16.56	39.5
State government				
Illinois State Board of Agriculture	4	5.5	2.58	6.1
N.Y. State Board of Agriculture	1	1.4	1.00	2.4
Pennsylvania St. Horticultural Soc.	1	1.4	0.50	1.2
Mass. State Cabinet Nat. History	1	1.4	0.50	1.2
Mass. State Board of Agriculture	1	1.4	0.25	0.6
Calif. State Board of Agriculture	1	1.4	1.00	2.4
Lancaster County Ag. Soc.	1	1.4	0.50	1.2
Maine State Board of Agriculture	1	1.4	0.25	0.6
Conn. State Board of Agriculture	1	1.4	0.25	0.6
Vermont State Board of Agriculture	1	1.4	0.33	0.8
N.Y. State Museum	1	1.4	1.00	2.4
Subtotal	14	19.2	8.16	19.5

Institution[a]	Individual affiliation			Weighted percent
	(no.)[b]	(%)	(weighted no.)[c]	
Federal government				
USDA	3	4.1	1.83	4.3
U.S. Entomological Commission	3	4.1	0.83	2.0
Hayden Survey	3	4.1	1.58	3.7
U.S. Survey Great Lakes	1	1.4	1.00	2.4
U.S. Coastal Survey	2	2.7	0.58	1.4
Smithsonian Institution	1	1.4	0.33	0.8
Wheeler Survey	1	1.4	0.33	0.8
U.S. North Pacific Survey	1	1.4	1.00	2.4
Subtotal	15	20.6	7.48	17.8
Private enterprise				
Western Union expedition	1	1.4	0.33	0.8
Private business	3	4.1	3.00	7.1
Subtotal	4	5.5	3.33	7.9
Total[a]	73	100.0	42.00	100.0

[a]44 institutions. [b]73 affiliations. [c]42 individuals.

exchange and curators of collections but also in the training and employment of entomologists. Considering the fact that many of these societies were founded as recently as the 1830s and 1840s, their effectiveness as multipliers of scientific activity and professionalism is quite impressive.[25]

Whereas the entomologists' affiliations with scientific societies stressed primarily systematics and curation of collections, their affiliations with state and federal government were linked primarily to agricultural concerns. Government and the scientific societies formed a close alliance in the scientific exploration of the continent.

In table 20, the emergent professionalism of entomology is indicated in terms of "titles" that were either assumed by the individuals themselves or assigned to them to indicate some professional activity in entomology, whether paid or not. About two-thirds of the entomologists for whom we have occupational information could be classified according to titles indicating some form of professional entomological activity, while at the same time only half of them were in fact receiving pay for their entomological or scientific services (see discussion of table 17, scientific income). Those in the intermediate range between the two-thirds with entomological "titles" and the half employed in entomology or other areas of science were in effect professionals searching for payment for their professional activities. To judge from titles alone, the most numerous were

Table 20. Functional Titles of Entomologists as Professionals

Number[a]	Title
11	Professor
9	State entomologist
8	Entomological museum curator
4	U.S. government entomologist
4	U.S. government scientist (with entomological functions)
3	Entomological editor
3	Insect collector and/or dealer
2	Secretary, entomological society
2	Zoological and entomological artist
1	State geologist (with entomological functions)
1	Entomological consultant
1	Librarian (with entomological functions)
1	Manufacturer, surgical (including entomological) forceps

[a]Total = 50.

professors, followed by state entomologists and by museum curators and government scientists (in both entomological and nonentomological positions). Here again, the number of professors is somewhat overstated in that many "professors" followed specialties in areas other than entomology, whereas those employed in government and in the scientific societies were engaged predominantly in entomology.

The foregoing review of the background, training, and employment of individuals represented in the *Record of American Entomology* makes it possible to construct a picture of the "type specimen" for American entomological authors about 1870. Born in New England, our entomologist was most likely a man, he spent his youth either in New England or in the mid-Atlantic states, and he resided in the mid-Atlantic states in the 1860s and 1870s. His parents were employed in a traditional profession, quite likely the Protestant ministry, or in commerce. He attended college and took studies that would normally lead to a career in medicine. He graduated from college, but he had little or no postgraduate formal education, and he received, overall, less formal education than his contemporaries in American science.

His entomological training took place primarily outside the formal educational system, in a self-taught apprenticeship relationship that very likely included a tutor with British background and entomological experience. He did not study in Europe, nor did he have direct access to European insect collections, but he was more likely than his fellow

American scientists to come into contact either with European immigrants who had studied science (including entomology) in Europe, or American born entomologists who had traveled to Europe to examine insect collections there.

He had an almost even chance of being employed in business, education, or government, but since his subgroup had a much stronger tendency toward entrepreneurship than other groups of American scientists (who preferred the professions), we will assume that he made his living in a retail or wholesale trade.

His interest in entomology led him to associate with one or more scientific institutions that promoted his specialty, and as of 1870, he may have had some source of scientific income from a college or university or from state or federal government. He actively sought full-time employment as a professional entomologist and probably entertained hopes of teaching in college, serving as a curator of insects in a museum, or working as a state or federal entomologist. In fact, he had an almost even chance of becoming a full- or part-time professional entomologist.

None of the entomological authors listed in the *Record of American Entomology* fits the description of the "type specimen" in every detail, but some do come close. Three individuals illustrate how the ideal appeared in the flesh.

George W. Peck, a descendant of *Mayflower* Pilgrims, was born in Boston in 1837. He had no college education but apparently exhibited an early interest in natural history, especially entomology and horticulture. He engaged in business in Boston for some years, then moved to New York, where he was a partner in a wholesale glass and tin supply company. He was a serious student of entomology, as evidenced by his election to the American Entomological Society in 1866 and his listing in the *Naturalists' Directory* for 1866 and 1878 with a specialty in North American Lepidoptera. As far as is known, he published only one article, in the *American Naturalist* (1869), on the habits of the *Lycosa* spider, which carries its young on its back. Later in life he moved to Roselle Park, New Jersey, where he assembled one of the best collections of Lepidoptera in the state. He died there in 1909.[26]

Oliver Spink Westcott was born in 1834 to American parents in Wickford, Rhode Island. He studied mathematics and foreign languages at nearby Brown University, apparently in preparation for a career in teaching, and graduated in 1857. We do not know who may have been instrumental in guiding his interests in natural history. In 1874 he resided in Chicago, Illinois, but he traveled to the AAAS meeting in Hartford, Connecticut, where he helped organize the Entomological Club of the

AAAS. In 1878 he resided in Racine, Wisconsin, where he apparently taught in public school. The *Naturalists' Directory* for that year lists his specialties as Lepidoptera and Coleoptera. His single known article, a note in the *American Naturalist* for 1873 on the habits of certain species of Lepidoptera, includes Westcott's account of having raised two species of parasites from eggs found in the cocoon. He settled in Oak Park, Illinois, about 1880 where he taught and served as principal of the Oak Park High School. He died there in 1919.[27]

Edward Norton was born in 1823 in Albany, New York. His father, John Treadwell Norton, was described as "honored," but nothing further is known about his parents. There is also no record of when or how he began the serious study of natural history. Following his graduation from Yale University in 1844, he traveled to Europe, but it is not known whether he visited European natural history collections. He settled in Farmington, Connecticut, where he achieved national prominence as a breeder of purebred cattle. He served for most of his life as secretary of the American Guernsey Cattle Club. In the 1860s and 1870s he pursued his studies in the order Hymenoptera, particularly the phytophagous (plant-eating) family of the order. His first paper was published in 1860, the same year in which he was elected a member of the Entomological Society of Philadelphia. By 1879, he had published fourteen separate titles in various entomological journals, the most important being an exhaustive monograph on the sawflies of North America, which appeared in the *Transactions of the American Entomological Society* (1867–1869). Altogether, Norton named and described approximately 250 new species of Hymenoptera. After 1880, his business obligations and the rapid development of entomological specialization discouraged him from publishing further, but he maintained an active interest in entomology. At the time of his death in 1894, he was a corresponding member of the American Entomological Society and several other scientific societies in the United States and Europe.[28]

How did the select group of entomological authors cited in the *Record of American Entomology* fit into the context of the larger entomological community? In order to gain a picture of the overall number of those with a scientific interest in North American insects about 1870, I compiled a master list from three sources: the *Naturalists' Directory* published in 1866 and 1878, the list of authors in the *Record of American Entomology*, and the list of members of the American Entomological Society for the years 1859 to 1878.[29] The total number on the master list came to 699 individuals. Addresses for 298 of these were listed for the period 1859–1866, and 486 were listed for 1878. The geographic location of only

98 individuals, however, or about one out of seven, could be traced from the 1860s through 1878.[30] This rather small percentage may be explained primarily in terms of the sources, in particular the unusual data available from the list of American Entomological Society members. During the decade following its founding in 1859, the society attracted a large number of entomologists to its membership who do not appear in other published sources. For example, of the 298 individuals for whom addresses are available for the period 1859–1866, 212 (or 71 percent) were members of the American Entomological Society. By 1878, when the next edition of the *Naturalists' Directory* appeared, the American Entomological Society was having serious financial and organizational problems.[31] As a result the number of members of the American Entomological Society for whom names and addresses were listed had *decreased* from 212 to 63. By far the largest decrease occurred in Philadelphia, the home of the American Entomological Society, where the number of known addresses decreased from 77 to 18. This decrease was in sharp contrast to the increase among those with entomological interests listed in the *Naturalists' Directory*, whose numbers grew from 298 in the 1860s to 486 in 1878.

The large drop in the known addresses of American Entomological Society members between the 1860s and 1870s, along with the relatively small percent of carryover between the two sets of addresses, makes a meaningful charting of the movement of the entomological population between the two decades problematical. (It is not reasonable to believe, for example, that the percent of entomologists in the mid-Atlantic states decreased by nearly fifteen percentage points from the 1860s to the 1870s.) On the other hand, the figures for 1878 alone do seem to project an accurate indication of the geographic location for the larger entomological community for that point in time (table 21). When one compares the figures for the larger entomological community with those for the entomological authors in the *Record of American Entomology*, one sees a higher percentage of entomological authors in the New England and mid-Atlantic regions. The Midwest, on the other hand, has three times as many entomologists in general as it has authors. Canada and the West have slightly more entomologists overall than authors. In the South the ratio of entomologists and entomological authors to the population is identical. These figures indicate that there were more entomological observers in relation to authors in the Midwest, West, and Canada, whereas the New England and mid-Atlantic regions had a higher concentration of entomological authors. This distribution is in line with the scientific leadership of the Northeast in all fields.

Table 21. Geographic Location of the General Entomological
Population and of the Entomological Authors in the 1870s

Region	Entomological authors		General entomological population	
	no.	%	no.	%
Canada	7	9.3	38	7.9
New England	23	30.7	90	18.6
Mid-Atlantic	31	41.3	163	33.7
Midwest	7	9.3	140	29.0
South	2	2.7	13	2.7
West	5	6.7	40	8.7
Total	75	100.0	484	100.0

Source: Naturalists' Directory (1866 and 1878). Of the 699 entomologists on the composite master list, the addresses of 484 are listed in the *Directory* for 1878.

These figures, combined with additional data, may be used to estimate the size of the overall entomological community about 1870. Three sets of figures are used here: the master list of 699, projections based on the comparison of known Entomological Society members in the 1859–1866 period with those in 1878, and the list of entomological correspondents of William Henry Edwards. The increase of known entomologists between 1859–1866 and 1878 comes to 61.3 percent. We can judge from the high dropout of American Entomological Society members that there existed a large number of serious students of entomology who do not appear in other published lists. Assuming that the number of society members had increased at the same rate as the overall entomological population (instead of dropping drastically), the number of American Entomological Society–affiliated entomologists would have increased to 341 instead of decreasing to 63.[32] This change would have resulted in a net increase of 278 American-Entomological-Society-related entomologists[33] for a total of 977, overall, in 1878.

Projecting the number of potential members of the American Entomological Society thus offers one approach to estimating the overall number of the American entomological community. Computations based on the number of Edwards's entomological correspondents, compiled from his manuscript letter file and citations in the *Butterflies of North America,* vols. 1 and 2 (1874 and 1884), offer confirmation of this estimate. The total number of Edwards's entomological correspondents comes to 151. Of these, 83 (or 55 percent) appear on the master list of 699, and 68 (or 45 percent) do not. In other words, almost half of

Edwards's entomological correspondents do not appear in published sources. Edwards's circle of correspondents comprised the specialists in the study of Lepidoptera. These, together with the coleopterists, formed the vast majority of the entomological community. One can assume that a comparable number of coleopterists who do not appear in published sources corresponded with LeConte and other experts in that field. Some of these would be the same individuals (i.e., many persons had interests in both Lepidoptera and Coleoptera as well as in other orders), but these duplicates would be offset by those who corresponded with experts in Diptera, Hymenoptera, and other orders. The total number of unlisted correspondents may be estimated conservatively at twice the number of Edwards's nonlisted correspondents, or 136. Adding the estimated number of unlisted correspondents to the master list of 699 results in an overall estimate of 835.

Through these two methods of estimating, one arrives at a low figure of 835 and a high figure of 977, or about 200 more than the number of entomologists known from published lists. The overall community of entomologists about 1870, then, may be estimated at about 900 individuals, of whom about 100 comprised the publishing elite. For every entomological author, there were nine knowledgeable entomological observers who supplied the authors with vital information.[34]

These little-known observers, collectors, and breeders of insects played a vital role in the entomological community. Osten-Sacken noted:

Entomologists often receive letters of enquiry from farmers, gardeners, mechanics and other persons, mostly deficient in a preparatory knowledge of natural history; and they generally have every reason to be astonished at the fulness and accuracy of the observations of these men of manual labor, as well as at the shrewdness displayed in the management of their experiments. Very often, an investigation is fully carried out by them, and all that they apply for to a scientific entomologist is, the scientific name of the specimen.[35]

From the private correspondence of prominent lepidopterists like Edwards, Lintner, and Scudder one gains a picture of this support system. Lintner, for example, turned often to Otto von Meske, a German immigrant neighbor of his, regarding questions about the Lepidoptera. When Lintner's fellow lepidopterist, Scudder, planned a trip to Europe, Lintner asked Meske to provide Scudder with introductions to entomologists in Germany.[36] Meske lived in Albany, was elected to the American Entomological Society in 1870, and was a valued colleague of Lintner, but nothing more is known of his life.

Dr. L. K. Hayhurst, of Sedalia, Missouri, assisted Edwards in similar

ways. Hayhurst was one of Edwards's most valuable sources of information for the butterflies in the Midwest and West. He worked as a physician for a midwestern railroad that sent him on frequent trips through
Missouri, Kansas, the Indian Nations (Oklahoma), and Texas. As he
traveled, he carefully observed the ranges and habits of the western butterflies. He was also expert in raising butterflies from eggs on the food
plant. In the *Butterflies of North America,* Edwards quotes Hayhurst at
length regarding the description, habits, and ranges of several species
about which Hayhurst had the most complete information.[37] Aside from
credits in Edwards's volumes, Hayhurst does not appear in the entomological literature, as an author, as a member of an entomological society,
or as a listing in the *Naturalists' Directory.*

Another little-known subgroup within the general American entomological community was the small but active group of women entomologists.[38] This group faced special problems because of the widespread
assumption that women should remain in "feminine," that is, nonintellectual, roles. If one began with this preconception, the idea of a
woman scientist was a contradiction in terms.[39]

Of the 699 on the master list, 24 (or 3.4 percent) were women who
embodied just such a contradiction. This percentage is confirmed in
Edwards's correspondence, where 5 out of 151 (or 3.3 percent) were
women. Two women, Mary Treat and Sara J. McBride, are cited as
authors in the *Record of American Entomology.* They comprise only 1.9
percent of the entomological authors, indicating that women tended to
occupy roles as observers, like entomologists in the Midwest and West.

In fact from the 1830s on, women had participated in American entomology in various roles: as independent investigators, publicists, illustrators, and amateur observers. The investigations of Margaretta Morris
into the habits of the Hessian fly and the seventeen-year locust in the
1840s led to her election as the first woman member of the Academy of
Natural Sciences of Philadelphia and (along with Maria Mitchell) as one
of the first two women members of the American Association for the
Advancement of Science.[40] In the 1870s, the tradition of women entomological investigators was carried on by Treat and McBride, both cited in
the *Record of American Entomology,* as well as by at least two others who
were not cited there.

In 1870, McBride took part in a debate among American dipterists,
initiated by a report by Riley that the larvae of *Simulium* caught young
trout by means of a web spun in the water. S. Green responded that the
threads were not strong enough to hold the young trout, but that they
filled the gills, thus interfering with the breathing, and suffocated the

fish. McBride disagreed with both Riley and Green, reporting her observations that the larvae, far from being dangerous to young fish, in fact served as an important food source for them.[41]

Treat, a native of Philadelphia, published a number of popular articles and books on entomology, botany, and natural history, the best known being *Chapters on Ants* (1879) and *Injurious Insects of the Farm and Garden* (1882). She developed a specialty in the study of insectivorous plants and challenged such authorities as Charles Darwin and Asa Gray on certain points.[42] In 1873, she published "Controlling Sex in Butterflies" in the *American Naturalist,* in which she reported experiments indicating that the proportion of male to female butterfly offspring in a single brood could be increased by reducing the amount of food fed to the larvae.[43] Conversely, well-fed broods of larvae were reported to produce a larger proportion of females. In a notice of Treat's article in the *Canadian Entomologist,* editor William Saunders expressed doubts about the validity of Treat's findings: "The authoress is unkind enough to suggest that male butterflies are produced only from half-starved larvae. . . . this new phase of 'women's rights,' though based on experiment, we cannot but regard as a fortuitous coincidence in the cases referred to and by no means a law of nature."[44] Later that year, Riley reported experiments of his own, which he said disproved Treat's theory. Riley concluded that in some cases the female larvae were larger and needed more food, and their mortality would therefore be somewhat greater, but their sex was not determined by the amount of food.[45]

Mary Murtfeldt and Emily A. Smith are not listed in the *Record of American Entomology,* but both participated as independent investigators in agricultural entomology in the 1870s. Murtfeldt, an amateur entomologist in St. Louis, Missouri, became an assistant to Riley, the Missouri state entomologist, in 1870. Riley wrote to Scudder that he encouraged Murtfeldt in her investigation of insects because she showed promise of developing into a good entomologist.[46] Obviously pleased with her performance, Riley took Murtfeldt with him to Washington, D.C., when he moved there to head the U.S. Entomological Commission. In the following years, Murtfeldt was recognized for her work as an entomologist with the Division of Entomology. She was one of only two or three women in the Department of Agriculture in the 1870s who was engaged in science (the other sixteen or so women were presumably clerks, librarians, and bookkeepers).[47] Murtfeldt later returned to Missouri, where she served as acting state entomologist from 1888 to 1896.[48] Not much is known of Murtfeldt's investigations in the 1870s, since she published nothing under her own name. This is not surprising, because

Riley followed the common practice of publishing the findings of his assistants under his own name. The fact that Murtfeldt continued to please the exacting and somewhat tyrannical Riley indicates that she was producing solid scientific work.

Emily A. Smith, a member of the Scientific Association of Peoria, Illinois, made her first appearance as an entomological investigator in 1878 when she presented her findings on the bark louse (*Lecanium acericorticis* Fitch) before the Entomological Club of the AAAS. That year, the bark louse had seriously damaged the maple trees in Illinois and adjacent states. Smith presented an account of the life history of the insect pest complete with microscopical preparations of the insect in various stages and of the parasitic chalcid she had discovered. She also described a control method she found effective that consisted of spraying the young insects with a solution of carbolic acid. Her presentation led to a "very fruitful" discussion of the pest, which was of particular interest to Thomas Basnett of Jacksonville, Florida, because a similar bark louse was threatening the orange groves in his state.[49] At another session of the Entomological Club in 1878, Smith described a collecting bottle she had invented that had a spring-operated lid, allowing the collector to have one hand free for netting insects.[50] In 1879, Thomas, the state entomologist of Illinois, hired Smith as his assistant. She and the two other assistants, George H. French and Daniel W. Coquillett, prepared much of Thomas's second annual report (1879) while he was busy with his duties on the Entomological Commission. A review of the volume noted Smith's original investigations on species of economic importance.[51] Smith apparently remained on the staff assigned to the Illinois state entomologist for several years.

A number of women entomologists wrote (or were perceived to write) as publicists for popular audiences rather than as scientists writing for scientific audiences. They fulfilled societal expectations of a distinct "woman's sphere" in science. In the 1850s, Charlotte DeBernier Scarbrough Taylor of Savannah, Georgia, published entomological works in *Harper's New Monthly Magazine*. Taylor's work illustrates the difficulties women faced in having their work taken seriously as science. Married to a wealthy merchant, Charlotte Taylor had begun in the 1830s to study the life histories of insects harmful to southern crops. A careful and accurate observer as well as a fine artist, Taylor produced some very good entomological work. Despite the accuracy of her observations, her work seems to have received little or no recognition from scientists or agriculturists. This lack of recognition was most likely due to the mode of publication in a popular magazine and to Taylor's lively, entertaining literary style. As a result, Taylor was viewed as a popularizer rather than a serious scientist.[52]

Another area well suited to the woman's sphere was the role of entomological illustrator. A number of women entomologists engaged in this very creative though little recognized activity. The job of illustrator fulfilled most satisfactorily the perceived role of women as supporters and nurturers of scientists as opposed to the masculine roles associated with more public scientific activities.[53] It also preserved their "modesty," which was considered the highest compliment one could offer a woman engaging in intellectual activity. The earliest known example of an American woman entomological illustrator is Lucy Way Sistare Say, the wife of Thomas Say. In the 1820s and 1830s, she provided some of the excellent drawings for Say's *American Entomology* and apparently all of the drawings for Say's *American Conchology* (which was Lucy Say's specialty).[54] Following her husband's death in 1834, Lucy Say continued to correspond with entomologists and other naturalists, and she was elected to the American Entomological Society in 1863, but she apparently did not publish on her own.

The most accomplished woman entomological illustrators in the 1870s were Mary Peart and her sister-in-law, Lydia Bowen, both of Philadelphia. Peart drew all the figures for the first two volumes of Edwards's *Butterflies of North America* and a portion of those for volume 3.[55] In order to draw accurately the life histories of the butterflies, including the preparatory stages, Peart raised hundreds of butterflies from eggs, which she used as models for illustrations. Bowen, who had worked on the plates for John James Audubon's birds, hand colored the drawings composed by Peart. The resulting plates were works of such rare beauty that some critics say they have never been surpassed.[56] Edwards once remarked that the American Entomological Society should make Peart an honorary life member for her contributions to the study of Lepidoptera, but the suggestion was not acted upon.[57]

The fact that certain roles, like illustrator or popularizer, were considered more appropriate for women than for men highlights the difficulties women scientists faced in overcoming societal barriers to their participation in science on an equal footing with men. One incident records what was probably the prevailing attitude toward women who expressed an interest in the study of insects. In 1852, Henry Goadby was engaged as instructor in entomology at the recently chartered Albany University. Prior to beginning the first lecture, Goadby noted that he was happy to see so many ladies present, but (he is quoted as saying) "he trusted they would remember that he was not placed here to deliver a 'popular' course of lectures. They would be different in character from those delivered in the Female Seminary. He was here to instruct *young men* who had come . . . to get an education, and he [would] endeavor to be true to his business. He should hope to say something which should be instructive and of practical value, if it did not please their fancy."[58] Goadby's comments clearly reflect the commonly held assumption that women would expect a popular course and also Goadby's resolve not to dilute a "serious" discussion of science for young men by lowering the discussion to the "popular" level appropriate for women.

In addition to the independent investigators, popularizers, and illustrators, numerous women served as entomological observers. For example, Edwards commended two women in New York for their exceptional skill as collectors. In a letter to Lintner he commented:

There are two collections of Lepidoptera in [New York], both belonging to ladies who have taken to collecting . . . for two or three seasons with such remarkable assiduity as to shame the best of us and whose success has been quite as remarkable. They have immense numbers of moths, especially nocturnidae . . . [taken at night with rum and molasses on trees]. Their system is to plaster about 20 trees in the orchards and woods nearby their houses, in the country of course. . . . they visit the trees at intervals of two hours with a lantern and tumbler. Finding any game [?] on the tree they clap the tumbler over it, introduce a can beneath [,] they then put a bit of sponge saturated with ether in the glass and generally the insect is killed. . . . They put a kerosene lamp in the open with a white cloth over the table and . . . on this they collect a great many [moths].[59]

Edwards added that the method developed by the two ladies was so successful that he planned to use it himself the next season. Another example was Mrs. Christina Ross, a frequent and valued correspondent of Edwards. Married to Bernard Ross, the chief trader for the Hudson's Bay Company, Mrs. Ross resided at many posts in western and northwestern Canada. She reported to Edwards on the Lepidoptera of the northern region, for which she is cited frequently in the *Butterflies of North America*.

Women thus comprised a small but active subgroup within the Brethren of the Net. They participated as observers, publicists, illustrators, and as independent investigators and authors. In the process, they surmounted not only the public ridicule ascribed to "bug hunters" in general but also the social stereotypes that linked their gender with nonintellectual pursuits. Those women entomologists who participated as investigators and authors of necessity relied more heavily than their male counterparts on male patrons for advancement in entomological careers.

In 1855, for example, E. O. Kendall wrote LeConte that a Miss Morris was extremely anxious to see him with reference to some of her bugs. Rather than writing LeConte directly, as male collectors did regularly, Morris considered it advisable to work through her friend Kendall. Her sponsor proposed that LeConte meet him for dinner, and afterward they would call on the "fair entomologists."[60]

In summary, the American entomological community about 1870 comprised about nine hundred serious entomological investigators of whom about a hundred served as the publishing elite. The nonpublishing entomologists were vital to the overall enterprise, supplying the information, contacts, and support routinely required by a discipline that was largely empirical in its methodology. Among the Brethren of the Net, women comprised a minority of 3 percent to 4 percent overall and about 2 percent of the publishing elite.

The analysis indicates that entomologists in general resembled the American scientific authors of a generation earlier in terms of their educational levels and their parental backgrounds. The entomologists as a group were distinguished by the unusually large percentage born and trained in Europe, especially England, and in Canada. In other respects, with regard to geographical background, entomologists resembled the general profile of American scientists for the antebellum period. New England states counted disproportionately as a place of origin, but the mid–Atlantic states soon became the residence of the majority of entomologists. In the Midwest and the South, entomologists in proportion to the total population deviated significantly from the proportion of scientists. The Midwest figured more prominently as a residence of entomologists than for scientists, whereas the South figured less prominently, both as a source of entomologists and a residence of entomologists than for scientists in general. California, meanwhile, developed into a thriving entomological center on the West Coast.

In comparison with other groups of scientists, entomologists came more frequently from families in commerce and trade, and the entomologists themselves tended to enter entrepreneurial pursuits. The parents and the entomologists themselves came chiefly from the traditional professions, primarily medicine, the clergy, and teaching. Despite the importance of agriculture to the growth and development of entomology, a very small proportion came from farm families.

Entomologists received their training in a combination of formal and nonformal contexts, the most important elements being informal apprenticeship under a knowledgeable entomologist, formal training in a college or university (especially under Harris and Agassiz at Harvard and the Museum of Comparative Zoology), and some exposure, whether direct or indirect, to European entomology, especially European insect collections. Formal education was overall less important in the entomologists' training than for contemporary groups of scientists. Entomologists seem to have maintained the amateur tradition longer than other scientists.

Scientific societies and government figure as exceptionally important sources of support in the early stages of the discipline, but by 1870 entomologists were moving into the universities. The pattern of affiliations indicates a conscious move toward professionalization, and the support of agriculture ensured that the entomologists' chances of finding paying positions were relatively good.

While quantitative analysis can lead to a fuller understanding of the conditions under which the American entomological community developed, such analysis cannot convey the growing sense of common purpose and camaraderie experienced by those personally involved. For qualitative appraisals of what the community meant to the participants, one must turn to the correspondence of the period. Two examples will suffice. In 1856, Fred[erick] W. Grant, a settler on the northern shore of Lake Huron near Coldwater, Canada West, wrote to LeConte, thanking him for the packet of books, which he said were "indeed a treasure of knowledge for one so entirely uninformed on the subject of American Coleoptera."[61] Grant was "particularly pleased" with LeConte's "Notes on the Classification of the *Carabidae* of the U.S.," which he said arranged these beetles according to "less arbitrary and artificial distinctions" than Latreille's classification. He ventured the opinion that [William Sharp] Macleay's system of classification (which by that time had generally lost favor) was the best guide to the true "affinities and analogies" in nature. Grant had already collected a considerable number of Coleoptera, many of which, he noted, differed from those of the European fauna. These excursions had cost him much time (which was a "dear article in the life of a Backwoodsman"), but he assured LeConte that he would "willingly spend three fourths of my time in the pursuit of my chief recreation and favorite study." He asked LeConte to send him a full list of desiderata so that he would know what to prepare for shipment. Grant closed by noting that he expected the arrival of his father, who was "by far a better Coleopterist than myself." When his father arrived, there would be a total of four entomologists in the remote setting, consisting of Grant, his brother and father, and Mr. Bush, who were "all pretty ardent though rather inexperienced: quite a large number for an unsettled district on Lake Huron . . . , is it not?"[62]

Plate 23, "Coleoptera," Townend Glover Notebooks, ca. 1850–1878, box 9, Record Unit 7126, Smithsonian Institution Archives, Washington, D.C.

In August 1868, Walsh wrote LeConte about the recent gathering of entomologists at the AAAS meeting in Chicago. He chided LeConte for not coming to the meeting, for he had missed seeing a "host of N[orth] A[merican] bug hunters," including Saunders, Hagen, Packard, Riley, and "that miserable old Illinois scalawag, Benj. D. Walsh."[63] While at the Chicago meeting, Hagen and his wife had stayed in the palatial residence of a Chicago millionaire, and following the meeting they visited the Walsh couple at their "humble cottage" near Rock Island. There the two entomologists and their wives "got along famously," the women discussing mutual interests and the men retiring to peruse Walsh's cabinet and to compare notes on their investigations of insects.

In these two letters one senses the growing solidarity of the American entomological fraternity. Representing a tiny minority, often isolated geographically, pursuing studies considered obscure and slightly ridiculous by the majority of their countrymen, individual entomologists were discovering to their relief and delight that they were not alone. They were indeed Brethren of the Net.

Acceptance and Implications of Evolution

9

The evolutionary theory proposed by Charles Darwin held a special significance for entomologists. Darwin considered himself an entomologist, among other specialties, and a considerable portion of his published work dealt with insects. As a young man, he had collected and studied insects; he helped organize and was a life member of the Entomological Society of London; on the voyage of the *Beagle*, he collected a large number of insect specimens, especially Coleoptera, and he drew on his observations of insects on that voyage in support of his theory.

Riley summarized Darwin's use of entomological data in the *Origin of Species* thus: Darwin declared that the preponderance of phytophagous over predaceous species in the tropics exemplified the relation of the insect and plant worlds; he showed that the insect faunas of Tierra del Fuego and Patagonia, though separated only by the Straits of Magellan, have nothing in common; he accounted for the paucity of insects on islands with the argument that the smaller the area, the less opportunity for the development of insect life; and he explained why insects on islands are apterous rather than winged: they would otherwise be blown out to sea.[1]

In the years immediately following the publication of the *Origin*, Darwin paid increasing attention to insects. He corresponded extensively with entomologists, asking questions, suggesting solutions, and encouraging them to pursue evolutionary themes in their own observations. In the 1860s, he used the meetings of the Entomological Society of London as a sounding board for his developing ideas on natural selection. The result of Darwin's intense interest in entomology in this period is apparent in *The Descent of Man* (1871), which draws upon entomological sources for much of its supporting data.[2]

American entomologists were keenly aware of the special relevance of Darwin's work to their discipline. Riley, writing only twelve years after the publication of the *Origin*, stated what had been apparent to many entomologists for some time: "next to plants . . . , insects offer . . . the best material for the inquiring mind. . . . Their rapid multiplication, the rapid manner in which one generation is often followed by another, the wonderful manner in which they are often affected by climate and food, especially during the preparatory . . . stages—all tend to furnish variation for Natural Selection to work upon, in a profusion unknown in the higher animals."[3] A decade after this statement, Riley expanded further on the relevance of entomology to evolutionary studies: "Indeed, the varying conditions of life in the same individual or species; the remarkable metamorphoses; the rapid development; the phenomena of dimorphism and heteromorphism; of phytophagic and sexual variation; the ready adaptation to changed conditions, and consequent rapid modification; the great prolificacy and immense number of individuals; the three distinctive stages of larva, pupa, and imago, susceptible to modification, as well as other characteristics in insects—render them particularly attractive and useful to the evolutionist."[4]

The intellectual revolution that Darwin initiated in American entomology complemented the growth of entomological institutions and the advances in systematics and practical applications in unexpected ways. Once considered as eccentric collectors, or adjuncts to geology or agriculture, entomologists suddenly emerged among those best qualified to judge the merits of the most momentous scientific questions of the age. Within their own specialty, problems that had been rendered increasingly perplexing—problems of classification, geographical distribution, variation, polymorphism, and ecological relations—now became susceptible of solutions within the context of an evolutionary framework. Evolutionary theory unified and strengthened the entomologists as a working community by furnishing direction to their investigations in systematics and practical applications, by strengthening their ties with European science, and by enhancing their status as investigators of fundamental biological questions.

American entomologists, like American scientists generally, tended to incorporate evolutionary concepts into their work rather than to engage in theoretical discussions. Among the various scientific disciplines, however, the entomologists demonstrated a greater willingness than others to discuss evolutionary theory openly in their professional writings. The debate that did occur among entomologists, both in public and private, may be divided into two phases. In the first phase, from about 1863 to 1867, Benjamin D. Walsh, with the encouragement of Darwin, challenged the views of prominent American opponents of Darwin, like Agassiz, James D. Dana, Scudder, and Packard. In the second phase, from about 1867 to 1870, either these opponents were won over to evolution, or, as with Agassiz, their views lost credibility among their scientific peers. During the latter period, the two leading opponents of Darwin among entomologists, Scudder and Packard, were converted to evolutionary theory. Indeed, Packard became the leading publicist for a distinct American school of evolution known eventually as Neo-Lamarckism. By the 1870s, almost every entomologist who expressed a view on the subject in public or private espoused some form of evolutionary theory, whether Darwinian or Neo-Lamarckian.[5]

Walsh initiated the public discussion of Darwinism among entomologists at a meeting of the Boston Society of Natural History in December 1863, where he defended the unorthodox view that varieties are incipient species. This discussion was followed by Walsh's article, "On Certain Entomological Speculations of the New England School of Naturalists," in which he challenged Agassiz's assertion (in a recently published volume on Lake Superior) that no insect species were common to both Europe and America.[6] Walsh pointed out that LeConte's list of Coleoptera (and Agassiz's own list of plants) in the same volume contradicted this statement. Drawing on other entomological authorities, Walsh compiled a list of some 360 identical or closely related species common to both continents, and he argued that Darwin's theory better accounted for these and other facts of geographical distribution than Agassiz's theory of special creation. Walsh indicated his intention to open the public discussion in a letter to Scudder in which he said his forthcoming article [probably the one on gall insects noted below] would "include 4 or 5 pages of horrible Darwinian heresies which will make Agassiz's hair stand on end."[7] In other articles in the *Proceedings of the Entomological Society of Philadelphia,* the *American Journal of Science,* and the *Proceedings of the Boston Society of Natural History,* Walsh demonstrated Agassiz's misrepresentation of Darwin's theory and his failure to consider seriously the evidence for natural selection. Walsh also challenged the views of Dana, Scudder, and Packard, all of whom had expressed their support of Agassiz or their opposition to Darwin's theory.[8]

Walsh sent copies of his articles to Darwin, reopening a correspondence between the two former classmates that lasted until Walsh's death five years later. Darwin replied that he was "very much pleased to see how boldly and clearly you [Walsh] speak out on the modification of species."[9] Darwin welcomed Walsh's forceful championship of evolutionary theory among American zoologists and employed his polemic abilities as he had already employed those of the botanist Asa Gray. "I am delighted at the manner in which you have bearded this lion [Agassiz] in his den . . . [and] attacked Dana's wild notions," Darwin wrote to Walsh.[10] Darwin expressed appreciation for Walsh's willingness to engage in public controversy, something that Darwin himself avoided.[11]

The main body of entomological data Walsh drew upon in support of Darwin was his investigation of the cynipid gall gnats that inhabited the willows in Illinois. The gall gnats comprised a group of species so closely resembling one another they were separable only on the basis of the galls that they formed. When Walsh began their study about 1860, only one species had been described. By 1864 Walsh had identified fifteen separate species, each of which was confined to a separate species of willow. The only reasonable explanation for the striking unity in structure, coloration, and life cycle of these species, and at the same time the distinct preference for different species of willow, said Walsh, was that they shared a common ancestor and that they had become separated as distinct species through the isolation of one factor—in this case the species of willow upon which they fed. To say that such phenomena were so "because the Great Author of Nature has willed them to be so," he concluded, "is no explanation at all." Only the "derivative" theory could adequately explain these facts.[12]

Having been placed on the defensive by Walsh's bold attack, Packard, Scudder, and Dana replied by pointing out what they considered to be factual errors in Walsh's entomological data. They offered only brief rebuttals on the main theoretical issue, namely Walsh's explanation of the data in terms of evolution. Agassiz did not reply to Walsh.[13] It is apparent from the exchange that a major difference separating Darwin's entomological supporters and opponents was what Ernst Mayr has called the dispute between those who were essentialist (or typological) thinkers and those who were population thinkers.[14] Essentialism is the notion that fixed, unchangeable ideas underlie observed variability and only they are permanent and real. This thinking, says Mayr, dominated the natural sciences from the seventeenth century up to the latter part of the nineteenth century. Essentialism was the primary obstacle to the acceptance of Darwinism, and in fact, Darwinism was the first effective challenge to its validity.[15]

Essentialist thinking among entomologists was closely bound to the concept of the "type." This concept had been the most brilliant contribution of Georges Cuvier and his students, among them Louis Agassiz.[16] Rejecting the idea that animals, and classes of animals, represented a gradual progression in the Great Chain of Being, and rejecting also Jean Baptiste Lamarck's concept of the plasticity of living forms, in which the environment could produce gradual and permanent changes from parent to offspring, Cuvier proposed that each animal, or class of animals, was a unique functional unity in which every organ was related in a specific way to every other organ in certain fixed relationships.[17] An animal's organs and their relationships to one another were based in turn on what Cuvier termed the "Conditions of Existence," or the environment. As functional units, animals could not vary significantly without abruptly perishing. Cuvier and his disciples thus perceived an innate tendency for animals to revert to the perfect type in which the total organism functioned perfectly, regulated by its own internal anatomy and the conditions of existence upon which it depended.[18]

Cuvier's rational approach to animal form led to great advances in comparative anatomy, paleontology, and systematics (all fields in which Agassiz made important contributions).[19] At the same time, Cuvier's interpretation of living forms in terms of stable functioning units slowed the acceptance of evolutionary theory because it reinforced essentialist, or typological, thinking. The tendency of animals to revert to their type left no foothold for variations to develop into new species. Cuvier's students, including such influential teachers as Agassiz and Richard Owen, never broke from his intellectual domination.[20]

Darwin's most original contribution, on the other hand, was to replace essentialism with population thinking. In Mayr's analysis, this initiated "perhaps the greatest revolution . . . in biology. . . . For the typologist, the type . . . is real and the variation is an illusion, while for the populationist the type (average) is an abstraction and only the variation is real. No two ways of looking at nature could be more different."[21] Mayr concludes that "virtually every major controversy in the field of evolution has been between a typologist and a populationist."[22]

Packard, Scudder, Thomas, and other opponents of Darwin frequently employed typological arguments. Packard summed up his technical refutation of Walsh's case by asserting that, despite some degree of variation in the animal kingdom, there is "a deeper-seated tendency of all . . . organisms towards a perfection of the *type* to which they belong."[23] Cyrus Thomas, an early opponent of Darwinism, argued that the classification of animals would collapse into chaos if the permanency of species were denied. "Starting with the admitted fact that species ever remain distinct, we next gather into a group those species having the parts of the body similarly arranged . . . , which . . . is termed a *genus*," and so forth through the family, order, and class.[24] Augustus R. Grote, another critic of evolutionary theory, also pointed to stability in animal forms as fundamental to classification. Grote insisted that "Even those Naturalists . . . who profess the Derivative Theory of Creation, require their followers first to understand the invariability of species" before moving on to the new theory.[25]

The conflict between typological and population thinking among the entomologists broke into open debate over the nature of the genus. Scudder and LeConte, following the teaching of Cuvier and Agassiz, maintained that each genus (as well as each species) was created independently. Walsh and Hagen disagreed: "allow me to ask you [Walsh wrote Scudder] whether you conceive that genera could have been created and had an actual existence before either species or individuals existed? I could as easily conceive of . . . virtue or patriotism having had an actual independent existence before matter and men were created."[26] Scudder also followed Cuvier in pointing to the gaps in the geological record as a proof of the stability of all taxonomic units. Walsh disagreed with this reading of the geological record also. Walsh held that genera did not exist in nature, but that "in geological time, there were intermediate forms between all allied genera, so they have no real existence, being dependent on the extinction of intermediate forms."[27]

The discovery of protective mimicry in butterflies by the English naturalist Henry Walter Bates widened the debate over evolutionary theory among entomologists in America and elsewhere. The evidence from mimicry convinced prominent American opponents that evolution was at work in the creation of new species, thus eliminating any significant opposition to evolutionary theory among American entomologists.

In the Amazon basin, Bates observed numerous instances in which species of one family of butterflies, the Pieridae, which were normally white, yellow, and orange, departed from the normal pattern to take on a close resemblance to species of another family, the Danaidae, which were marked in bright red, yellow, blue, and orange. Bates reasoned that the Danaidae "models" were unpalatable to predators and that any Pieridae that were modified in that direction would tend to survive. The closer the mimic resembled the model, the better its chances for survival, and the same selection process that favored the mimic pattern would operate to keep it stable.[28] Here was dramatic evidence of natural selection not in the geological record but in living populations.

Bates presented his findings in a paper before the Linnaean Society of London in 1862. He considered the mimetic butterflies "a most beautiful proof of the theory of natural selection."[29] Upon reading the paper, Darwin wrote to Bates, "The mimetic cases are truly marvelous. . . . I am rejoiced that I passed over the whole subject in the 'Origin,' for I should have made a precious mess of it. . . . one feels present at the creation of new forms."[30] Darwin circulated Bates's paper among other entomologists, and in 1865 he arranged for Bates to have a copy sent to Walsh.[31] Similar cases from various parts of the world were soon reported by entomologists such as Alfred Russel Wallace in Malaysia, Thomas Belt in Nicaragua, Roland Trimen in South Africa, and Walsh and Riley in the United States.[32]

The case reported by Walsh and Riley was the now well-known mimicry of the monarch butterfly by the viceroy. The Danais family of butterflies, which Bates found so abundant in the tropics, was represented in North America by only two species, *Danaus archippus* Fabr., the monarch, which was common in much of the United States and Canada, and *D. berenice* Cram., which replaced *archippus* in the southern states. Walsh and Riley found among the Nymphalis family (which were normally blue-black or black and white), one species, *N. disippus* Godt., that was almost an exact counterpart of *D. archippus*. The mimicry was so close that some authorities had confused the two.[33] Walsh and Riley, who were the first to identify this as a case of mimicry, pointed to details of the case that supported a Darwinist explanation. For one thing, the model and the mimic were always found to inhabit the same region at the same season. For another, the mimic *N. disippus* was abundant in its range, whereas all other species of the Nymphalis family were relatively rare. Since the larvae and pupae stages of *N. disippus* were almost indistinguishable from other species in the Nymphalis family and would therefore be thinned out at about the same rate by predators, Walsh and Riley reasoned that only protection by mimicry in the adult stage could account for the abundance of *disippus*.[34] Upon further investigation, Riley cited the even more conclusive evidence (ironically first reported by antievolutionist Scudder) that in the south, where *D. archippus* was replaced by a darker relative, *D. berenice,* the mimetic *disippus,* was darkened to the same degree.[35] Riley pointed out that the striking resemblances of these butterflies were formerly regarded as "curious analogies in nature, intended to carry out the . . . plan of the Creator," but viewed in light of the Darwinian hypothesis, they now "acquired an immense significance."[36]

Riley cited this case in an exchange of letters on mimicry that ran through several issues of *Nature* (London) in 1870–1871. He refuted the objection of Scudder (among others) that predation takes place primarily in the larva stage and that therefore mimicry in adults could not be effective in the selection of the adult form.[37] Riley also answered the objections of Scudder and others with experiments and observations he had made in the monarch case. While admitting that natural selection could not account for the variations themselves, Riley maintained that evolution provided the best explanation for mimicry.[38]

The debate over mimicry apparently played a significant part in Scudder's conversion to evolutionary theory. In 1869, his fellow antievolutionist Grote wrote to Scudder, "I hope you will not adhere to Mr. Bates's theories."[39] As noted above, Scudder opposed the Batesian explanation in 1870. By 1872, however, Scudder had changed his mind. Grote chided Scudder for "first abusing and then starting Natural Selection schools of your own."[40] By 1876, Scudder was publishing papers that demonstrated how mimicry could be explained in terms of natural selection, and by 1880, he was proudly citing the contributions of American entomologists to evolutionary studies.[41]

The conversion of Packard, the other leading opponent of evolution, is associated with the rise of a distinct American school of evolutionary theory, called Neo-Lamarckism. This alternative to Darwinian theory was developed independently in 1865 and 1866 by Alpheus Hyatt, professor of anatomy at Harvard University, and the paleontologist Edward Drinker Cope. Though convinced of the general validity of evolution, Hyatt and Cope were dissatisfied with Darwin's concept of natural selection because it failed to account for the occurrence of variation in living forms. Drawing on the speculations of Lamarck, they proposed that environmental influences—especially geography and climate—could act as primary agents of change.[42]

Cope's original contribution to the theoretical formulation was the "law of acceleration and retardation," which related the rate of change in the species to the embryonic development in the individual. In 1871, Cope explained the Lamarckian mechanism as meaning that the embryo's development was influenced through the parent's body. If the environment were to change so that the parent would have need for increased use of a particular organ, the increased use of that organ would direct the "growth force" of its offspring toward that part, thus causing variation in the developing embryo.[43] Neo-Lamarckians like Cope held that superficial characters could be formed by natural selection, but fundamental characters, those that led to the formation of new species, were formed by the influence of the environment.[44]

Packard, who was raised in a devout Christian household and believed firmly in the special creation of species, responded to the challenge of Darwinism by reformulating his thinking about the origin of species according to naturalistic explanations. In the years from 1867 to 1870, he adopted evolutionary explanations of descent. Eventually rejecting both Darwin's emphasis on natural selection and the law of acceleration and retardation, he came to view the direct influence of the environment as the ultimate cause of evolution.[45] Apparently his researches into the development and anatomy of the horseshoe crab and the study of arthropod ancestry played a role in his conversion to evolutionary explanations. In 1874, Packard and Frederick Ward Putnam investigated the insect fauna of Mammoth Cave, Kentucky, where strange underground creatures seemed to confirm Lamarckian evolutionary development. In the cave insects Packard found evidence for "reverse evolution," that is, the loss of organs such as eyes, a change that he concluded had been induced through the absence of light.[46] Packard's travels in the West in the 1870s exposed him to a vast laboratory of environmental diversity and to the adaptation of plants and animals to these environmental influences. The idea that climate caused variation guided Packard in his study of geometrid moths, in which he took measurements of species of moths inhabiting different climatic regions from the Atlantic coast to the Pacific. His study was published as a monograph in Hayden's annual report for 1876.[47]

Packard soon became the leading publicist for the Lamarckians. He traveled to Europe in 1872, where he gathered material for a biography of Lamarck; he wrote popular articles explaining the theory that rivaled Darwin's; and in 1884 he proposed the term "Neo-Lamarckism," by which the school was known thereafter.[48]

Lamarckian theory received wide support from the 1870s until the 1890s because it offered solutions to problems that troubled Darwinists and non-Darwinists alike. Lamarckism supplied an explanation for the cause of variations that Darwinism did not. It postulated rapid evolutionary change, through environmentally induced variations, that accommodated the objections of physicists like Lord Kelvin that the earth was too young for natural selection to have produced the tremendous variety of living forms. By emphasizing a vital life force upon which the environment acted, Lamarckism maintained a clear distinction between living and nonliving forms. This belief in a life force, as opposed to Darwin's mechanistic version of natural selection, made Lamarckism more compatible with prevailing religious views.[49]

American entomologists' accommodation to religion is evident in two addresses before the AAAS by LeConte and Riley. In his 1875 presidential address before the AAAS, LeConte argued for the essential difference between the biological and the physical sciences. Precise measurement was not possible in biology, he maintained, because the logic of biology was "not formal, but perceptive. . . . The aesthetic character of Natural History . . . prevents the results . . . from being worked out with the precision of a logical machine." Physical scientists could not see the work of the Creator as clearly as biologists, he argued further. He held that biology could indicate the existence of the Creator but that natural theology could prove nothing further in the realm of religion.[50] Riley, defending the Lamarckian tenet that acquired characteristics could be inherited, expressed the view that the production of new forms through the use of organs implied a Designer. This view, he said, provided a new and hopeful view of nature and of man.[51]

The shift toward Lamarckism in the 1870s took place among naturalists in Europe and America, but it was in America that Neo-Lamarckism gained its widest acceptance. Lamarckism achieved its greatest acceptance in the 1870s, a period in which Darwin gave increasing weight to the Lamarckian explanations of variation in succeeding editions of the *Origin of Species*. By 1880, the Neo-Lamarckians outnumbered the Darwinists among American naturalists.[52] The interest in environmental influences in America was the natural result of the vast extent and diversity of the continent, which included a wide variety of geographic and climatic zones made available to the American naturalists through western exploration and the spread of rail transportation.[53]

As a group, American entomologists seem to have been attracted to Lamarckian positions even more than their fellow naturalists. The rapid reproduction among invertebrates and their relatively greater plasticity of forms seemingly fit the Lamarckian hypothesis. Riley and other entomologists expected to observe the rapid evolution of new insect species within a few years' time. In 1888, Riley attempted a general synthesis of Darwinian and Lamarckian theory.[54] In the 1880s, following August Weismann's distinction between the germ cells (which pass along unchanged by environmental influences on the parent) and somatic cells (which can be modified but are not inherited), the pendulum began to swing back from Lamarckian environmental explanations of evolution to the Darwinist emphasis on natural selection.[55]

The conversion of Scudder and Packard, the two leading opponents of evolution, meant that American entomological leaders were now virtually unanimous in their support of evolution. How less prominent entomologists responded to evolutionary theory is more difficult to assess. Occasional references in the entomological journals and in correspondence indicate that the "amateurs" adjusted to evolutionary theory in the same way as the leaders. The treatment of evolution in the entomological societies and journals, including explicit and implicit editorial policy as well as printed correspondence from the members, serves to indicate their thinking.

The Cambridge Entomological Club is of particular interest in this regard because Agassiz's influence continued, though he died in 1873, the year before the club was founded. Here, if anywhere, one might expect to find entomological opposition to evolution. Scudder served as first president, and he was a dominant influence in the first twenty years of the club. In his first presidential address, Scudder explained, "We have favored the biological side of our science . . . as the most important . . . and yet . . . the least known. . . . We have desired . . . to uphold the superior value of questions which have a more direct philosophical bearing."[56] The emphasis on higher "philosophical" aspects of entomology was reiterated often.[57] The fact that Scudder and other Agassiz students had broken with Agassiz and accepted evolution explains the lack of open opposition to evolutionary theory in the club's journal, *Psyche*. In many ways, however, the policy established for *Psyche* followed the intellectual tradition of Agassiz and Cuvier. George Dimmock, the first editor, explained that *Psyche* would be "devoted principally . . . to . . . the habits, general anatomy and physiology of Arthropoda."[58] Descriptions of new species were expressly left to other journals, except where necessary to establish more general principles. Agricultural entomology, consisting of "measures of defense against insects" was also excluded; however, life histories and accounts of the biological relations of insects were welcomed.[59]

The ease with which the Cambridge Entomological Club accommodated the tradition of Agassiz and Cuvier to evolutionary theory is a striking confirmation of how both scientists had brought their studies to the very threshold of evolutionary thought, without actually crossing over themselves.[60] Neither Cuvier nor Agassiz were able, or willing, to see that the morphological similarities they used in comparison existed because of phylogenetic relationships. For them, similarity was explanation enough because it fulfilled their teleological quest for the rule of law in the natural world. For them, species were by definition communities of resemblance, whereas for the evolutionist, species were ultimately communities of reproduction.[61]

Agassiz's students and heirs inherited his comparative methodology and his quest for generalization ("Facts are stupid things, until brought into connection with some general law," he once told Scudder), but they overcame his philosophical objections to evolution.[62] They continued to search for "philosophical" generalizations in the life histories, habits, and environmental relationships of insects, but they were not averse to interpreting these in terms of evolutionary theory. The extent to which the club's membership had become accommodated to evolutionary theory is indicated by the remarks of its president, W. P. Austin, in 1879, who noted that evolution had created widespread interest in the study of geographical distribution: "The increasing popularity of the theory of the continuity of organic life . . . has given a great impetus to the study of questions of distribution, while at the same time every new fact in regard to distribution [tends to either] confirm or throw doubt on the theory."[63] Austin concluded that "so long as every species was supposed to be due to a special creative act questions of distribution were of very little interest," but now that it was accepted that species evolved, the data of geographical distribution took on great significance.[64]

In the case of the Entomological Society of Philadelphia (later the American Entomological Society), the most striking feature was the society's vigorous sponsorship of Walsh's exposition of Darwinism as noted already in the Walsh-Packard-Dana exchange. Overt references to Darwinian concepts in scientific journals were unusual during the Civil War. For example, the papers published by the Philadelphia Academy of Natural Sciences, the parent group to the entomological society, during the same period contained no open references to Darwinism. When the academy editors did begin publishing articles with references to evolution, these articles simply incorporated evolutionary hypotheses and supporting data in scientific reports rather than advocating the acceptance of evolutionary theory in the polemical style used by Walsh.[65]

LeConte, the first president of the Entomological Society of Philadelphia, was converted to evolutionary ideas sometime in the late 1860s. Thereafter, he proceeded to revise his classifications in accordance with evolutionary concepts, "sinking" many of his early species that he felt were founded on minor variations.[66] In 1875, when the disagreement involving the Pearsall collection had blown over (see Chapter 3) and he again served as entomological society president, LeConte was elected to the presidency of the American Association for the Advancement of Science as well. For his presidential address before the AAAS he chose the topic of the evolutionary descent of beetles, which he traced through a study of their present geographical distribution.[67] From the lack of any objection to the views expressed by Walsh or LeConte, or indeed any open references to evolution in the society's journal or minutes, one may infer that the members concurred with evolutionary theory.

The editors of the *Canadian Entomologist*, C. J. S. Bethune and William Saunders, refrained from editorializing either for or against evolutionary theory in the pages of the journal. Judging by the contents of the *Cana-*

dian Entomologist from its founding in 1868 through 1880, however, it may be inferred that the editors and those who read the journal were tacit, if not outright, supporters of evolution. For example, William Henry Edwards reported numerous observations and experiments on polymorphism in butterflies and on classification that he explained on the basis of evolution. Other references to evolution occurred frequently in the 1870s. In November 1872, Theodore L. Mead, a collaborator of Edwards, reported finding a variety of the butterfly *Limenitis misippus* that, like *Nymphalis disippus,* resembled *Danaus archippus,* the well-known unpalatable model. Mead speculated that if he had not collected the butterfly and thus intervened in the "struggle for existence" the variety "might have given rise to a new species."[68] In the same year, George John Bowles, an entomologist from Montreal, Canada, discussed his observations of an unusual variety of the European cabbage butterfly that he had first seen in 1863, and he speculated that, since he was able to rear a specimen in captivity, the variety might be well on the way to becoming a distinct species. Noting that the variety occurred in England also but that it was more common in America, Bowles speculated that the tendency to diverge had been accelerated by conditions in the New World.[69] The single antievolutionary reference in the *Canadian Entomologist* during the 1870s was written by the British entomologist, Francis Walker. In a review of various explanations for the geographical distribution of insects, Walker denied that any clear evidence of a line of descent from previous extinct species to present species had ever been traced. He favored the explanation that insects had been created as distinct species, had then been forced to the tropics during the ice age, and subsequently spread north and south to their present distribution.[70]

The Brooklyn Entomological Society, founded in 1872 by George D. Hulst and four other lepidopterists and coleopterists, began publishing its *Bulletin* in 1878.[71] Devoting primary attention to collecting and raising butterflies and beetles for display and exchange, the members seemed content to accept the consensus on evolution that entomologists had reached by the 1870s. In 1885, the *Bulletin of the Brooklyn Entomological Society* was merged with *Papilio,* another New York entomological publication. John B. Smith, editor of the new journal *Entomologica Americana,* noted that although some members might be disinclined to accept evolutionary theory, the best classifications were based on evolutionary premises.[72]

Other periodical publications in which entomologists regularly expressed their views and in which one might expect to find any opposition to evolution, such as the report of the Entomological Club of the

AAAS, or the *Proceedings of the Buffalo Society of Natural History* (edited by Grote), confirm the conclusion of this survey: American entomologists quickly and with near unanimity accepted evolutionary theory, Darwinian or Neo-Lamarckian, and incorporated this theory in their work.

The ready acceptance of evolution by the American entomological community contrasts sharply with the reception of evolution by Darwin's colleagues in the Entomological Society of London. From the time of publication of the *Origin of Species,* the London society members were divided on the issue of Darwin's theory.[73] The initial response to *Origin* was hostile, and though there were some converts during the next few years, in the mid-1860s the majority remained anti-Darwinians. In the late 1860s and early 1870s, the presidency was held alternately by prominent Darwinists and anti-Darwinists (Bates and Wallace followed by John O. Westwood and Sidney Smith Saunders). During these years, evolutionary theory was debated frequently at the meetings, apparently in an open and friendly atmosphere.[74] Opposition to Darwin came chiefly from the taxonomists, who feared that their work would be disrupted, and from champions of natural theology, like Westwood, who charged that Darwin's theory was not supported by facts and that it threatened established religious views. Collectors may also have feared the loss of status that comparative work implied.[75] By 1875, after fifteen years of debate, positions within the society remained essentially unchanged, with members divided between Darwinists and anti-Darwinists.[76] The hesitancy, if not hostility, with which the London society viewed evolutionary theory as contrasted with the thoroughgoing acceptance of evolution by the American entomological societies indicates a fundamental divergence between the entomological communities on either side of the Atlantic.

The American acceptance was based on several interrelated factors. First, Americans were more typically field-oriented naturalists who had the opportunity to observe the tremendous climatic and geographical variety of the North American continent and who were conditioned to seek external or environmental relationships among living organisms.[77] Second, their emphasis on agricultural entomology reinforced their appreciation of the biological and ecological aspects of insects. The relatively greater interest in climatic and geographic variations and the emphasis on biological and ecological aspects of entomology in the investigation of insect pests of agriculture reinforced one another. Third, the fluid, dynamic character of American social and political institutions and practice caused Americans to think of change—whether in society, politics, or nature—in terms of aggregates and dynamic populations rather than in hierarchies and stable patterns. Nineteenth-century Americans, observing the filling of the North American continent with a restless, mobile population and the creation of a political structure comprised of multiple local governments, were prone to think in terms of populations rather than typologies.[78] Whatever the explanation, the openness of American entomologists to evolutionary theory merits attention in the overall context of evolutionary debate in the Atlantic scientific community.

The entomologists, in their near unanimous support of evolutionary theory, played a significant role in the dissemination of evolutionary theory among American intellectual leaders, such as the clergy, doctors, lawyers, and teachers. The orthodox clergy at first expected the scientists to reject Darwin as they had rejected the evolutionary theory in Robert Chambers's *Vestiges of the Natural History of Creation* in the 1840s. Disappointed by the scientists' support for Darwin, the opponents of Darwin, made up mostly of the clergy with the assistance of some distinguished scientists and educators, like Agassiz, Edward Hitchcock (president of Amherst College), and F. A. P. Barnard (president of Columbia College), began a delayed campaign against evolutionary theory.[79] The publication of *The Descent of Man* in 1871, which outraged orthodox church people in particular, marked a broadening of the controversy over evolution that few literate people could avoid.[80]

Outspoken support of Darwin by entomologists like Walsh and Riley helped force the issue. Prior to Darwin, entomologists had stressed the balance of nature or natural theology (or both) and had thus served as faithful allies of the clergy in support of the argument from design. Fitch, a solid churchman, once wrote to Harris, the son of a minister, concerning his observations of some projections used by certain insects when breaking out of the cocoon. He then added, "here is . . . a most interesting illustration of the mode in which nature produces a remarkable modification of a particular part, where special office is to be fulfilled by it."[81] The rapid conversion of American entomologists to evolution reversed the role of this important community in subsequent debates. The Darwinian explanation of mimicry among insects was central to the entomologists' new role as apologists for evolution. Opponents of evolution had challenged evolutionists to produce evidence of speciation among living organisms. Skeptics considered such evidence more important than the interpretation of the geological and paleontological record such as developed by Edward Drinker Cope and Othaniel Charles Marsh.[82] Leading entomologists interpreted mimicry and its evidence as proof of the continuous production of new species. The production of new species was precisely what many entomologists in the 1870s expected to observe.

Within the context of Neo-Lamarckism, such phenomena as widespread variation and polymorphism were seen as evidence of rapid evolution. Riley, for example, noted that though the formation of species in other classes may never be observed because of the time required, it was probable that this could be observed in insects.[83]

American entomologists were thus among the earliest and most prominent advocates of evolutionary theory. Walsh and Riley published their views in agricultural papers and reports, and they anticipated and met the expected objections from some readers. They noted, for example, that the great objection to the evolutionary explanation of mimicry was not based on scientific reasoning, but on the fact that it proceeded from Darwinian principles, "and Darwin . . . is a horrible and pernicious monster who holds that Man is nothing but a Gorilla."[84] Riley rejected the notion that a belief in evolution was heretical; on the contrary, he held that the conception of creation by natural process was a much higher vision than one of instantaneous creation of each species.[85]

Riley's advocacy of evolution in the St. Louis Academy of Science shows the influence entomologists had on other intellectual leaders. Dr. George Engelmann, a founder of the St. Louis Academy, physician, botanist, and collaborator of Asa Gray and other prominent scientists, regarded Darwin's theory with distaste. At a society meeting in 1871, Riley confronted Engelmann with the evidence for evolution. He argued from the basis of his and Walsh's observations that many European species of plants and animals flourished in America, but few American species became pests in Europe. The Walsh-Riley explanation involved a curious recapitulation of the debate between Jefferson and Buffon three-quarters of a century earlier (see Chapter 1). Reversing Jefferson's argument that New World animals were more robust than those of the Old World, Riley and Walsh now argued that the Old World was in fact geologically younger and had newer, more highly improved plants and animals that were better equipped in the struggle for existence than the "old-fashioned" American species. Old World species were consequently hardier than those of the New World! A few months after this exchange Riley presented the evidence for mimicry and its Darwinian explanation to the academy. Over the next several years Riley and others spoke on evolutionary topics, most arguing for, but some against, evolution. By 1876, Engelmann had partly accepted evolution, and by 1877, when Riley made evolution the subject of his presidential address before the academy, Engelmann agreed with him.[86]

The most significant effect of the Darwinian revolution for the entomologists was the central position it gave them within the new field of evolutionary biology. Edward S. Morse, in an 1876 review of American contributions to evolutionary theory, noted that "the study of animals has been raised to [a new] dignity" as a result of evolutionary theory. Prior to Darwin, the disciplines of chemistry and geology had enjoyed the highest prestige because they produced wealth, while zoology was "a mere adjunct to geology, or a means to thwart the ravages of insects." With the advent of evolutionary theory, however, zoology had become "the pivot on which the doctrine of man's origins hinges."[87] Among American contributors to evolutionary theory, Morse singled out the entomologists for their outstanding achievements, citing the work of Walsh, Riley, Scudder, Packard, Edwards, Putnam, and Grote.[88] These American entomologists, in different ways, had established significant evolutionary principles or supplied important confirmation of evolutionary hypotheses. Through their association with famous European naturalists like Darwin, Wallace, and Bates, in the investigation of evolutionary questions, the American entomologists gained enhanced prestige. The investigation of evolutionary questions by American entomologists went hand in hand with the discoveries of paleontologists and other American scientists to forge a new relationship between American and European science in which Americans were no longer content to be in a supporting role.[89] Indeed, in evolutionary investigations, as in agricultural applications, American entomologists were assuming leadership roles. For the entomologists, whose chosen objects of study had once seemed at best insignificant and at worst repugnant, evolution meant a rise in prestige unparalleled in the history of the discipline. As Francis M. Webster, an economic entomologist in the 1890s, noted, the entomologist was now valued not only because he enabled the farmer to grow "two blades of grass" but also because he was "solving the great problem of life and its diffusion over the . . . globe."[90]

The study of insects was now recognized as important both for its practical benefits to agriculture and for its contributions to the great scientific and philosophical questions of the day, and those who studied insects benefited from this new recognition.

10 William Henry Edwards and Polymorphism in Butterflies

The entomologists' most striking post-Darwinian application of natural selection was their demonstration of the utility of seemingly insignificant insect forms like the design and coloration in butterfly wings. What had once been admired as beautiful indications of Divine Creation were now shown to be the results of natural processes. The evolutionist Alfred Russel Wallace cited the usefulness of all forms and relations among organisms, either past or present, no matter how insignificant they may seem, as the most important principle set forth in the theory of natural selection.[1]

Mimicry among the Lepidoptera was the first and most famous interpretation of insect forms according to evolutionary theory. Polymorphism among the Lepidoptera and other insect orders also drew the attention of entomologists, who were eager to discover evolutionary explanations for other patterns they observed among insect forms. Prior to the theory of natural selection, entomologists had considered the regular occurrence of two or more quite distinct forms within a single butterfly population (polymorphism) as a curious anomaly in the Divine plan, whose significance was not yet clear. In the context of natural selection, however, polymorphism assumed significance both as evidence for natural selection and as a means for studying evolution in process.[2]

The American lepidopterist William Henry Edwards played a central role in discovering the extent of polymorphism among North American butterflies. He also developed new techniques for raising and experimenting on polymorphic species as subjects for studying natural selection. Edwards came from a distinguished and successful American family whose members included the Puritan minister Jonathan Edwards (his great-great-grandfather) and the inventor Colonel William Edwards (his grandfather). The latter developed an improved method for tanning hides that allowed Americans to compete successfully with the British in this commodity.[3]

Born in New York in 1822 to William W. and Helen Ann Mann Edwards, William Henry Edwards was raised on an estate in the Catskill mountains near Hunter, New York. As a boy, William Henry roamed through the woods and meadows of the district. There he developed a love of nature and a zeal for the study of natural history that remained with him all of his life. From the village school in Hunter he went to Williams College, where he graduated in 1842, then to law school in New York City in preparation for joining his father's New York business, which had branches in banking, insurance, and European imports.[4]

Edwards's scientific interests turned first to ornithology. Upon graduation from Williams, he sought employment as the mounter of bird specimens for the U.S. Exploring Expedition, which had just returned from the South Pacific with thousands of specimens. In the confusion and rivalries that plagued the disposition of the expedition's specimens, Edwards's application led nowhere.[5] Four years later, however, at the close of his law studies, Edwards and his uncle, Amory Edwards, who had served as consul at Buenos Aires, embarked on an expedition up the Amazon River, where they explored the main river from Pará to Manaos and the great delta islands of Maraja and Mexiana. Upon his return, Edwards wrote a small book about his explorations entitled *A Voyage up the River Amazon* (1847). This book was the first of several travel narratives appearing over a decade and a half that documented the initial scientific exploration of the Amazon valley.[6] Edwards's book was important more for its depiction of the luxuriant and exotic flora and fauna in the distant Amazon jungle than for its science. This popular travel account appeared just as two English naturalists, Henry Walter Bates and Alfred Russel Wallace, were planning a joint collecting trip somewhere in the tropics. Upon reading Edwards's account they were persuaded to make the Amazon their destination. By coincidence, Edwards happened to be in London in 1848, just as Bates and Wallace were preparing to leave, and he advised them on conditions and gave them letters of introduction to foreign merchants in Pará.[7] Edwards's initial contact with Wallace and Bates marked the first of many instances in which he played a significant role in the scientific revolution of which natural selection was to become the centerpiece.

Following his Amazon adventure, Edwards turned his attention from birds to insects, and he came into contact with John Akhurst of Brooklyn, New York, a taxidermist and collector of Coleoptera and Lepidoptera who guided the efforts of many aspiring entomologists in those years. Akhurst showed Edwards how to mount and store insects, and he supplied him with cork and pins. In 1856, Edwards commenced the systematic collection of butterflies from successive summer residences in New York and Massachusetts, especially at Newburgh, New York. In the 1850s he met the entomologists who organized the Entomological Society of Philadelphia (later the American Entomological Society) as well as entomologists from other parts of the country. Baird supported Edwards's investigations by sending him Lepidoptera from governmental expeditions for evaluation and publication of new species. The British Museum and the Museum of Natural History at St. Petersburg, Russia, also sent Edwards their butterflies from British and Russian America for the same purpose. In 1861, Edwards began publishing descriptions of new lepidopterous species in the *Proceedings of the Academy of Natural Sciences.* With the appearance of specialized American entomological journals, he published regularly in the *Proceedings of the Entomological Society of Philadelphia,* the *Canadian Entomologist,* and other journals.[8]

In 1864, Edwards traveled to the Kanawha Valley in West Virginia to assess the feasibility of mining coal in a district that was being opened up by the Chesapeake and Ohio Railway. Four years later he moved to Coalburgh, about twenty miles north of Charleston, West Virginia, where he founded the Ohio and Kanawha Coal Company. As president of the company, Edwards found ample time to pursue his entomological interests. In 1868, Edwards published the first installment of *The Butterflies of North America,* the work for which he is most famous. Intended eventually to encompass the description and illustration of all North American species in all of their phases, the publication appeared as a series of pamphlets, which were then combined into three volumes, the first in 1874, the second in 1884, and the third in 1897. Upon this publication rested Edwards's reputation as one of America's greatest lepidopterists. Edwards is regarded as the most innovative nineteenth-century American lepidopterist in terms of breeding butterflies for the study of their evolutionary development. Scudder, the other great nineteenth-century American lepidopterist, is considered to be preeminent in terms of precise descriptions and in systematics.[9]

In his work with butterflies, Edwards enlisted the assistance of hundreds of people, three of whom deserve to be mentioned because of their special contributions. Theodore L. Mead was a young and enthusiastic collector who, beginning in 1869, spent much time with Edwards at

Coalburgh. There the two collected, reared, and later experimented with butterflies in all their stages. Mead also made collecting expeditions to the Rocky Mountains and the Pacific Coast, and the fruits of these journeys supplied Edwards with important information about the western butterflies. Mead published little under his own name, but he is often credited by Edwards with important discoveries. In 1882, Mead married Edwards's eldest daughter.[10]

Edwards's illustrators, Mary Peart and Lydia Bowen, have been noted for their outstanding illustrations (see Chapter 8). It should be emphasized here that the favorable reception of Edwards's *Butterflies of North America* rested in no small part on the beautiful illustrations produced by these two artists. Incidentally, the expense of producing and printing illustrations of such quality added greatly to the cost of the publications, and Edwards reported that he personally had to subsidize their publication.[11]

Edwards's interest in polymorphism came from two sources: (1) the traditional concern with classification and (2) the more recent demonstration of the significance of polymorphism in natural selection. The concern with classification arose from his goal of establishing an accurate classification and life history of all American butterflies. Entomologists had discovered that one problem in the classification of butterflies was the occurrence, or suspected occurrence, of polymorphic forms. In some cases, for example, two or more alternating forms occurred within one interbreeding species, and these were sometimes so different in appearance that they were classified as separate species. When their full history was known, however, they were shown to be alternating forms of one species. Seasonal dimorphism, or the appearance of two or more distinct seasonal forms, had been known since about 1830, when it was demonstrated that the European butterflies *Araschnia prorsa* and *A. levana* were in fact the winter and summer forms of the same species.[12]

Another kind of polymorphism, known as sexual dimorphism, was pointed out by Walsh in 1865 in the case of the tiger swallowtail butterfly, which was classified under two names, *Papilio turnus* and *P. glaucus.* For over a century *P. turnus* and *P. glaucus* had been considered separate species, *turnus* being yellow and *glaucus* black. Walsh contended that *glaucus* was merely a smaller and darker form of *turnus,* that occurred only in the female. This conclusion was supported by subsequent observation and testimony. Research revealed that no one had ever reported a male *glaucus;* furthermore, in 1832 James Ridings of Philadelphia had captured the two forms while in copulation, and George Newman, also of Philadelphia, had raised both forms from the same batch of eggs. The identity of *turnus* and *glaucus* had in fact been proved numerous times by breeding, but its significance was lost on the pre-Darwinians.[13]

The second source of Edwards's interest was the demonstration by Bates and Wallace that polymorphism was important in the study of natural selection. We may be certain that Edwards read and digested Bates's paper on mimicry (1862) and Wallace's articles "The Malayan Papilionidae" (1864) and "Mimicry, and other Protective Resemblances among Animals" (1867).[14] In "The Malayan Papilionidae," Wallace gave a lengthy discussion of polymorphism in the context of natural selection. He explained that polymorphism was produced when extreme forms were better suited to the "conditions of existence" than the connecting forms. When the connecting forms became extinct, the forms at either extreme survived as distinct forms within the same insect population.[15] Following this line of evolutionary reasoning, Wallace defined polymorphism as "the co-existence in the same locality of two or more distinct forms, not connected by intermediate gradations, and all of which are occasionally produced from common parents."[16]

Wallace explained why butterflies served as ideal subjects for the study of natural selection: they had been studied intensively by naturalists for over a century so that their species, variations, and distribution were well documented; they were remarkably uniform in general structure and habits (the larvae ate vegetation and the adults sucked juices), yet they were as numerous as species as the other orders, indicating a high degree of specific modification; and within this general uniformity they had developed a peculiar wing structure that acted as a register of the minutest details of organization and modification. Wallace cited Bates, who said the wings of butterflies "serve as a tablet on which Nature writes the story of the modifications of species."[17] Add to this the fact that butterflies exhibited a remarkable tendency toward polymorphism, and one may well understand the entomologists' excitement in the search for and discovery of evolutionary explanations of polymorphism among butterflies.

Edwards's contributions to the study of polymorphism fall into three areas: his discovery of numerous examples of the phenomenon among North American butterflies, his development of a method of raising butterflies that provided accurate information about the relationship between different broods of dimorphic species, and his experiments involving the effects of temperature on polymorphic species. The discoveries and documentation of polymorphism in the first and third areas both depended on the development of a reliable method of rearing butterflies

from the egg to the adult; therefore, Edwards's method of raising butterflies was in some respects his most important achievement.

For many centuries, people had raised butterflies by confining larvae or
chrysalids in containers, and the practice became widespread among collectors from about the mideighteenth century.[18] Collectors were naturally curious about the process of development from larva to adult. But a
more important reason for raising butterflies seems to have been the
desire to obtain perfect specimens for their collections, with wings un-
scathed by the elements or by the collector's net. Some collectors raised
rare species for sale or exchange as a way of enlarging their own collections. The sharing of information about how to raise butterflies was an
important function of entomological societies. For example, the lepidopterists and coleopterists who organized the Brooklyn Entomological Society in 1872 (Edwards was a founding member) devoted much attention
to methods of raising butterflies for mounting and exchange.[19]

By the 1860s, as certain species of Lepidoptera like the European cabbage butterfly became more destructive, American entomologists began
raising insects more systematically than formerly (when perfect specimens for the collection had been the main objective) in order to develop
effective control measures. Raising specimens from eggs also helped
American entomologists solve difficult problems posed by the complex
geographic distribution of American butterfly species.[20] The Canadian
entomologist William Saunders, who was concerned with agricultural
pests, reported raising diurnal Lepidoptera in order to trace their life
cycles. He found that by confining pregnant females of the family Hes-
peridae in separate boxes and by varying the amount of light admitted to
the boxes, he was able to obtain eggs, which he then raised to adults.[21]
Other lepidopterists, such as Scudder and Lintner, quickly adopted Saunders's method. Scudder, who was preparing a volume on the butterflies of
New England, requested that entomologists send him eggs, larvae, and
chrysalids obtained by rearing for his research.[22] Though the method
developed by Saunders and others represented a definite advance, it still
had one serious limitation: the females of many species could not be
induced to deposit eggs in confinement.

Edwards faced this difficulty in his attempts to unravel the polymorphic forms of certain swallowtail butterflies, known as *Papilio ajax*
and *P. marcellus*. By 1864, Edwards's observations had convinced him that
there were two or three forms of one species that had been classified as
separate species. Edwards experimented by raising many broods from
eggs, but he was unable to prove that one form came from another

because the females would not lay their eggs in captivity.[23] Finally, in the spring of 1870, Edwards experimented by placing a female *ajax* in a nail keg with a papaw bush, the food plant of *ajax* larvae. The female proceeded to lay eggs, which Edwards then kept in confinement on the food plant through the larval and chrysalid stages. "I expect to prove by this brood," he announced, "that *marcellus* and *ajax* are but different broods of the same insect."[24] In August, the chrysalids which had come from the *ajax* eggs all produced *Papilio marcellus* adults, thus proving the identity of *marcellus* and *ajax* as one species, and at the same time providing the first demonstration of the Edwards method of raising butterflies. That same season, Edwards reported similar successes in having the females of three other species deposit eggs when confined with the food plant of the larvae. He expressed his confidence that this method of taking eggs would always succeed.[25]

Edwards's first success left many questions unanswered with respect to *Papilio ajax-marcellus,* but the investigative method for this and other species had been established. During the next several years he made many refinements of the system, such as using bags instead of kegs to confine the females and developing special boxes designed to provide proper temperature and air circulation for the developing larvae and chrysalids. The essential feature, however,—confining the female on the food plant of the larvae—remained the key element of his method.

Edwards's method of rearing butterflies spread quickly among American lepidopterists, many of whom collaborated with Edwards in the investigation of polymorphic species.[26] In their zeal for rearing butterflies by this method, Edwards and his fellow American lepidopterists excelled in unraveling the early stages of diurnal Lepidoptera, and in this area of research they took a decided lead over European lepidopterists.[27] In 1884, after a decade of success with his method, Edwards summed up the results of his innovation in these words:

For a century collectors have amused themselves in rearing caterpillars found on the food plant, or from eggs gathered here and there on various plants; but who knew what type of butterfly laid those eggs? It is not ten years since the fortunate discovery was made in this country that the female might be induced to lay her eggs readily in confinement, so that breeding could be conducted with certainty; and in these years I do not hesitate to say, more has been learned of the life-history of American butterflies than is today known of European, though not a district of Europe, but has had a long succession of active lepidopterists and diligent students of dried butterflies.[28]

Citing the Americans' successes in this field, Edwards concluded that "sitting in one's closet and speculating on dried butterflies will not do."[29]

Having developed his method in 1870, Edwards was able within two seasons to work out the complex polymorphism of *Papilio ajax*. From careful observation each season at Coalburgh, Edwards postulated the presence of three distinct forms: *walshii* (which appeared about March 15 and disappeared about June 1; this was a new name given by Edwards); *telamonides* (which appeared a few weeks later than *walshii* and remained until the end of June; this corresponded to the form previously called *ajax*); and *marcellus* (which appeared about June 1 and disappeared at the end of October).[30] In 1870, Edwards established that *telamonides (ajax)* produced *marcellus* later in the summer. In addition, one of the chrysalids from the *telamonides* brood lived over the winter and produced a *telamonides* in the spring, further confirming that these two regularly occurred as spring and summer forms. This left the status of *walshii* unclear, but working on the hypothesis that this was an earlier spring form of the same species, Edwards confined several *walshii* females in April 1871. By August, the brood from these *walshii* eggs had produced all three varieties. Summarizing the results of these experiments, Edwards stated that *walshii* from overwintering chrysalids produced *telamonides* and *marcellus* in the same season (and occasionally *walshii*); *telamonides* from overwintering chrysalids produced *marcellus* the same season and *telamonides* the following spring; and *marcellus* produced several successive broods of *marcellus* the same season (and occasionally a *telamonides*). The last fall brood of *marcellus* overwintered to produce *walshii* and *telamonides* the following spring.[31] In the course of two seasons, Edwards had established the polymorphic relationship of three forms of *ajax*, each of which had been considered a separate species.[32]

Edwards, Mead, and others began investigations of several other species for which data on the early stages were lacking, including several suspected polymorphic species. Three additional examples of these early discoveries will be given, two among the genus *Grapta* and one from the genus *Phyciodes*. Lepidopterists had long been uncertain how to classify certain *Grapta* that appeared consistently in the same geographical location at the same seasons. Some of them had black wings, and some had red wings. They had traditionally been classified as two variations of one species, but in 1869 Lintner separated them into *Grapta interrogationis* (black winged) and *G. fabricii* (red winged).[33] In June 1871, in an attempt to clarify the status and history of the *Grapta* species, Edwards confined two female *interrogationis* on the food plant, the hop vine, and from these he obtained thirty-eight larvae. Within a month these larvae developed into adults of both *interrogationis* and *fabricii,* of which there were about twice as many of the former as there were of the latter.[34] Even though Edwards had not yet proved the full cycle (which would have required producing *interrogationis* from eggs laid by *fabricii* females), he reasoned from the analogous case of *ajax* that the two should be considered dimorphic forms of one species rather than separate species or varieties.[35] So remarkable were Edwards's findings that Lintner at first expressed doubts that the two *Grapta* could be the same species because they were so different in appearance.[36]

Having demonstrated the dimorphism of *interrogationis,* Edwards suggested that *Grapta comma* and *G. dryas,* which he had cited as separate species in volume 1 of the *Butterflies of North America,* would also prove to be dimorphic forms of a single species.[37] In 1873, Mead succeeded in confining two female *G. dryas* on the food plant of the larvae, and the resulting eggs developed into both *dryas* and *comma.* In volume 2, Edwards changed the names to *Grapta comma* with two varieties, harrisii and dryas. In *G. comma* the seasonal dimorphism was more marked than in *G. interrogationis,* the *dryas* form being the prevailing summer form and *comma* being the autumnal form.[38]

The third example was the clarification by Edwards and Mead of one of the most variable American butterflies, *Phyciodes tharos.* For a century, since *tharos* was first described by the English entomologist Dru Drury, there had been confusion in the classification of *tharos* in its many forms and between *tharos* and its relatives. In 1864, Edwards separated a species he called *Melitaea [Phyciodes] phaon* from *tharos* and shortly thereafter the Philadelphia collector Tryon Reakirt described *batesii* as a separate species. In 1868, Edwards described yet another new species, *Melitaea [Phyciodes] marcia,* which had been considered a variation of *tharos.*[39] Many questions remained concerning this complex group. Following the success with rearing *ajax* in 1870, both Edwards and Mead tried unsuccessfully for five years to raise larvae from the eggs of *tharos.* Finally, in 1875, Mead discovered the food plant of *tharos,* the aster, and both he and Edwards commenced raising larvae of *tharos* and *marcia* to determine their relationship.[40] The resulting offspring proved the dimorphic relationship, *tharos* being the summer form and *marcia* being the winter form with four distinct variations (A, B, C, and D).[41]

Until 1875, Edwards interpreted the significance of his rearing experiments primarily in terms of classification and the elucidation of the early stages in the life history of American butterflies. In the first volume of the *Butterflies of North America,* Edwards noted how the findings on *ajax* and *interrogationis* illustrated the weakness of the current classification system, which was founded on the markings and coloration in the adult stage only.[42] Though Edwards expressed his early findings in terms of classification, he was aware that his experiments had implications for broader questions of natural selection and evolution. His success at breeding polymorphic forms drew the attention of Bates, Wallace, and the German naturalist August Weismann. Bates wrote to Edwards in 1871 that the discovery of polymorphism in the cases of *Papilio ajax* and *Grapta interrogationis* were of "the greatest interest [with regard to] the great question of the origin of species [and that they opened up] a long vista of possible consequences."[43] Bates suggested that an alteration in the climate had apparently changed the number of broods in many species, and he speculated that such broods might become separate species if the North American climate again became tropical.[44] Wallace wrote in 1873 that Edwards's experiments on dimorphic forms demonstrated one way in which species were formed.[45] Weismann wrote to Edwards in 1872, congratulated him on his success with *P. ajax,* and expressed regret that he had not known of this work earlier, because Edwards's findings had direct relevance to similar experiments he (Weismann) was carrying out on European species. Weismann also sent a copy of his recently completed monograph *On the Effect of Isolation in the Formation of Species.*[46]

Weismann taught zoology and comparative anatomy at the University of Freiburg, Germany. A convinced Darwinist from the time he read the *Origin* in 1861, he had given an important defense of Darwinism in his inaugural lecture at Freiburg in 1868, and from that time he was the leading Darwinist in Germany.[47] Weismann's studies of dimorphic butterflies began in 1868 as part of a series of studies of metamorphosis and variation in butterflies that was published as part of his *Studien zur Descendenztheorie* (1875–1876).[48] Convinced that variation in butterflies touched significant questions of evolutionary descent, Weismann began experiments on *Araschnia levana-prorsa,* the first species in which seasonal dimorphism had been demonstrated. *Levana* is the winter form, which is produced from overwintering larvae, and *prorsa* is the summer form, which is produced from summer larvae. In his experiments, Weismann attempted to induce dimorphism artificially in order to postulate its historic origin.[49] After examining and rejecting several possible explanations for the origin of dimorphism, Weismann hypothesized that dimorphic forms might have been formed by the direct influence of temperature changes. He tested this hypothesis by subjecting *levana* chrysalids to cold. Instead of the summer form (*prorsa*), which under normal conditions would have followed, these chrysalids produced an intermediate form, *porima,* which had the design of the summer form but the coloration of the winter form. Through the repeated application of cold, Weismann succeeded in producing some complete reversions to *levana.* Further experiments demonstrated, however, that the process could not be reversed; the application of heat to *prorsa* larvae could not produce *levana.*[50]

Weismann explained these findings in relation to the probable history of the species. He reasoned that the winter form, *levana,* was the original, or primary form that had existed at the time of the last glaciation. As the climate warmed, there was time for a second (summer) generation, *prorsa.* Because *prorsa* had originated as a variation of *levana,* the artificial application of the original, colder, conditions could bring a reversion to the original form; however, the opposite result, that is, the production of the secondary form by heat, could not be obtained.[51] Experiments on other dimorphic species supported Weismann's hypothesis. In the case of *Pieris napi, P. napeae,* and *P. bryoniae,* Weismann succeeded through the application of cold in forcing a complete reversion of all the secondary forms to the primary form (though the opposite could not be produced). Weismann theorized that the dimorphism of this species was not as old and therefore not as well established as *Araschnia,* and therefore a complete reversion was possible.[52]

From these experiments Weismann concluded that, although the exact physical-chemical process of variation was not yet known, significant variations could be induced by direct external conditions, like the application of cold. He reasoned, therefore, that each major alteration of the climate, such as the glacial epochs, must have given rise to variations and ultimately new species. He concluded further that because each species had been produced by a unique set of climatic influences that had formed the basic ingredients for natural selection, each species could thereafter be modified only in certain limited directions that depended on the physical nature of the particular species. Changed external conditions did not produce identical reactions among living organisms, like chemical changes registered on litmus paper. The effect of changing external conditions on a living organism depended on its unique evolutionary history. For Weismann, the one-way reversibility of dimorphism provided strong evidence for the stability of the organism, which set limits on its further direction of development.[53] This emphasis on the stability of organisms, including their power to pass on precise structural features to the offspring, became a central theme in Weismann's Neo-Darwinist formulation of evolutionary theory. Reversion to latent hereditary forms, such as the temperature-induced reversions in butterflies, could be interpreted both as evidence for the cause of variation (external conditions) and as evidence for the stability of hereditary material in the organism. In the 1870s, naturalists viewed variation and heredity as parts of the same process: variation worked for change in organisms, while heredity served to maintain their stability. Weismann emphasized the stability of organisms in his search for keys to the hereditary process.[54]

Edwards read Weismann's "Über den Saison-Dimorphismus der Schmetterlinge" (in translation) shortly after it appeared in 1875. Though he had known that his own breeding of dimorphic forms might have relevance for evolutionary questions, it was not until he learned of Weismann's experiments that he possessed a theoretical framework and an experimental method for defining this relevance. He immediately began similar experiments in the effects of cold on *ajax*.[55] Edwards obtained eggs from the winter (*telamonides*) form and placed them on ice. As in the case of Weismann's winter brood (*levana*), most of these reverted to *telamonides* instead of producing the summer *marcellus* form, which under normal conditions would have appeared. Edwards then applied cold to the first summer generation of *marcellus* and found again that most of these also reverted to *telamonides* (though normally *marcellus* produced two more *marcellus* generations before the chrysalids of the last one overwintered to produce *telamonides*). The first experiment was not entirely successful because Edwards was away from home for some days and the box was mistakenly removed from the ice. Edwards often had problems in attempting to conduct experiments under controlled conditions (he regularly carried boxes of iced insects on the train from Coalburgh to the Catskills and back). Weismann also recalled how, when he carried experimental butterfly pupae on the train, the mechanical action apparently had an effect on the developing insects that was similar to the application of cold.[56]

Reporting his results to Weismann, Edwards asked why he had not been able to produce the other winter form, *walshii*, by the application of cold. Weismann suggested that *telamonides* was the primary form (even though *walshii* appeared earlier in the spring), and that *walshii* and *marcellus* had developed as dimorphic forms of *telamonides*. Of particular interest to Weismann was the finding that a portion of all the summer (*marcellus*) generations remained in the chrysalid stage to overwinter, emerging in the spring as one or the other of the winter forms. This phenomenon did not occur in any of the European species known to Weismann. He explained the meaning in evolutionary terms: the summer form, *marcellus*, had been established more recently than had the summer form, *Araschnia prorsa*, and was therefore less stable. As the North

American climate had warmed, two summer generations had been added. Because they were so recent, and not yet firmly established, a portion of each brood reverted to the original summer brood in which the chrysalids had overwintered.[57] In subsequent experiments on *ajax,* Edwards found that the summer form could not be induced by the application of heat, thus confirming Weismann's hypothesis that the winter forms were primary.[58] In 1875, Edwards published an abstract of Weismann's paper on seasonal dimorphism in the *Canadian Entomologist* along with the results of his own experiments.[59]

In his further researches, Edwards combined Weismann's experimental method with field studies and geographic distribution to formulate hypotheses about the evolution of American butterfly species. He began experiments on *Phyciodes tharos* in 1875, the same year in which he and Mead proved its seasonal dimorphism with *marcia.* Edwards possessed over five hundred specimens of *tharos* representing all its forms for its geographical range. In *tharos,* it will be recalled, Edwards had established the winter form, *marcia,* which had four distinct varieties (A, B, C, and D), and a summer form, *tharos.* Edwards found that the application of cold caused the summer generation to revert to the winter generation completely in the females and partially in the males. This finding was similar to Weismann's results with *Araschnia levana.* Edwards concluded that *P. marcia,* variation B, was the nearest to the primitive type because it was the most frequent form to appear among those exposed to cold. Also, its coloration occurred in the allied species, *Phyciodes phaon* (in the Gulf states) and *P. vesta* (in Texas), both of which were seasonally dimorphic and both of which were restricted in their winter broods to the form corresponding to *marcia* B. Edwards explained the significance thus: *tharos, phaon,* and *vesta* were once varieties of one species that had the color still common to all three. At the time they became permanent species, *tharos* was giving rise to several subvarieties and as the climate became warmer each of these (along with *phaon* and *vesta*) developed a summer form. As the climate warmed more, the winter form (*marcia*) colonized to the north and the summer form to the south to their present distributions. In the Catskills, the species had two generations, one *marcia* and one *tharos;* in Coalburgh, there were four generations, *marcia,* followed by three generations of *tharos.* This evolutionary history accounted for the extreme variability of the species, which had caused confusion in the attempts to classify *tharos.*[60]

A final example will demonstrate Edwards's sophisticated combination of theory, field observation, and experimentation to trace the probable rise of butterfly species from dimorphic forms. Among the *Satyrus* butterflies, Edwards suspected a dimorphic relationship between *Satyrus alope* and *S. nephele,* whose ranges overlapped.[61] The geographical range of *S. alope* extended from North Carolina to New York and west to Texas, while in the northwestern United States from Indiana to the Pacific, and in Canada, *S. nephele* was the sole representative.[62] In New York and New England both *alope* and *nephele* were found, including a series of intergrades between the two. *Alope* and *nephele* were considered two separate species and were figured by Edwards as such in 1866; however, by 1876, Edwards suspected a dimorphic relationship between the two in the intervening belt. Calling upon lepidopterists throughout the range of both *alope* and *nephele* to send eggs for rearing, Edwards and Mead found that the broods raised from eggs in the exclusively *alope* district produced *alope,* while the broods from the exclusively *nephele* district produced *nephele.* The broods from eggs in the overlapping area where both *alope* and *nephele* occurred produced varying offspring. There, the *nephele* eggs produced either *alope* or intergrades between the two, whereas *alope* eggs produced only intergrades. *Alope* and *nephele* thus proved to be dimorphic forms of the same species but only in the overlapping belt.[63]

Edwards then compared *alope-nephele* with *Satyrus pegala,* a closely allied species that occurred in the Gulf states but whose range was separated from *alope* by a break in Georgia and Mississippi. The *pegala* larvae fed on coarse sea grasses, whereas *alope* larvae fed on soft meadow grasses. In reconstructing the history of the *Satyrus* group, Edwards postulated that in former geologic times, the range of *pegala* may have been more extensive and *alope* may have arisen as a dimorphic form more suited to the upland meadows. When the sea grass and meadow environments became separated, each form retreated to its own habitat, and the two became separate species. Subsequently, *alope* gave rise to *nephele.* Edwards concluded that if conditions were to change so that the dimorphic forms in the overlapping belt died out, *alope* and *nephele* would also become separate species, and only a reconstruction of their evolutionary history could show how any of these species had arisen as dimorphic forms of the other. Edwards believed a similar process had given rise to *alope* var. *texana,* a related variety that occupied the Southwest. He noted that the belt of intergrade was approximately the same as several other dimorphic species, notably *Limenitis arthemis-proserpina* and *Papilio turnus-glaucus,* and he speculated that the geologic history of the region may account for these facts.[64]

Edwards's explanation of speciation in *Satyrus* is a striking combination of experiment, hypothesis, and field studies by an American naturalist in the 1870s that shows a total rejection of typological concepts in favor of explanations based on populations. Though Edwards had used the breeding and field methods earlier, his use of experiment and theory came directly from Weismann. Edwards's explanation of speciation in the *Satyrus* butterflies, through seasonal polymorphism and geographic isolation, is essentially the same explanation given by modern biologists.[65]

Weismann and Edwards, along with other Darwinians (including Darwin himself) probably overestimated the importance of this kind of evolutionary process in relation to other ways in which species had arisen. Lacking an explanation for the origin of variations and the way in which variations were inherited, they tended to think in terms of the direct action of the environment-producing variations that in turn were filtered through natural selection to produce new species. The "blending" of individually induced variations in individuals thus "pushed" the population in a direction more adapted to changing conditions. Modern genetics postulates a different origin for variations and a slower rate of variability than was supposed in the 1870s, but neither of these advances changes the fundamental concept of speciation held by Edwards and those who thought like him.[66]

Edwards's work on polymorphic butterflies was important both in systematics and in evolutionary theory. Prior to 1852 there were only 137 butterfly species known and described for the United States and Canada; between 1852 and 1860, 61 new species were described. From 1860 to 1874, 311 additional new species were described, bringing the total to 509.[67] Even more striking was the advance in knowledge of early stages of butterflies. In 1868, when the first issue of *Butterflies of North America* was published, Edwards noted that little more was known about butterfly larvae than had been known by John Abbot seventy years earlier. In 1884, with the issue of volume 2, Edwards could state, "All that has now changed, and today it can be said that the preparatory stages of North American butterflies as a whole are better known than are those of Europe; and so many zealous workers are now busy . . . that another period of sixteen years may leave comparatively little to be done in these investigations."[68]

Edwards's contributions in the field of evolutionary studies were also substantial. While he followed leads suggested by theorists like Darwin, Bates, Wallace, and Weismann, he also produced a considerable amount of original work in both experimentation and theory. Above all, he led the way in the sheer volume of new information about polymorphic butterflies.

A fuller picture of his contributions is apparent when one examines his relationship with Bates, Wallace, and Weismann, three towering figures in the early development of evolutionary biology. Edwards's influence on the decision of Bates and Wallace to go to the Amazon has long been known. What is not so well known is that Edwards remained in contact with Bates and Wallace after both had discontinued the active pursuit of Lepidoptera, while at the same time Edwards entered the period of his greatest productivity.

Bates's paper on mimicry (1862) was followed a year later by *The Naturalist on the River Amazons,* which explained mimicry in accurate and vivid language that pleased both scientific and popular readers. These two publications represented in many ways the high point of his career. Despite his fame as a naturalist and author, Bates found it difficult to find a position in London, where the entomological establishment was dominated by anti-Darwinists.[69] In 1864, he settled for a position as assistant secretary of the Royal Geographic Society. There he continued to write on the systematics of the Lepidoptera and other insect groups and in the process supplied Darwin with valuable information on mimicry as it related to evolutionary theory, which bolstered later editions of the *Origin* as well as the *Descent of Man* (1871). In the 1870s, Bates turned from the Lepidoptera, even selling his butterfly collection, and concentrated his attention on the Coleoptera, becoming a world-renowned authority on their systematics.[70] During that period, Bates praised Edwards's work on polymorphism and expressed envy at Edwards's opportunity to engage in the rearing and study of butterflies. In 1875, he wrote to Edwards, "How interesting it is to get so near a glimpse . . . of the actual creation of a species by nature!"[71] Though Bates suggested some things to watch for in the experiments, it was Edwards who carried out the experiments and observations in the 1870s.

Wallace likewise had given up active study of the Lepidoptera by the time Edwards began his major contributions. Like Bates, he gave suggestions, offered encouragement, and praised Edwards for the contributions he was making, but he did not enter into the experiments or field studies himself. In 1887, while on an American tour, Wallace visited Edwards in Coalburgh, where they compared views on the advances in the study of Lepidoptera in the forty years since they had first met. It is clear therefore that Edwards, Bates, and Wallace stimulated and encouraged one another in their thinking and work in evolutionary studies over a period of forty years. This shared endeavor helped each in his own way to make important contributions.[72]

Edwards's relations to Weismann are somewhat easier to chart. The stimulus supplied to Edwards by Weismann's experiments with cold and his evolutionary explanation for the rise of seasonal dimorphism are clear. Significant also is Edwards's close collaboration with Weismann in the years 1875 to 1878, when they wrote to each other frequently about polymorphic species. In those years, Edwards was the leading experimenter in the field. Entomologists in Germany, Great Britain, and elsewhere in Europe were not pursuing the study of polymorphism with the same zeal as Edwards and the Americans. Weismann praised Edwards's work on *Papilio ajax* as a demonstration of an unusually complex example of polymorphism and a more conclusive proof of his theory than had been obtained in the European species. He expressed regret that he had not had Edwards's results when he had first begun publishing on seasonal dimorphism. In 1882, when the British entomologist Raphael Meldola prepared an English edition of Weismann's *Studies in the Theory of Descent,* Weismann revised the section on seasonal dimorphism to include Edwards's experiments, and he added an appendix showing Edwards's experimental results.[73]

Edwards's collaboration with Weismann in the years 1875 to 1878 and the inclusion of his findings in the 1882 edition of *Studies in the Theory of Descent* represent the summation of their direct contact and the peak of Edwards's contributions to the development of evolutionary theory. After 1878, each person pursued separate lines of investigation. During the period of their collaboration, Weismann's experimental findings with respect to seasonal dimorphism forged an important link in his reasoning about the nature of variation and inheritance. Noting that the exact cause of variation—how cold or warmth affected the color of a butterfly wing, for example—remained a mystery of chemical-physiological process, Weismann nevertheless became convinced through his studies on dimorphism and other work that variations did not arise from changed environmental conditions alone but depended on the nature of the organism and how it had evolved. Change in a given species, Weismann concluded, depends both on the external conditions and on the "physical nature" of the varying organism, that is, on the internal physiology of the organism undergoing variation.[74] Weismann was moving toward the distinction between phenotype and genotype, between somatic cells and the germ plasma, which he soon developed into a theory of particulate inheritance.[75]

Weismann, like Darwin, developed his evolutionary theory through a comprehensive examination of all available evidence, and it is therefore difficult to establish precise causes for developments in his thinking. It is nevertheless suggestive that his rejection of Lamarckian (environmentally induced) causes of evolution occurred during the years he was in contact

"Polymorphism in Butterflies: Winter and Summer form of *Papilio Ajax,* Showing the winter form var. *Telamonides* (16) and the summer form var. *Marcellus* (17) of *Papilio Ajax,*" from August Weismann, *Studies in the Theory of Descent; with Notes and Additions by the Author,* translated and edited by Raphael Meldola with a prefatory notice by Charles Darwin (2 vols., London: Sampson, Low, Marston, Searle, and Rivington, 1882), vol. 2, plate 2 with Edwards. In the first edition of his *Studien zur Descendenztheorie* (1875–1876), he allowed for environmental causes in species change, including the possibility for the inheritance of acquired characteristics, but by the time the English translation *Studies in the Theory of Descent* appeared (1882), he had departed radically from this position and had rejected any possibility of Lamarckian-induced variation.[76] It is clear from the prominent place given Edwards's findings on polymorphism in butterflies in the revised edition that this evidence played an important role in Weismann's thinking in those years, and it is therefore likely that this evidence influenced his total rejection of Lamarckian evolution.

In 1878 Weismann turned from entomology to the study of the reproductive development in the cells of hydrozoa in his search for an explanation of the nature of the hereditary material. In the meantime, in 1875–1879, a series of dramatic observations from microscopists showed the combined presence of male and female nuclei in the fertilized egg cell and cell division by mitosis. Weismann thereafter focused his attention more specifically on the cell nucleus as the probable carrier of hereditary material.[77] By 1883 he made the crucial distinction between the somatic cell and the life of the germ plasma, and by 1886 he challenged the view that sexual reproduction was a blending of two parental germ plasmas. In 1892 he published a full theory of germ plasm and a particulate hypothesis of heredity. These theoretical advances consummated years of study, experimentation, and reflection, including the study of seasonal dimorphism, and Edwards shares a notable role in a crucial phase of his theoretical development.[78]

On the continent, the tradition of lepidopteran research, which involved the manipulation of environmental factors on the preparatory stages of butterflies to produce changes in color and wing patterns in the adults, has continued to the present. In the decades around the turn of the century, this line of experimentation again revolved around the question of the inheritance of acquired characteristics. The definitive rejection of the inheritance of acquired characteristics received a consensus by the 1940s.[79] More recent research has demonstrated that the length of daylight is the primary factor triggering the development of one or the other of dimorphic lepidopteran forms, though in some special cases—like *Araschnia* (and presumably *Papilio ajax* and other American species)—temperature acts as a secondary factor. Today the primary significance of such dimorphism is seen as ecological. It helps to ensure that the various forms of the adult insect will have the best chance for survival at different seasons of the year.[80]

Meanwhile, Edwards continued his program of researching the history of North American butterflies, which he completed with the publication of volume 3 in 1897. In the course of gathering material for volume 3, he established the presence of seasonal dimorphism in numerous additional species, but these discoveries, though necessary for his taxonomic work, did not attract the same attention from evolutionary biologists as his first discoveries in the early 1870s.[81] It was Edwards's fascination with butterflies and his continued occupation with description, illustration, and classification at a time when many biologists were turning to such fields as cytology that explain the relative lack of recognition of his earlier original contributions to evolutionary biology.

Edwards's achievements as a lepidopterist and an evolutionist relied on the lepidopterists and collectors who by the 1870s had developed a widespread network across North America. These observers, collectors, correspondents, and experimenters provided the support necessary for Edwards and others who led the way in evolutionary studies. These "zealous workers" benefited in turn from the extension of rail transportation across the continent in the period following the Civil War. In 1878, for example, Edwards wrote to his fellow lepidopterist Lintner that the rapid progress in construction of the Southern Pacific Railroad across Arizona would soon open that province fully to collecting. Edwards and his network of collectors also benefited from the growth of towns and cities, often with resident members of the entomological fraternity.[82] By the 1870s, Albany and New York City had joined Philadelphia and Boston as thriving centers for the study of Lepidoptera, complete with specialized societies and journals.[83] Edwards's circle of correspondents included lepidopterists along the Atlantic seaboard from Canada to Florida, the Midwest (Cincinnati, St. Louis), the West (San Francisco), and the Southwest (Texas, Arizona, and New Mexico). Collectors sent him specimens and information from such remote outposts as Elko, Nevada, Vancouver Island, Fort Simpson in Northwestern Canada, and St. Michael, Alaska.[84]

Edwards's investigations profited from the wide geographical and climatic diversity of the area covered by his circle of correspondents. As questions pertaining to dimorphism arose, Edwards was able to inquire of collectors in the areas where the dimorphic forms overlapped and where they were isolated to answer these questions.[85] In this quest, he sometimes gathered as many as five hundred specimens representing forms from the known range of a species.[86] European lepidopterists like Bates and Zeller frequently pointed out Edwards's "great advantage" in having access to such a vast series of specimens from the relatively undisturbed insect fauna of North America.[87]

Edwards's studies of polymorphism among the Lepidoptera represent another instance in which the American entomological community by the 1870s exerted its leadership in entomological science. The interest in variation and polymorphism, and the bearing of these phenomena on evolution, broadened and deepened the expertise and competence of American entomologists, adding a new dimension to a community already invigorated by the search for solutions to destructive insects. The capture and raising of butterflies, once considered an idle avocation, by the 1870s had become a tool used by lepidopterists to delve into questions of the origin of species. In the 1870s, Edwards and the American lepidopterists led the world in this endeavor.[88] The stimulus of evolution thus strengthened and enriched the intellectual context of the American entomological community.

The Yucca Moth   11

In addition to mimicry and polymorphism in insects, Darwin's writings
on natural selection sparked widespread interest in the biological and
ecological relations of insects and plants.[1] The groundwork for this new
area of study was laid in the *Origin of Species* (1859) and was developed at
length in Darwin's treatise *On the Various Contrivances by which British and
Foreign Orchids are Fertilized by Insects* (1862). Later works filled out various aspects of this theme, but the orchid book kindled the most sustained
interest among botanists and entomologists.[2] Michael Ghiselin has
pointed out that Darwin's orchid book may be read on several different
levels.[3] Superficially it appears to be an entertaining account of the remarkable adaptations of the orchid plant that encourage fertilization by
insects. On a deeper level it explains the enormous importance of cross-
fertilization. It appears, however, that Darwin's primary interest in writing the book was neither of these but rather a deliberate challenge to the
philosophical and metaphysical basis of the argument from design.

Prior to the publication of the *Origin of Species,* a major argument for
the existence of God was the appeal to the self-evident rationality and
plan of the universe. Natural theology played a key role in the support of
the argument. For example, Bishop William Paley, a leading exponent of
the doctrine in the early nineteenth century, cited the orderly arrangement of the parts of the flower and the "contrivances" that allowed
insects to pollinate them as evidence for the existence of a "contriver."
Asa Gray, Darwin's foremost advocate in America, also used the word
"contrivance" to describe the various adaptations of plants. Gray refused
to abandon the argument from design altogether, substituting for it what

John Dewey later called "design on the installment plan." Darwin, with fine irony, adopted the language of Paley and Gray but turned the argument on end, so as to hold up the idea of biological design to ridicule.

Like natural theologians before him, Darwin explained how the various parts of the orchid are arranged to facilitate pollination by insects, but he then went on to demonstrate how such adaptations arose through the modification of preexisting organs that had had quite different functions. In other words, he showed that the present structures were not designed with their present functions in mind at all but were ad hoc adaptations of organs "designed" for quite different "purposes." By reducing the argument to absurdity, Darwin dealt a crushing blow to the notion of design in biology. The rapid decline of its use in theology and its elimination from biology altogether may be attributed primarily to him.[4]

What captured the immediate attention of most readers of Darwin's orchid book, however, was not its philosophical attack on the argument from design but its revelation of the complex and vital relationship between flowering plants and the insects that cross-fertilize them. In a letter to Gray, Darwin complained, "No one else has perceived that my chief interest in my orchid book has been that it was a 'flank movement' on the enemy."[5] Despite Darwin's intentions, the orchid book was used primarily by zoologists and botanists to explore this new area of relationships in the plant and animal kingdoms rather than by protagonists in the evolution debates. For example, the German botanist Hermann Müller, following Darwin's lead, developed the study of the interrelationships of plants and insects into a major field of study.

One of the most startling developments in the field of insect-plant relationships was the discovery that various yucca species and species of *Pronuba* moths mutually depend upon each other for their own propagation and survival. Significantly, this discovery resulted from the collaboration of a botanist, Dr. George Engelmann, and an entomologist, Riley.

Engelmann was educated in medicine in Germany before immigrating to America, finally settling in St. Louis in 1832. There he practiced medicine and carried on scientific studies in botany, meteorology, geology, and zoology that gained him a reputation as the leading scientific personality in the American West. Engelmann led in the founding of the St. Louis Academy of Science in 1856 and the Missouri Botanical Garden in 1859. He was a key correspondent and authority on western plants for Gray and other botanists in America and Europe.[6]

Engelmann retired from active medical practice to devote his energies to scientific pursuits in 1869, the same year Riley came to St. Louis as the first state entomologist. Although they disagreed initially on the subject of evolution, the two scientists had many interests in common, and they compared notes and exchanged views at the St. Louis Academy of Science and elsewhere. Engelmann had first become familiar with the yuccas on a trip through the Southwest shortly after arriving in America. After years of study in the field and in his garden, he published his observations on the entire genus. In 1872, he called Riley's attention to some unusual aspects of the yucca flower, especially its elaborate adaptation for fertilization, which he suggested was accomplished by some as yet unspecified insect.[7]

Engelmann's observations were not guided by Darwin's writings; in fact, at that time he was an anti-Darwinian. Riley, however, as a staunch supporter of Darwin's views, immediately interpreted the yucca flower structure in terms developed by Darwin in the orchid book and elsewhere.[8] Riley soon found the insects that pollinated the yucca. They all belonged to a genus of tineid moth that he named *Pronuba yuccasella,* or in popular language the yucca moth. The remarkable thing about the relationship between the moth and the yucca, Riley pointed out, was that in contrast to the orchids, which could be pollinated by many different lepidopterous insects, the yucca depended on moths of this single genus.

The method of pollination was equally unusual. The yucca flower was constructed in such a way that the stamens with their pollen-producing anthers were located some distance below the top of the pistil, that is, below where they needed to be placed in order to fertilize the ovules. The nectary, located just under the pistil, and distant from the pollen on the anthers, afforded no particular aid in the usual manner of inducing insects to cross-fertilize. Given this situation, Riley explained, the female pronuba moth acted as a "foster mother" to the yucca. Flying from plant to plant, always at night, she collected large pellets of pollen from the anthers, utilizing maxillary tenacles, or mouth parts, wonderfully "contrived" for this purpose, and then rolled the pollen into a ball under her neck. Finally settling on one plant she apparently deemed suitable for the rearing of her young, she clung to the top of the pistil and forced the pollen into the stigmatic tube. Only in this way could the yucca be assured of cross-fertilization and thus ensure the propagation of the species.

The yucca in turn served a vital role in the life history of the moth, for the pronuba larvae could only develop in the fertilized ovaries of the growing seeds of the yucca. Riley felt that the moth's actions in pollinating the yucca were "purposeful" in that they ensured food for the developing larvae. During the first season, Riley was unable to observe exactly how and when the female pronuba deposited the eggs. He speculated that she thrust the eggs into the ovaries, perhaps through the pollen tubes, at the time she pollinated the stigma, but he had not seen this. He was certain that the moth eggs were placed in the fertilized yucca ovaries, for he had dissected the young fruit and found the developing larvae there.[9]

Riley reported these remarkable findings to the St. Louis Academy of Science and to the AAAS in September 1872. His paper at the AAAS elicited considerable interest and discussion on the part of Gray, Edward S. Morse, and others who remarked on the striking example of mutual dependence between an insect and plant species.[10] Upon learning of these findings, Hermann Müller wrote to Riley that this was the "most wonderful instance of mutual adaptation" yet discovered.[11] Such absolute interdependence was proof that the plant and animal species had evolved together, said Riley. As a practical demonstration of the accuracy of his statements, Riley predicted that the range of the pronuba moth would correspond to the range of the yucca and that the yucca would not produce seed where the moth was not present.[12]

Riley publicized his findings in the meetings and publications of scientific societies and in his annual reports to alert other naturalists to the facts he had uncovered and to encourage them to add their observations during future seasons. During the course of many years, Riley, with the help of others, was able to fill in most of the details of the life history of the pronuba moth and its method of pollinating the yucca. He found that the sequences of ovipositing and pollination were even more apparently "purposeful" than he had at first suspected.

The female moth, having collected pollen from several yucca flowers, then flew to a plant that she evidently judged to be exactly at the right stage for pollination. These were invariably new flowers that had been open for only one or two nights, the ovules of older flowers not being susceptible to fertilization. There she thrust the eggs directly into the yucca ovules by means of an exceptionally long ovipositor. Following this, in what seemed to be a deliberately calculated action, she ran up to the top of the style, where she thrust the pollen down into the stigmatic opening, working her head vigorously for several seconds. The sequence of activities, first ovapositing, then pollinating, differed from the usual method of cross-fertilization by insects, where the insect carried pollen from plant to plant in a "non-purposeful" manner strictly incidental to the search for nectar. Through the entire sequence of ovapositing and pollinating, the yucca moth paid no attention whatever to the nectary, which at any rate was located in such a place that it served no apparent "purpose" in pollination. Biologists now speculate, in a manner worthy of Darwin's best irony, that the yucca nectary actually serves to *prevent* pollination by insects other than the pronuba by luring them away from the pollen-bearing anthers.

By using an ink dye, Riley was able to trace the path of the ovipositor into the ovules. The path of the dye showed that on the fourth or fifth day following ovipositing by the female, the larvae hatched and fed on the ovules. Riley found that only a small portion of the ovules of each plant contained larvae, thus allowing the propagation of most of the yucca seeds while at the same time supporting the larvae. Upon reaching full size, the larvae descended to the ground, where they burrowed beneath the surface, waiting for the future season's blooming. It is now known that a single brood emerges from the ground over a period of three years, thus ensuring continuation of the species even if, as occasionally happens, the yucca fails to bloom in a particular year. About two weeks before the yuccas bloomed, the larvae assumed chrysalid form, emerging as adults just as the yucca came into bloom. Riley attempted to force the early emergence of adult moths by means of hothouse conditions but failed. He concluded that because the blooming period of the yucca was so short (during the one or two nights when the flower was fully open) that the moth's habit of developing at the proper season was strongly fixed.[13] Riley also found further examples of pronuba moths and the yuccas they pollinated, some being restricted to one species and others pollinating several different yucca species.[14]

Riley's observations and conclusions regarding the evolutionary relationship between the yucca and the pronuba moth met some spirited objections. From the ranks of the American entomologists two individuals, Jacob Böll and Vactor T. Chambers, questioned Riley's observations and conclusions. Eventually Riley demonstrated that the differences in observations resulted from confusion between two species, the *Pronuba yuccasella,* which fertilized the yucca, and the *Prodoxus decipiens,* which did not. Riley then went on to state that the two moths had no doubt descended from a common ancestor, which accounted for the close family resemblance between the adults. The structure and life history, particularly in the larval stage, however, clearly demonstrated the evolutionary adaptation of pronuba for fertilizing the yucca. The pronuba larva, which left the yucca plant to burrow into the soil, had developed legs, whereas the prodoxus larva, which remained in the yucca, was legless. The pronuba emerged as adult moth when the yucca bloomed, whereas the prodoxus moth appeared earlier, as its larvae fed on the stem as well as the seeds and food for its larvae was available earlier. Riley's "discovery" (actually a renaming) of the bogus yucca moth ended any public objections to his evolutionary explanation of the yucca moth from within the entomologists' ranks.[15]

On the whole, botanists rather than entomologists seem to have been more disturbed by the philosophical challenge to the argument from design inherent in Riley's evolutionary explanation. Thomas Meehan, editor of the *Gardener's Monthly,* presented a paper at the Saratoga AAAS meeting in 1879 in which he challenged Riley's contention that the yucca depended on pronuba for fertilization, offering as evidence yucca seed capsules that had no pronuba larvae.[16] Gardeners frequently objected to Riley's claim that the range of the yucca and the pronuba moth must of necessity be the same.[17] A gardener in England voiced this objection in openly anti-Darwinian terms: "I fancy insect agency is talked of in a very unscientific way by too enthusiastic followers of Mr. Darwin. I remember reading Professor Riley's dogma that the Yucca could only possibly be fertilized by a certain American insect, and being amused at it because I had seen it fruiting well at the south of Europe."[18] Riley replied by saying that, on infrequent occasions, the yucca may be fertilized by other insects and at times a deformity in the flower allowed self-fertilization. Neither of these eventualities, however, occurred frequently enough to ensure the perpetuation of the species. Experiments by Riley and others confirmed that where the moth was excluded, the yucca did not produce seed. Also, where yuccas grew in the open, one could never be certain that the pronuba moth was absent, for it remained hidden in the half-closed flowers during the day and could be observed for only a few hours in the evening and night. There was a high probability that the moth had become established in England and southern Europe, especially since Riley had sent pronuba larvae to correspondents in those areas.[19]

The relationship between the yucca and the pronuba moth provided dramatic evidence that plants and animals had evolved together. This finding expanded the entomologists' range of evolutionary investigations beyond the areas of insect mimicry and polymorphism to include the evolution of whole biotic communities. The use of evidence from the yucca-moth relationship in the debate over evolution was no doubt important, as in, for example, the conversion of Engelmann to evolutionary views. By the 1870s, however, almost all American entomologists were evolutionists, so this aspect of the evolutionary debate had limited relevance within their circle. The central importance of the yucca moth investigations among American entomologists was the increased emphasis on biological studies, in particular the study of functional relations among plants and animals, which now assumed an equal or greater importance to their classification. The concept of natural selection proved to be a more productive guide to observation and experimentation than natural theology or the concept of the balance of nature.

It is instructive that Riley, an agricultural entomologist, used Darwinian theory to investigate the biological and evolutionary relationships of two "noneconomic" species, the yucca and the pronuba moth. The case of the pronuba moth thus demonstrates a unique blend of practice and theory that by the 1870s characterized American entomology.

12 The Debate over Entomological Nomenclature

By the 1870s, the various strands of American entomology had coalesced into a solid institutional structure, with the kinds of internal strengths, problems, and controversies that characterize mature scientific disciplines. The drive to describe and systematize the American insect fauna, combined with the patriotic assembling of type specimens in American collections, had produced a richness of insect collections and a magnitude of printed descriptions of American insects worthy of an established entomological community. As a result, this community now faced a crisis in the nomenclature of its science. The immediate problem was one of synonymy, that is, the existence of two or more names for the same insect. The advent of Darwinism intensified this crisis by introducing radical new principles to the definition of species and their systematic arrangement.

The American debate over entomological nomenclature began in 1872, when Scudder published a "Systematic Revision of American Butterflies" in advance of his book, *The Butterflies of New England*. Scudder and Edwards were regarded as the two leading authorities on the American Lepidoptera, and it is therefore easy to understand the consternation felt by lepidopterists as they reviewed the sweeping changes Scudder proposed for American species. Scudder engaged in a radical "splitting" of genera on the basis of minute differences in the genitalia and other organs, and he renamed scores of familiar species by giving priority to previously unrecognized authorities. According to Bethune, editor of the *Canadian Entomologist*, Scudder's revision contained 96 new genera, only 16 of which retained the names in common usage. Among the new genera proposed by Scudder, 42 were given entirely new names, and 39 others were given obsolete names from Jacob Hübner, a German entomologist of the early nineteenth century whose names had previously not been widely acknowledged. In addition to the introduction of new names in a wholesale manner, Scudder's radical splitting of genera resulted in a classification in which 96 genera contained only 228 species, which amounted to almost one genus for every two species, and in fact 45 genera contained only a single species.[1]

The "sweeping, revolutionary, [and] radical" changes in nomenclature proposed by Scudder became a major item of discussion among the entomologists at the AAAS meeting at Dubuque, Iowa, in September 1872, who were discussing the organization of an Entomological Club. Among those assembled in Dubuque there was near unanimous disapproval of the changes proposed by Scudder. Many of those present expressed the hope that he would reconsider the changes before publishing his book.[2] Bethune suggested that if Scudder persisted in his revision, other lepidopterists should agree among themselves to ignore the proposed names until such a time as the matter could be generally settled.[3] When the entomologists met again at the Portland, Maine, meeting of the AAAS in 1873, they voted to establish a committee on nomenclature that should consider all such questions and report at the next meeting. The aging Reverend Morris, in effect America's emeritus lepidopterist, was chairman of the meeting. He appointed a committee consisting of Riley (chairman), Scudder, Bethune, and LeConte, all of whom, with the exception of Scudder, favored retaining traditional nomenclature when possible. Morris, himself a critic of Scudder's revisions, later added Edwards, another Scudder opponent, to the committee, in the hopes of having the matter "cut and dry" by the next meeting.[4] In subsequent years, the membership of the committee changed, the most important being the substitution of Grote, a Scudder supporter, for Edwards. The key figures on the committee were Riley, Scudder, and LeConte.

Morris charged the committee on nomenclature to study the matter and propose a code of nomenclature for all American entomologists. This was a bold step for a scientific discipline just forming its own subsection within the AAAS, especially in light of the fact that the AAAS was considering a general revision of the rules of nomenclature.

The code under which the AAAS nominally operated had been drawn up by the British Association for the Advancement of Science in 1842, adopted by the American Association of Geologists and Naturalists in 1845, and revised by the British association in 1865. In the latter year, the AAAS had appointed a committee on nomenclature, but nothing definite had been proposed and no formal action had been taken on the British revisions. Many AAAS members considered the code inadequate. Scudder, among others, had suggested that the Committee on Nomenclature be given new instructions so that explicit rules could be proposed and adopted.[5] Edwards argued that entomologists faced more pressing needs than other zoologists in the clarification of the rules of nomenclature; therefore, they should assume the initiative in amending the rules to meet their special needs and then propose these as the basis for general adoption by the AAAS.[6] It was finally agreed that the entomological committee on nomenclature would give guidelines for the entomologists alone but would wait for the full report of the association's Committee on Nomenclature before adopting binding rules.[7] It was with this understanding that the debate was carried on during the years from 1873 to 1876 when the committee produced a model code for the guidance of entomologists. While Scudder's proposed changes in the Lepidoptera occupied the center of attention, the issues affected all of entomological nomenclature and indeed all of zoological nomenclature. For example, not long after Scudder's revisions appeared, a similar revision for the Coleoptera appeared in George R. Crotch's "Checklist of North American Coleoptera." LeConte objected to these changes in the nomenclature for Coleoptera on much the same grounds that most lepidopterists objected to Scudder's changes in the Lepidoptera.[8]

Objections to the kind of sweeping revisions proposed by Scudder and Crotch were based on four arguments. First, it was held that the revival of antiquated names (like those of Jacob Hübner) that had never received recognition in the scientific literature was a misuse of the rule of priority. Second, the abandonment of familiar names by recognized authorities in favor of unfamiliar names by obscure authors was considered unjust to the great entomologists of the previous generation who had laid the foundations for recent progress in systematics and who had established the nomenclature in current usage. Third, the frequent changing of names was considered injurious to the growth of entomological science because it discouraged beginners, from whose ranks the next generation of entomologists would come, and because it placed an unnecessary burden on amateur collectors, who would face difficulties in constantly renaming and rearranging their specimens. Fourth, American converts to Darwinism opposed what they saw as excessive splitting of genera and species on the basis of nonbiological criteria because it ran counter to principles of classification provided by evolutionary theory.

By the midnineteenth century, opinions varied widely as to how rigorous one should be in applying the rule of priority when naming species or genera. The "rule" of priority had developed subsequent to Linnaeus, the founder of the binomial system, and achieved widespread acceptance only gradually by different disciplines and in different countries. Furthermore, the "type" concept, which became a cardinal principle of organization of the early nineteenth century, was unknown to Linnaeus and the eighteenth-century founders of various disciplines. In the midnineteenth century there was not even agreement on a common beginning point for the binomial system. In 1842, the British Association for the Advancement of Science adopted the twelfth edition of Linnaeus (1766), and in 1845 the American Association followed its lead, but most naturalists in northern Europe preferred the tenth edition (1758), and the botanists went back even further.[9]

Even more troublesome than the uncertainty of a beginning point (which for an early discipline like entomology made a great difference) was the existence of multiple authors in the post-Linnaean period. These authors, often working in isolation from each other, supplied thousands of names and built classification systems on scanty data, and their works were regularly discovered and revived by zealous systematists like Scudder.[10] Such a case was the German lepidopterist Jacob Hübner, a designer at a cotton factory in Augsburg, Germany, who amassed a large collection and who published his names and classifications in various places in the years 1796 to 1828. Edwards charged that Hübner had never published his system and names in a monograph, as specified by the rules of the British and American associations, and that his names had been published in privately printed pamphlets and sale catalogs of insects.[11] Edwards contended that Hübner had never been regarded as an authority by German lepidopterists and had only lately been revived by some British lepidopterists and by Scudder and Grote in America.[12] Two of Scudder's main sources for proposed changes were Hübner's *Verzeichniss bekannter Schmetterlinge* (1816) and *Tentamen* (1806). The latter, a manuscript checklist in Latin, has been a continuing source of controversy among systematists.[13] Most entomologists who expressed an opinion on this point agreed with Edwards. For example, Saunders, who succeeded Bethune as editor of the *Canadian Entomologist,* objected to the practice of "digging about in the old bones of nomenclature."[14]

Many entomologists who objected to Scudder's zealous application of priority held that such practice did a grave injustice to recognized authorities of the previous generation who had introduced scientifically accurate descriptions and classifications to the science.[15] Edwards and some others even suggested that those who engaged in wholesale revisions on the basis of long-forgotten works did so out of personal vanity, so that they could attach their own names to newly created genera. On this issue personalities inevitably came into play. Edwards, who had recently clashed with Grote in the process of organizing the Brooklyn Entomological Society, wrote to Lintner: "I knew Grote when he was young entomologically and I know him now, and if his whole ento-mological work has not been for the 'assertion of ourselves' first and the good of science last, I do not understand him. It is characteristic that he chops up genera so as to either get his own or Hübner's name after most of them—and then tack 'Grote' after the species named because he cre-ated the genus."[16] In response to these charges, Scudder and Grote in-sisted that the pursuit of pure science required a strict application of the rule of priority. Questions of "justice" had no place in the discussion.[17] Grote pointed out that Hübner was post-Linnaean and (assuming one accepted the mode of his publications) that he had clear priority for many of the species and genera that had been revised and changed by later authors.[18]

Many lepidopterists objected to Scudder's revisions on the grounds that they created unnecessary labor and confusion among those who would have to learn new names and rearrange their cabinets.[19] In 1877 Saunders admitted that the uncertainty resulting from Scudder's revisions had so discouraged him that he had done nothing in his cabinet for several years.[20] Others objected that the frequent revisions by a handful of experts worked a hardship on "those who had comparatively little time to devote to Entomology," including the many amateurs who sup-plied leading figures with vital information. At the meeting of the Ento-mological Club in 1876, Edmund B. Reed, who served variously as secretary-treasurer, curator, and librarian of the London branch of the Entomological Society of Ontario (formerly Canada), urged that the interests of this group be considered. It was from the present novices in entomology that the future leaders of entomological science would be drawn, he maintained, and the frequent revisions were an unnecessary discouragement to beginners.[21] James Behrens, a collector in San Fran-cisco whose species had been described by LeConte, Packard, Grote, and Edwards, among others, argued that collectors would be deterred in their pursuit of new captures.[22] Edwards added that American ento-mology relied heavily on these zealous workers because the discipline was growing rapidly through popularization and because it was closely tied to agriculture, with its numerous nontechnical reports. He argued that the nomenclature should be kept simple and understandable to the layman and to the agriculturist.[23] Not all entomologists agreed. Mor-rison, the beau ideal of insect collectors, supported Scudder in most matters. He noted that the present confusion had arisen precisely because collectors had often assigned names to their captures without considering the rule of priority. The only solution, he argued, was a strict adherence to the rule of priority, no matter how inconvenient it might be at first.[24]

The other great objection to Scudder's revision, voiced by Edwards, Riley, and other supporters of Darwin, was that Scudder's revisions were based on trivial, or nonbiological, distinctions not in harmony with evolutionary criteria. American entomologists, having rapidly applied Darwinian theory in their various specialties, comprehended the need to revise their classifications accordingly, though the actual working out of new systems came only slowly.[25] Darwin had recognized that one of the greatest strengths of his theory lay in its explanation of the "natural" classification of organisms naturalists now universally recognized. He proposed a genealogical standard of classification in which taxa were treated as communities of descent. Darwin held that in nature there was (or once had been) a continuous series between organisms, and the only reason that specific or other distinctions could be made at all was because the intervening forms had been eliminated.[26] He also held that there was no essential difference between species and varieties, since a variety was an incipient species.[27] American entomologists like Walsh, Riley, and Edwards supported Darwin's genealogical standard in classification. Walsh wrote to Edwards: "The closet naturalists are the mischief. I am more convinced every day that a large percentage of species cannot be distinguished by the mere comparison . . . of dried specimens of the imago. I have no doubt whatever that many so-called species are in reality an aggregation of several distinct species, and that on the other hand many so-called species (often described from single specimens) are nothing but varieties."[28] As a consequence of their Darwinian persuasion, Edwards, Riley, and other evolutionists tended toward "lumping" species rather than "splitting" them on fine distinctions, particularly where these distinctions were not based on what they considered patterns of descent or real biological distinctions.

The strict Darwinists among American entomologists disagreed fundamentally with Scudder's concept of species. Upon receipt of Scudder's revision, Riley replied: "Starting on a Darwinian basis, and knowing, from experience, how variable species are . . . I feel convinced . . . that the more we understand our species, the more we become acquainted with their biology, . . . the more we shall find they run into one another. . . . they are linked together by intermediate species, in such a manner that in a *natural system* we shall never be able to draw arbitrary lines."[29] Riley felt that Scudder's genera were based on distinctions that were too trivial, such as the relative length of tibia or tarsus, or the position of a minor wing vein.[30] Edwards heaped ridicule on what he considered Scudder's nonbiological classifications borrowed from Hübner: "Hübner . . . amused himself with assorting the known butterflies into batches or parcels, as a child would sort his alleys and taws, by color, stripes, and shape, putting blues into one lot, browns into another, one-striped into a third, two-striped into a fourth, regardless of characters which would be generic, that is, *which would indicate blood relationship or common descent.*" Edwards concluded that it was rare to find one of his batches or "coitus" that coincided with a genus.[31] To Lintner he wrote, "Hübner did not divide his butterflies into genera but into 'herds and hordes.' . . . One horde will sometimes cover species which are alike in color, but stand in . . . distinct genera."[32]

Edwards cited his findings on dimorphism in butterflies to underscore the need for a biological basis for classification. Such a system, he said, should include attention to each stage of the insect's development. Summing up his disagreement with Scudder on this point, he wrote to Lintner:

Scudder makes up his mind about species from his examination of their tails, genitalia, which cannot . . . be decisive as to specific differences. Why did he not find in that way that [*Grapta interrogationis,* var. *umbrosa* and var. *fabricii*] were all the same? Why that the var. of Ajax were not all same? . . . He and I doubtless differ radically as to what is a species. He would include all the vars. of a species with the species itself even though such var. may be breeding true each to itself and may have got to be as distinct as G. [Album ?] and Comma for instance.[33]

Scudder, who at that stage only grudgingly admitted evolutionary explanations for morphological characters, apparently did not admit dimorphism at all in his schemes of classification.[34]

While Edwards, Scudder, LeConte, Grote, Mead, and others exchanged fire through the pages of the entomological journals, the Committee on Nomenclature, likewise divided between Scudder and his critics, worked slowly toward preparing a code for consideration by the Entomological Club.[35] In 1876, the committee submitted the proposed code to the Entomological Club at the AAAS meeting in Buffalo, New York. Scudder had opposed discussing the code at the AAAS meeting, because, he said, the committee had never met as a whole, and the members had not compared their views with the zoologists in other departments. The fact that the rules were presented in 1876 represented a victory for Scudder's critics.[36]

The committee submitted a list of eleven rules with notations indicating which had the unanimous approval of the committee and which had a divided opinion. Predictably, the questions of priority raised by Scudder's revisions carried divided opinions. The committee recommended (with a divided vote) that where a specific name had been accepted and in use for a period of twenty years, the name should not be changed in favor of a name from an author of prior date (rule 2). They proposed (with a divided vote) that the personal name that followed the species should be the name of the person who first proposed the species name and not that of a later author who rearranged the generic combination (rule 8). They agreed (unanimously) that the name following the genus should be the author who established the genus, but that the name of the author who had first proposed the term would be cited in brackets (rule 3).

Each of these rules was directed in some measure against Scudder's revisions. After a lively debate, the Entomological Club voted to adopt the rules for the guidance of American entomologists. They also passed a resolution that these rules could not be changed unless the changes were announced during the club's annual meeting at the AAAS and approved by vote at the annual meeting the following year.[37] Despite the fact that a vote had been taken and the rules approved by a majority, uncertainty remained as to whether the rules were binding for all American entomologists and, if so, how they could be enforced. Those rules that had received a divided vote were especially open to question. Most entomologists seem to have regarded the rules as advisory in nature. They depended for their authority on a general agreement to follow the advisory.[38] Scudder himself made no move to change what he had published, a position Edwards charged amounted to a willful refusal to honor the action of the Entomological Club. For lack of any mechanism to deal with recalcitrants, no further direct move against Scudder's revisions was made.[39]

While the rules adopted in Buffalo addressed the question of priority in various ways, they did not address the question whether evolutionary theory should serve as a basis for classification and the definition of species. This was an area that probably could not be legislated, but it could be influenced in other ways. And while Edwards and others were pleased that the authority of Scudder's revisions had been undermined, they felt something more was needed to set entomological nomenclature back on track.[40] Shortly after adoption of the rules at the Buffalo meeting in September 1876, Hagen, curator of insects at the Museum of Comparative Zoology, called upon Edwards to publish a checklist of American butterflies using the traditional nomenclature and citing standard authorities. This strategy, Hagen argued, would furnish an alternative to Scudder's nomenclature and would provide a powerful impetus for Americans to put the adopted rules into practice.[41] Edwards had already begun gathering material for such a checklist and had solicited advice in the undertaking from Philipp C. Zeller, a German schoolmaster and renowned lepidopterist. Though not a Darwinian, Zeller agreed with Edwards on the need to consider the total biology of insects in matters of classification. With such backing, Edwards welcomed Hagen's invitation and began work on the checklist at once.[42]

In preparing the checklist, Edwards had the active collaboration of Hagen, Lintner, and Adolph Speyer, another leading German lepidopterist. Speyer's support was critical, for he had not been involved in the debate, and his authority could offset Scudder's. "Publishing this under Speyer's auspices will completely block out Scudder if he now comes forward to claim his genera," Edwards wrote to Lintner.[43] Edwards sent specimens representing entire groups of Lepidoptera to Speyer, who then used these in suggesting arrangements for the checklist.[44]

Edwards's *Catalogue of the Lepidoptera of America North of Mexico* appeared in March 1877, with ample time for distribution prior to the fall meeting of the AAAS. In the introduction, Edwards explained that the arrangement was based on a Darwinian conception of species. He drew heavily upon American authors who, following Darwin, considered the preparatory stages as well as the adult in their classifications. In addition he cited European authors who, unlike Scudder, retained later well-known authorities in their nomenclature. Prominent among these were Speyer and Edward Doubleday, a British lepidopterist who in the 1830s and 1840s had acquired American species through Harris and others and who had collected extensively in the United States.[45] Edwards did not list a single genus based on Hübner. "I couldn't admit Hübner anywhere," he wrote to Lintner. "That is a case of give an inch, take an ell. If Hübner is an authority at all, he is almost everywhere."[46]

Cresson, the secretary of the American Entomological Society, distributed copies of Edwards's *Catalogue* to all members of the society and to entomologists listed in the *Naturalists' Directory* with the express purpose of countering Scudder's revisions.[47] In a review of the *Catalogue* for the *Canadian Entomologist,* Saunders praised Edwards's work for its "conservative" character, predicting that lepidopterists could once again get on with their labors without having to rename their specimens and rearrange their cabinets. Saunders noted with approval that in crediting genera, Edwards followed the rules adopted by American entomologists at Buffalo. For example, Edwards appended the author who first gave the genus a proper definition and not the name of the later reviser of the genus. Saunders applauded the exclusion of Hübner's genera, but he considered the exclusion of two of Scudder's own genera unfortunate. Even though he considered Scudder's *New England Butterflies* to be a hindrance to the orderly progress of taxonomy in the Lepidoptera, he felt Scudder was entitled to consideration where his genera were based on solid distinctions.[48]

When the AAAS met in Nashville, Tennessee, in September 1877, William H. Dall, chairman of the AAAS Committee on Nomenclature, presented the committee recommendations for consideration by the membership. Dall, a federal scientist who had collected for Edwards in Alaska and a zoologist who specialized in malacology, had tried in vain to avoid being drawn into what he considered "unseemly squabbles" among American lepidopterists.[49] Constant badgering by Edwards, Riley, LeConte, and others, however, convinced Dall and the committee that entomologists faced critical problems in the matter of nomenclature. The rules Dall's committee proposed to the AAAS in 1877 reflect to a considerable degree the desires of the entomologists, particularly those of the anti-Scudder camp.[50] The AAAS committee explicitly rejected Scudder's attempt to retain old forms of zoological families, subfamilies, and tribes in zoological nomenclature, ruling that names higher than genera were not subject to the rule of priority.[51] The committee endorsed LeConte's suggestion that the twelfth edition of Linnaeus be recognized as the beginning point for zoological nomenclature.[52] At Riley's suggestion, the committee recognized an exception to the rule that restricted specific names to only one word. In the case of gall insects, he convinced them that the name of the plant upon which the insect lived should also be included.[53] Apparently, the biggest concession to the entomologists was the recognition that each scientific specialty required the flexibility to decide among its own members on how to treat obsolete authors for purposes of priority. The full committee thus gave its sanction to the independent action taken by the entomologists a year earlier, implicitly recognizing the Entomological Club as an independent tribunal capable of ruling on entomological nomenclature. LeConte, citing recommendations of Alfred R. Wallace, president of the Entomological Society of London, had urged that such a system of referral to a standing committee be established.[54] Finally, the full committee noted with approval the suggestion of Riley and LeConte that a name that had been standard in the zoological nomenclature for twenty-five years should remain in preference to revisions based on older authors. American entomologists, now well established, were now reversing the position held by Harris in the 1830s, when he insisted on the right of priority even of descriptions located in obscure publications (see Chapter 1). The committee, probably wisely, left such decisions up to the majority in each branch.[55]

How much impact the proposed or adopted rules of nomenclature had on the actual practice of entomologists is questionable. Systematists and taxonomists are notorious individualists relatively immune to outside influence or criticism. In 1878, for example, Scudder proposed yet another revision that brought howls of protest from Edwards. Denouncing Scudder's disregard for rulings by the Entomological Club and the AAAS, Edwards charged that he stuck to his system out of stubbornness: "If he [Scudder] said the horse was 16 feet high he sticks to it. And as he chooses to stick to his blunder of naming species by their tails, so be it. . . . His descriptions based on tails giving no further indications of the color, size, form or nature of the species, no one in the world could ever know what he was talking about, or tell one of his pseudo-species from another."[56]

At any rate, the matter of compliance is not central to our concern. What is more important to note is how the debate reflected the maturity of American entomological science by the 1870s. The very occurrence of the debate over nomenclature is a sign of the rapid growth of the American entomological community and its increasing scientific sophistication. The referral of the problem to an entomological organization that represented all American entomologists, the debate in entomological journals and elsewhere, and the resolution of the debate in the AAAS mark striking advances in entomological institutions compared to a generation earlier, when American entomologists first organized. Indeed by the 1870s, the American entomologists represented a major segment of the international scientific community, and they assumed leading roles in such matters as the revisions of internationally recognized rules of nomenclature. In such matters, Americans often gave more weight to evolutionary theory and biological considerations in questions of nomenclature than their colleagues in Europe. In all these ways the debate over nomenclature reflected the maturity of the American entomological community.

Conclusion

At the Entomological Club meeting of the AAAS in 1878, Lintner recounted the progress American entomology had made over the previous four decades. He pointed with pride to the increase in the number of entomologists, the advances in entomological knowledge, the expanding number of entomological publications, the generous support of government, and the American contributions to systematics and evolutionary studies. Lintner's talk before the AAAS was one of many self-congratulatory addresses by American entomologists around 1880. His account provides an opportunity to review not only the advances themselves but the reasons for them.[1]

The increase in numbers of American entomologists was indeed impressive. Whereas in the 1830s, there had been less than a half dozen individuals pursuing entomology seriously enough to publish on the subject, Lintner estimated that by 1878 there were about 30 American entomologists who published regularly. Lintner's estimate was conservative compared with others: that same year, Scudder counted 41 publishing American entomologists, and in 1869, Packard had listed 52. As we have seen, the total number of publishing entomological authors in about 1870 was closer to 100, but the estimates of 40 to 50 who published regularly are in line with the data cited in Chapter 8. In 1879, Lintner estimated the total number of American entomologists at 835, a figure that corresponds to the figure of approximately 800 based on data presented above (Chapter 8).[2] By comparison, we may recall that Kohlstedt calculated there were 2,068 participants in the AAAS in the period from 1848 to 1860.[3] In 1880, entomologists comprised the largest

253

single group of zoologists in North America. Among American naturalists in general, they were second only to the botanists when viewed in terms of sheer numbers of participants.

The advance in knowledge was equally impressive. In 1835, Harris had listed a total of only 994 described species of Coleoptera and 140 Lepidoptera. By the 1870s, standard checklists for Coleoptera contained 7,450 species, and checklists for Lepidoptera contained 1,132 moths and 506 butterflies. In very rough figures the number of American insects described in the forty-year period increased about ten times. With respect to the study of insect life histories, the Americans had no rivals. At the beginning of the period, fewer than a dozen life histories were known, and many of them were incomplete. By the 1870s, hundreds of life histories were known and those of the most injurious species, like the Rocky Mountain locust, were documented in thousands of publications.[4]

The growth of American entomological institutions in general was considered a cause for celebration. In the 1840s, Americans had complained of the paucity of American entomological literature and the lack of entomological societies, and they worried about the plundering of the American insect fauna to the advantage of European collections and publications. The few private insect collections (like Melsheimer's) did not reflect the entomological wealth of the continent, and (as the fate of Say's collection demonstrated) these collections lacked institutional stability.

Forty years later the situation had changed. Whereas in the 1840s, Americans had only a few general scientific publications, like the *American Journal of Science,* American entomological authors in the 1860s and 1870s could publish in a variety of specialized entomological publications. Monographs published by the Smithsonian Institution included an expanding series dedicated to the various insect orders, and agricultural entomology reports sponsored by state and federal governments were so numerous that separate reference works were required to list them.[5]

The organization of American entomologists into scientific societies showed a similar improvement. By the 1870s, the entomologists had founded a multiplicity of organized groups in different geographical locations offering a variety of activities for those interested in the collection and exchange of insects, systematics, agricultural applications, and evolutionary studies. In addition, scientific societies with general memberships, like the AAAS, provided opportunities for entomologists to meet and share information.

By the 1880s, American entomological collections had advanced from their precarious existence of the 1840s to a position of stability and permanence in world science. Central entomological collections in Philadelphia, Boston, Cambridge, and Washington, D.C., were now the definitive sources of information for the American insect fauna. These collections had served to reverse the flow of American specimens to Europe, and they ensured the continued advancement of American entomology.

More than anything else, generous public support for entomology in the United States provided a yardstick by which to measure the progress of the science. There was nothing comparable in Europe or elsewhere to the volume and the quality of government-sponsored entomology produced each year by entomologists on the payroll of the states and the federal government. By the 1870s American entomologists were setting the standard in government-sponsored entomology for other nations to follow.[6] In the area of practical applications and control measures, the American entomologists had no rivals. Their achievements were evident in a whole range of control measures that used chemical pesticides, meteorological data, quantitative studies of the contents of predators' stomachs, locust warning systems, and other investigative approaches and measures that they had pioneered.

The advances in American entomology extended beyond institution building, systematics, and practical applications to encompass theory at the highest level. In 1876, when Edward S. Morse summarized the contributions of American zoologists to evolutionary studies, he pointed to the special contributions made by American entomologists.[7] The entomologists had been quick to accept the evolutionary explanations offered by Darwin, they had played a leading role in the studies of insect adaptations that supported Darwin's theory, and they had led in the development of a distinctive American evolutionary school of Neo-Lamarckism.[8] Walsh, Riley, Packard, Scudder, Edwards, and Hagen had developed evidence for evolution in studies of insect eating habits, gall formation, geographic distribution of insects, insect mimicry and polymorphism, plant and insect interdependencies, and the influences of climate and temperature on insect development in various stages. They had published support for Darwinian or Lamarckian evolution in government publications, scientific journals, and the popular press. In the debate over entomological nomenclature in the 1870s, entomologists led the American scientific community in the first stages of the revision of taxonomy according to evolutionary principles. In this process, they set precedents that influenced the rules for American and European scientists regarding scientific nomenclature.

From the 1840s to the 1870s, the American entomological community developed into a community of distinctive composition and style. Its composition derived from a mixture of individuals from European and American backgrounds. More than one-fourth of the entomological elite in the 1860s and 1870s were immigrants from Europe, primarily Great Britain. The "American" interest in the insect fauna of North America was to a considerable extent imported, in the form of individuals who had European training in entomology. This European, particularly British, presence in the formative stages of American entomology was a decisive influence even though it is sometimes difficult to say exactly how this influence operated in the American context.

The most prominent feature of the "American style" was its field orientation, in particular its concern with the ecology of plant and animal communities. This orientation evolved from the combination of interests in taxonomy, geographical distribution, the preparatory stages of injurious species, the search for control measures, and the evolution of varieties and species. A second distinctive feature was the tendency of American entomologists to view insects in terms of populations rather than typologies. This thinking in turn predisposed Americans favorably toward acceptance of evolutionary theory and enhanced their field investigations. One outstanding exponent of this style was Edwards, who with a network of collectors pursued evolutionary investigations into preparatory stages and polymorphism in butterflies. Through the combination of systematic, geographic, agricultural, and evolutionary concerns, American entomologists developed an early and sustained competence in ecological studies. In fact, by the 1880s, the Americans comprised the largest and most productive group of practicing ecological investigators in the world.[9]

As a result of these developments, the overall prestige of American entomologists rose enormously. By the 1870s, entomologists no longer felt they needed to apologize for the supposedly trivial importance of their subject. Few could doubt the great destructive power of insects and the necessity for entomologists' efforts in the control of injurious species. In addition to their function as agricultural advisers, entomologists took a leading position in the discussion of evolution, the most engaging intellectual topic of the era. By the 1870s, some agricultural entomologists had already begun referring to themselves as "professionals" with concomitant status.

The rapid development of American entomology around midcentury was the result of three social developments: (1) the striving of Americans to forge scientific institutions equivalent to those in Europe, reinforced by the infusion of European—mainly British—personnel into the American context, (2) rapid and massive agricultural change in the northern states and Canada, and (3) investigations into the evolutionary origins of insect forms. Of the three, agriculture was the most decisive. Agriculturists persuaded legislatures to appropriate funds for entomological investigations. This infusion of funds was effective, however, only because of the presence of individuals who, under the influence of European precedents, had begun the pursuit of entomology largely independent of agricultural concerns. Entomologists with agricultural and nonagricultural orientations worked together to develop the institutions that were essentials for their discipline. In time, evolutionary theory provided American entomologists with a basic theoretical foundation that informed their investigations in all areas, from taxonomy to insect control.

Though contemporaries saw the rise of American entomology as an almost unblemished success story, critics a century later pointed to a darker side, the indications of which were only faintly visible by 1880. The entomologists' successes in controlling injurious insects such as the Colorado potato beetle encouraged them to advocate the widespread use of arsenic-based insecticides, the first of a long line of chemical insecticides to pose serious problems for public health and ecological systems.[10] In addition, the entomologists' success with agricultural pests led to an alliance between the entomologists and the more successful farmers and their supporting institutions, an alliance that eventually operated to the disadvantage of small farmers and landless agricultural workers. This alliance raised the old question of elitism in a democracy in a new guise: for whom should science be performed, and who should benefit? In 1869, for example, Walsh and Riley advocated a federal study of insects affecting the cotton crop of the South with the argument that two good crops would do more good for the region than any other policy.[11] With hindsight, one might ask whether support for cotton production in the South at that time might also have reinforced the economic, racial, and social inequities that plagued the region.

If the successful promotion of agricultural entomology set the stage for future problems in the areas of public health and special interest politics, the drive toward professional status that was just emerging in the 1870s had the unfortunate effect of excluding women from positions of leadership within the profession. Until the 1870s, women comprised a small but active minority within the American entomological community. Investigators like Margaretta Morris and Mary Treat were regarded as exceptions. By the late 1870s, however, there were enough women entomologists and other women with professional aspirations to prompt their male colleagues to question their status within the professional and scientific community. When, in the 1880s and later, American scientists resolved this question generally, women entomologists, like women scientists in other fields, were relegated to "women's work," for example as scientific assistants, and were excluded from full participation in scientific and professional organizations and from scientific advancement.[12]

These issues of public health, ecological balance, and discrimination became public only as American agricultural entomologists developed into a fully "professional" group in the period after 1880, and they approached these matters with the self-assurance, even arrogance, of a newly successful professional group. The generation of entomologists under study here, however, was too involved in forging an effective community and institutions to give much thought to these issues.

The North American entomological community drew its characteristics from a unique combination of personnel, resources, and ideas, at a particular period in American history. The community, and the science they practiced, inevitably shared the character of America in those years. The Brethren of the Net represented a tiny elite that had its origins in the late Jacksonian era and developed its mature characteristics during the Civil War and the first years of the Gilded Age. During those years the United States and Canada were transformed from loose confederations of states and provinces with continent-sized hinterlands into centralized nation-states extending from the Atlantic to the Pacific. Between 1840 and 1880, the United States fought the Mexican War and the Civil War and added California, Oregon, Texas, the Southwest, and Alaska to its continental domains. In 1867, the British Canadian provinces became the Dominion of Canada. The United States was linked by rail from coast to coast by 1869, and Canada followed suit in 1885. The drive toward central organization and the rational management of natural and human resources that characterized the period was evident also in the organization and institution-building activities of the entomologists. Indeed the metaphor of "warfare" against the insect foes of agriculture came naturally to agricultural entomologists of that generation.

As they pursued their musings about insects, the entomologists of midcentury forged a science that was thoroughly identified with continental expansion across North America. Natural history, more than any other science, is shaped by the cultural and natural resources of national units. North America in the nineteenth century was about as close to the ideal setting as one can imagine for the rapid development of entomological science. Drawing on the established scientific traditions of various European countries and America itself and working within stable political systems expanding continentwide, the entomologists were well positioned to take advantage of the rich harvest of specimens from western exploration, to forge alliances with horticulturists and agriculturists facing new problems of the agricultural revolution, and to seek out stable institutional bases for their discipline within expanding and multifaceted educational and governmental structures.

The members of this elite were unified by their aesthetic and intellectual interest in insects rather than by any common socioeconomic background. In comparison with the majority of their contemporaries, who were involved in the rough-and-tumble struggle for economic advancement, they marched to the beat of a different drummer. Despite the fact that many entomologists came from families in trade and commerce, they chose to expend their talents and energies in intellectual pursuits with only modest prospects of monetary reward. Small wonder that they appeared strange in a generation caught up in the fever of land speculation, gold rushes, and railroad stock manipulation. As Zimmermann wrote to LeConte, "What genuine Yankee would travel to the Rocky Mountains to catch nothing but bugs?"[13]

The entomologists of that era took full advantage of "amateur" talent available in a democratic system at a time when "amateurs" could and did make solid contributions to entomology. By the next generation, from 1880 to 1900, when entomology became more firmly institutionalized in university programs requiring advanced degrees, and when biological investigations moved increasingly into laboratory settings, contributions by amateurs were no longer as important.

The generation of midcentury entomologists thus achieved and even surpassed the wish often expressed by American scientists of the Jeffersonian and Jacksonian eras that America produce a science fully equivalent to that of Europe. These midcentury entomologists were the first to combine a viable entomological community with vigorous institutions. They pioneered in reorienting the study of insects from natural theology to evolution. They represented the last generation that was trained primarily in the self-taught/apprentice tradition. In these ways, they experienced, more than any generation before or since, what it meant to be part of the Brethren of the Net.

Appendix I

Entomological Authors Cited in the *Record of American Entomology*

1 Adair, D. L.
2 Andrews, W. V.
3 Angus, J.
4 Bassett, H. F.
5 Beal, W. J.
6 Behr, H.
7 Bethune, C. J. S.
8 Billings, B.
9 Blake, C. A.
10 Blake, James
11 Borden, B.
12 Bowles, G. J.
13 Boutell, E. G.
14 Brown, A. M.
15 Burgess, E.
16 Byers, W. N.
17 Cabot, L.
18 Caulfield, F. B.
19 Chambers, V. T.
20 Clemens, J. B.
21 Coleman, N.
22 Cope, E. D.
23 Couper, W.
24 Cox, E. T.
25 Cresson, E. T.
26 Crotch, G. R.
27 Dall, W. H.
28 Devinny, V.

29 Dodge, C. R.
30 Edwards, W. H.
31 Emerton, J. H.
32 Fish, W. C.
33 Fitch, A.
34 Gentry, T. G.
35 Gillman, H.
36 Glover, T.
37 Green, S.
38 Grote, A. R.
39 Hagen, H. A.
40 Haldeman, S. S.
41 Harris, T. W.
42 Hays, W. J.
43 Herrick, E. C.
44 Higginson, T. W.
45 Horn, G. H.
46 Jones, J. M.
47 Julich, W.
48 LeBaron, W.
49 LeConte, J. L.
50 Leidy, J.
51 Lincecum, G.
52 Lintner, J. A.
53 Lockwood, S.
54 Mann, B. P.
55 Maynard, C. J.
56 McBride, S. J.

57 Mead, T. L.
58 Melsheimer, F. E.
59 Merrill, J. C., Jr.
60 Minot, C. S.
61 Mitchell, L.
62 Morehouse, G. W.
63 Morrison, H. K.
64 Nichols, G. W.
65 Nickerson, M. C.
66 Norton, E.
67 Orton, J.
68 Osten-Sacken, C. R.
69 Packard, A. S.
70 Parker, H. W.
71 Peck, G. W.
72 Perkins, G. H.
73 Pettit, J.
74 Philbrook, P. H.
75 Provancher, L.
76 Rathvon, S. S.
77 Reakirt, T.
78 Reed, E. B.
79 Riley, C. V.
80 Ritchie, A. S.
81 Robinson, C. T.
82 S [J.P.S.]
83 Sanborn, F. G.
84 Saunders, W.

85 Scudder, S. H.

86 Shimer, H.

87 Smith, G. D.

88 Smith, S. I.

89 Sprague, P. L.

90 Strecker, F. H.

91 Stretch, R. H.

92 Summers, S. V.

93 T [E.H.T.]

94 Thomas, C.

95 Treat, M.

96 Townsend, B. R.

97 Trouvelot, L.

98 Uhler, P. R.

99 Wagner, S.

100 Walsh, B. D.

101 Wescott, O. S.

102 Whitney, C. P.

103 Wilder, B. G.

104 Williams, J. B.

105 Wood, H. C.

106 Wright, C.

107 Young, C. A.

108 Zimmermann, C.

Appendix 2

Entomological Authors Cited in the *Record of American Entomology* Ranked According to Priority

1 Riley, C. V.
2 Packard, A. S.
3 Walsh, B. D.
4 LeConte, J. L.
5 Thomas, C.
6 Scudder, S. H.
7 Saunders, W.
8 Osten-Sacken, C. R.
9 Fitch, A.
10 Hagen, H. A.
11 Lintner, J. A.
12 Uhler, P. R.
13 LeBaron, W.
14 Harris, T. W.
15 Edwards, W. H.
16 Strecker, F. H.
17 Horn, G. H.
18 Bethune, C. J. S.
19 Grote, A. R.
20 Glover, T.
21 Crotch, G. R.
22 Sanborn, F. G.
23 Cresson, E. T.
24 Haldeman, S. S.
25 Emerton, J. H.
26 Sprague, P. L.
27 Melsheimer, F. E.

28 Rathvon, S. S.
29 Smith, G. D.
30 Clemens, J. B.
31 Norton, E.
32 Dodge, C. R.
33 Behr, H.
34 Whitney, C. P.
35 Mead, T. L.
36 Burgess, E.
37 Perkins, G. H.
38 Stretch, R. H.
39 Mann, B. P.
40 Maynard, C. J.
41 Reed, E. B.
42 Shimer, H.
43 Morrison, H. K.
44 Wood, H. C.
45 Bassett, H. F.
46 Chambers, V. T.
47 Minot, C. S.
48 Blake, C. A.
49 Couper, W.
50 Lincecum, G.
51 Smith, S. I.
52 Merrill, J. C.
53 Zimmermann, C.
54 Wescott, O. S.

55 Billings, B.
56 Wagern, S.
57 Leidy, J.
58 Peck, G. W.
59 Provancher, L.
60 Reakirt, T.
61 Andrews, W. V.
62 Bowles, G. J.
63 Herrick, E. C.
64 Cope, E. D.
65 Wright, C.
66 Pettit, J.
67 Orton, J.
68 Beal, W. J.
69 Angus, J.
70 Parker, H. W.
71 Julich, W.
72 Jones, J. M.
73 Ritchie, A. S.
74 Williams, J. B.
75 Nichols, G. W.
76 Lockwood, S.
77 Treat, M.
78 Trouvelot, L.
79 Robinson, C. T.
80 Higginson, T. W
81 Dall, W. H.

82 Cox, E. T.

83 Coleman, N.

84 Caulfield, F. B.

85 Wilder, B. G.

86 Gentry, T. G.

87 Gillman, H.

88 Blake, J.

89 Byers, W. N.

90 Fish, W. C.

91 S [S.P.S.]

92 Green, S.

93 T. [E.H.T.]

94 Morehouse, G. W.

95 McBride, S. J.

96 Townsend, B. R.

97 Brown, A. M.

98 Devinny, D.

99 Boutell, E. G.

100 Philbrook, P. H.

101 Nickerson, M. C.

102 Mitchell, L.

103 Borden, B.

104 Hays, W. J.

105 Cabot, L.

106 Summers, S. W.

107 Adair, D. L.

108 Young, C. A.

Note: No data for the last thirteen individuals, numbers 96–108.

Abbreviations

AAAS	American Association for the Advancement of Science
AAEE	American Association of Economic Entomologists
AE	*American Entomologist*
AES	American Entomological Society
AgH	*Agricultural History*
AJS	*American Journal of Science*
AN	*American Naturalist*
ANSP	Academy of Natural Sciences of Philadelphia
APS	American Philosophical Society
AR	*Annual Review of Entomology*
BES	Brooklyn Entomological Society
BSNH	Boston Society of Natural History
BSW	Biological Society of Washington, D.C.
Bull.	Bulletin
CDAB	*Concise Dictionary of American Biography*, 1964
CE	*Canadian Entomologist*
DAB	*Dictionary of American Biography*, 1959
DSB	*Dictionary of Scientific Biography*, 1970
EA	*Entomologica Americana*
EMM	*Entomologist's Monthly Magazine*
ER	*Environmental Review*
ESA	Entomological Society of America
ESP	Entomological Society of Philadelphia
ESW	Entomological Society of Washington, D.C.
ISAS	Illinois State Agricultural Society
ISHS	Illinois State Horticultural Society
JHB	*Journal of the History of Biology*
JHI	*Journal of the History of Ideas*
JNYES	*Journal of the New York Entomological Society*
JSBNH	*Journal of the Society for the Bibliography of Natural History*

MCZ	Museum of Comparative Zoology
MSB	Museum of Science, Boston
MSBA	Missouri State Board of Agriculture
MSHS	Missouri State Horticultural Society
NA	U.S. National Archives
NYES	New York Entomological Society
NYSA	New York State Archives
NYSAS	New York State Agricultural Society
NYSL	New York State Library
OIAS, SIA	Official Incoming Assistant Secretary Correspondence, Smithsonian Institution Archives
OIS, SIA	Official Incoming Secretary Correspondence, Smithsonian Institution Archives
PE	*Practical Entomologist*
PF	*Prairie Farmer*
Proc.	Proceedings
QRB	*Quarterly Review of Biology*
RAE	*Record of American Entomology*
RG	Record group
SIA	Smithsonian Institution Archives
SLAS	St. Louis Academy of Science
Trans.	Transactions
USNM	U.S. National Museum
WVSA	West Virginia State Archives
WVUL	West Virginia University Library
Zool. Rec.	*Zoological Record*

Notes

1. Entomology in the American Context to 1840

1. John C. Greene, *American Science in the Age of Jefferson* (Ames: Iowa State University Press, 1984), xiv, 3, and 12.
2. This background is discussed more fully in my dissertation, Conner Sorensen, "Brethren of the Net: American Entomology, 1840–1880" (Ph.D. diss., University of California, Davis, 1984), chap. 1. See Max Beier, "The Early Naturalists and Anatomists During the Renaissance and Seventeenth Century," in Ray F. Smith, Thomas E. Mittler, and Carroll N. Smith, eds., *History of Entomology* (Palo Alto, Calif.: Annual Reviews, 1973), 85–89; S. L. Tuxen, "Entomology Systematizes and Describes, 1700–1815," in ibid., 102–105; Jacob Lorch, "The Discovery of Nectar and Nectaries and Its Relation to Views on Flowers and Insects," *Isis* 69 (December 1978), 526–31.
3. Tuxen, "Entomology Systematizes," 97–102.
4. Ibid., 108; R. L. Usinger, "Prefatory Chapter: The Role of Linnaeus in the Advancement of Entomology," *AR* 9 (1964), 4, 8–9, and 13.
5. Mary P. Winsor, "The Development of Linnaean Insect Classification," *Taxon* 25 (February 1976), 62.
6. Ibid., 62–63.
7. Claude Dupuis, "Pierre André Latreille (1762–1833): The Foremost Entomologist of His Time," *AR* 19 (1974), 7–8; Tuxen, "Entomology Systematizes," 112–13; and Carl H. Lindroth, "Systematics Specializes Between Fabricius and Darwin, 1800–1859," in Smith, Mittler, and Smith, eds., *History of Entomology,* 121–23.
8. Since Linnaeus, the number of orders has increased from seven to twenty-five, and the relative positions of the orders have changed. See Url Lanham, *The Insects* (New York: Columbia University Press, 1964), 33–39.
9. Greene, *American Science in the Age of Jefferson,* 6, 10–11, 188, and 253; George H. Daniels, *American Science in the Age of Jackson* (New York: Columbia University Press, 1968), 42–46 and 192.
10. Greene, *American Science in the Age of Jefferson,* 318–19.

11. Ibid., 29–30 and 227; Brooke Hindle, *The Pursuit of Science in Revolutionary America* (1956; repr., New York: W. W. Norton, 1974), 322–23.
12. Greene, *American Science in the Age of Jefferson*, 385–91.
13. Ibid., 318–19.
14. Ibid., 53–55, 277, and 281; Patsy A. Gerstner, "The Academy of Natural Sciences of Philadelphia, 1812–1850," in Alexandra Oleson and Sanborn C. Brown, eds., *The Pursuit of Knowledge in the Early American Republic: American Scientific and Learned Societies from Colonial Times to the Civil War* (Baltimore, Md.: Johns Hopkins University Press, 1976), 180–84.
15. Greene, *American Science in the Age of Jefferson*, 70–71, 82.
16. Ibid., 74–75; Sally Gregory Kohlstedt, "The Nineteenth-Century Amateur Tradition: The Case of the Boston Society of Natural History," in Gerald Holton and William A. Blanpied, eds., *Science and Its Public: The Changing Relationship*, Boston Studies in the Philosophy of Science 33 (Boston: D. Reidel Publishing, 1976), 178–82; Thomas T. Bouvé, *Historical Sketch of the Boston Society of Natural History; with a Notice of the Linnaean Society, which Preceded it,* Anniversary Memoirs of the BSNH, (Boston: BSNH, 1880), 3–6, 15–16, 21–24. In 1967, the Boston Society of Natural History became the Museum of Science, Boston.
17. See Chapter 8.
18. Greene, *American Science in the Age of Jefferson*, 103–105; Kenneth R. Nodyne, "The Founding of the Lyceum of Natural History," *Annals of the New York Academy of Sciences* 172 (1970), 141–49.
19. I. Bernard Cohen, "The New World as a Source of Science for Europe," *IX Congreso Internacional de Historia de la Ciencia* (Barcelona-Madrid, 1959), 98.
20. Arnold Mallis, *American Entomologists* (New Brunswick, N.J.: Rutgers University Press, 1971), 3–9 and 244; Greene, *American Science in the Age of Jefferson*, 294–95. On the tradition of rearing insects, see especially Chapter 10, on William Henry Edwards.
21. Mallis, *American Entomologists,* 13–16; Greene, *American Science in the Age of Jefferson*, 70–71, 82–84, and 301; Peck is credited with the first American paper on systematic zoology, a description of a new species of fish in the *Transactions of the American Philosophical Society* (1794). See Daniels, *American Science in the Age of Jackson*, 16.
22. Thomas Say, *American Entomology; or, Descriptions of the Insects of North America,* 3 vols. (Philadelphia: S. A. Mitchell, 1824–28); Harry B. Weiss and Grace Ziegler, *Thomas Say: Early American Naturalist* (Springfield, Ill.: Charles C. Thomas, 1931), 24–38, 88, 114–30, and 200; Greene, *American Science in the Age of Jefferson*, 281, 306–11; Simon Baatz, "Patronage, Science, and Ideology in an American City: Patrician Philadelphia, 1800–1869" (Ph.D. diss., University of Pennsylvania, 1986), 74–99; Charlotte M. Porter, *The Eagle's Nest: Natural History and American Ideas, 1812–1842* (University: University of Alabama Press, 1986), 4–5 and 134; "Thomas Say," in Clark A. Elliott, ed., *Biographical Dictionary of American Science: The Seventeenth Through the Nineteenth Centuries* (Westport, Conn.: Greenwood Press, 1979), 228; Jeffrey K. Barnes, "Insects in the New Nation: A Cultural Context for the Emergence of American Entomology," *Bull., ESA* 31:1 (1985), 24.
23. Mallis, *American Entomologists,* 243–44; Richard Adicks, ed., *LeConte's Report on East Florida* (Orlando: University Presses of Florida, 1978), 6–8; Lindroth,

"Systematics Specializes," 127 and 135; C. F. Cowan, "Boisduval and Le-Conte: *Histoire générale et iconographie des lépidoptères et des chenilles de l'Amérique septentrionale,*" *JSBNH* 5 (1969), 132; Cyril F. Dos Passos, "The Dates of Publication of the *Histoire générale et iconographie des lépidoptères et des chenilles de l'Amérique septentrionale,* by Boisduval and LeConte, 1829–1833[–1834]," *JSBNH* 4 (1962), 48–56; C. F. Cowan, "Boisduval's *Species général des lépidoptères,*" *JSBNH* 5 (1969), 121–22; LeConte-Dejean correspondence in Le-Conte Collection, APS.

24. Mallis, *American Entomologists,* 12 and 20; Robert Snetsinger, *Frederick Valentine Melsheimer, Parent of American Entomology* (University Park: Entomological Society of Pennsylvania, 1973), 1–18; Say to J. F. Melsheimer, April 27, 1817, in William J. Fox, ed., "Letters from Thomas Say to John F. Melsheimer, 1816–1825," *Entomological News* 12 (June 1901), 173.

25. Mallis, *American Entomologists,* 25–33; Thaddeus W. Harris, *A Report on the Insects of Massachusetts, Injurious to Vegetation* (1841; repr., New York: Arno Press, American Environmental Studies, 1970).

26. Say to John F. Melsheimer, May 24, 1816, in Fox, ed., "Letters from Thomas Say to J. F. Melsheimer," *Entomological News* 13 (January 1902), 11.

27. Weiss and Ziegler, *Thomas Say,* 171–74.

28. Harris to Say, November 18, 1824, in Samuel H. Scudder, ed., "Some Old Correspondence Between Harris, Say and Pickering," *Psyche* 6 (September 1891), 140.

29. Harris to LeConte, December 17, 1829, John Eatton LeConte–Thaddeus William Harris Correspondence, Microfilm no. 580.1, APS (hereafter cited as APS Film 580.1).

30. Harris to LeConte [1830?], John Eatton LeConte–Thaddeus William Harris Correspondence, Microfilm no. 554.1–2, APS.

31. Asa Fitch to Harris, December 30, 1846, MCZ Archives, Harvard University, Cambridge, Massachusetts.

32. In 1848, Samuel Stehman Haldeman complained that he could not find a copy of J. T. C. Ratzeburg's volumes on forest entomology, though they were standard items in European libraries. Haldeman to John L. LeConte, May 27, 1848, LeConte Coll., APS.

33. LeConte to Harris, January 12, 1830, APS film 580.1; Harris to Hentz, April 3, 1830, in Thaddeus William Harris, *Entomological Correspondence of Thaddeus William Harris, M.D.,* edited by S. H. Scudder, Occasional Paper of the BSNH 1 (Boston: BSNH, 1869), 95.

34. Thomas Say, "On a South American Species . . . ," in Thomas Say, *American Entomology: A Description of the Insects of North America,* ed. John L. LeConte (New York: J. W. Bouthon, 1869), vol. 2, p. 32.

35. Quoted in Harry B. Weiss, *The Pioneer Century of American Entomology* (New Brunswick, N.J.: Author, 1936), 110.

36. LeConte to Harris, March 31, 1830, APS film 580.1; see also Harris's proposal for LeConte to conduct an exchange between him (Harris) and Dejean and also his request that LeConte inform him which of his beetles had been named by Dejean. LeConte to Harris, ibid., and Harris to LeConte, February 19, 1830, LeConte Coll., APS.

37. Cowan, "Boisduval and LeConte," 126.

38. Harris to LeConte, December 17, 1829, APS film 580.1.

39. The Congress adopted the rule that only names published after January 1, 1758, would be recognized and that the first occasion upon which the name was used would be the accepted one. In 1958, the rule was amended so that any name that had been in common usage for fifty years would remain the officially recognized scientific name. See Kjell B. Sandved and Michael G. Emsley, *Insect Magic* (New York: Viking Press, 1978), 28.

40. Frederick Valentine Melsheimer, "A Catalogue of the Insects of Pennsylvania," reprinted in Snetsinger, *Frederick Valentine Melsheimer*, 21–22. On the patriotic urge of Americans to publish their own scientific discoveries, see George Basalla, "The Spread of Western Science," *Science* 156 (May 5, 1967), 614–17.

41. Weiss and Ziegler, *Thomas Say*, 61–78, 93–94, and 200–201.

42. Hindle, *Science in Revolutionary America*, 12–33.

43. See E. A. Schwarz, "Some Notes of Melsheimer's Catalogue of the Coleoptera of Pennsylvania," *Proc., ESW* 3 (1895), 135.

2. "A Few Literary Gentlemen": The Entomological Society of Pennsylvania, 1842–1853

1. Morris to Harris, July 22, 1839, MCZ Archives.

2. On the organization of American science in this period, see Sally Gregory Kohlstedt, *The Formation of the American Scientific Community: The American Association for the Advancement of Science, 1848–1860* (Urbana: University of Illinois Press, 1976); Kohlstedt, "The Geologists' Model for National Science, 1840–1847," *Proc., APS* 118 (April 1974), 179–94; A. Hunter Dupree, *Science in the Federal Government: A History of Policies and Activities to 1940* (1957; repr., New York: Arno Press, 1980), chaps. 3–6; and Robert Bruce, *The Launching of Modern American Science* (New York: Alfred A. Knopf, 1987).

3. Melsheimer to Harris, November 24, 1842, Thaddeus William Harris Notebooks 12, BSNH Records, MSB; "Entomological Society of Pennsylvania," *AJS* 44 (1843), 199–200; A. G. Wheeler, Jr., and Karl Valley, "A History of the Entomological Society of Pennsylvania, 1842–1844 and 1924–Present," *Melsheimer Entomological Series* 24 (1978), 16–17.

4. LeConte to Harris, August 4, 1840, APS film 554.1–2.

5. Harris to Samuel S. Haldeman, November 15, 1842, LeConte Coll., APS.

6. Harris to LeConte, November 16, 1842, in ibid.

7. See Diane Lindstrom, *Economic Development in the Philadelphia Region, 1810–1850* (New York: Columbia University Press, 1978), for an account of the region of most importance to the entomological society.

8. The influence of transportation is apparent in letters of the members found in the Samuel S. Haldeman Collection, Library, ANSP. See also George Rogers Taylor, *The Transportation Revolution, 1815–1860* (New York: Holt, Rinehart and Winston, 1951), 141–44 (passengers), 149–51 (mail), and 139–40 (express).

9. George Daniels notes that cultural patriotism reached its highest peak of excitement in this period. See Daniels, *American Science in the Age of Jackson*, 42; Bruce, *Launching of Modern American Science*, 25–27.

10. Wheeler and Valley, "A History," 16.

11. Haldeman to Harris, October 31, 1842, Harris Notebooks 14, MSB.
12. Ibid. In 1847, Haldeman stated that the entomological society was meeting irregularly. Haldeman to LeConte, December 3, 1847, LeConte Coll., APS. Although there seem to be no references to meetings after 1847, the main project of the society, the Melsheimer catalog, was in progress until the early 1850s.
13. Mallis, *American Entomologists*, 34–35.
14. Haldeman to Harris, October 31, 1842, Harris Notebooks 14, MSB.
15. Mallis, *American Entomologists*, 34–35.
16. Melsheimer to [Haldeman], June 24, 1842, Haldeman Coll., ANSP.
17. John G. Morris, "Contributions Toward a History of Entomology in the United States," *AJS* 51, n.s. 1 (January 1846), 27; Mallis, *American Entomologists*, 12–13.
18. Mallis, *American Entomologists*, 284–85; Morris to Baird, Friday [?], 1859, OIAS, SIA.
19. Morris, "History of Entomology"; Morris to Professor Johnson, March 22, 1844, Official Correspondence, ANSP (hereafter cited as Off. Corr., ANSP).
20. Morris to Haldeman, Melsheimer, Ziegler, and Baird, [July 1846], Haldeman Coll., ANSP.
21. A. G. Wheeler, Jr., "Rev. Daniel Ziegler, D.D.: A Promoter of Entomology in Mid-Nineteenth Century America," *Melsheimer Entomological Series* 28 (1980), 21–26.
22. Ibid., 24.
23. Ziegler letters in Haldeman Coll., ANSP; D. Ziegler, "Descriptions of New North American Coleoptera," *Proc., ANSP* 2 (1844–1845), 43–47, 266–72; Director's Report, *Sixth Annual Report, Museum of Comparative Zoology* in *Harvard Museum of Comparative Zoology Reports, 1861–1874* (Cambridge, Mass.: Welsh, Bigelow, 1871–1873), 30 and 35–36 (hereafter cited as *MCZ Reports*).
24. A. G. Wheeler, Jr., "The Rev. Solomon Oswald, Forgotten Member of the Original Entomological Society of Pennsylvania," *Melsheimer Entomological Series* 30 (1981), 43–46.
25. Except where noted, the account of LeConte is based on Samuel H. Scudder, "John Lawrence LeConte, 1825–1883," National Academy of Sciences, *Biographical Memoirs*, vol. 2 (Washington, D.C.: National Academy of Sciences, 1886), 261–93, and Mallis, *American Entomologists*, 242–48.
26. LeConte to Haldeman, July 31, 1844; October 24, 1846; January 2, 1849, and July 21, 1851, Haldeman Coll., ANSP; LeConte to Joseph Henry, August 18, 1852 [not sent?], LeConte Coll., APS. In 1844, he planned to go to Oregon and in 1845 to the upper Missouri by means of a Chouteau fur trading boat, but apparently neither plan was carried out. LeConte to Haldeman, April [?] 1844 and March 14, 1845, in Haldeman Coll., ANSP.
27. Morris to Melsheimer, May 14, 1844, MCZ Archives.
28. Harris to Melsheimer, November 26, 1835, MCZ Archives.
29. Haldeman to Harris, October 31, 1842, Harris Notebooks 14, MSB.
30. Melsheimer to Harris, December 7, 1844, Harris Notebooks 12, MSB.
31. Herman A. Hagen, "Dr. Christian Zimmermann," *CE* 21 (1889), 54.
32. Hagen, "Zimmermann," *CE* 21 (1889), 54.
33. Pickering to Harris, February 18, 1836, Harris Notebooks 12, MSB.

34. Morris to Harris, February 17, 1840, and February 13, 1841, MCZ Archives; Zimmermann to J. E. LeConte, April 5, 1843, LeConte Coll., APS.
35. Harris to J. E. LeConte, April 5, 1843, LeConte Coll., APS.
36. Morris to Harris, February 13, 1841, MCZ Archives.
37. LeConte to Haldeman, May 20, 1849, Haldeman Coll., ANSP.
38. Schaum to Haldeman, January 11, 1848, in ibid.
39. Mallis, *American Entomologists*, 405–408.
40. See Hentz correspondence in Harris, *Entomological Correspondence*.
41. The publication of Hentz's *The Spiders of the United States* was delayed until 1875 because of the expense of printing the illustrations. Mallis, *American Entomologists*, 408.
42. Hentz to Haldeman, October 22, 1842, and January 27, 1842, Haldeman Coll., ANSP. The attitude of entomologists like Zimmermann and Hentz, who felt isolated from their entomological brethren while living in the South, should be balanced against those of other scientists, particularly geologists and botanists, who found the South more congenial to their interests. See James X. Corgan, ed., *The Geological Sciences in the Antebellum South* (University: University of Alabama Press, 1982), especially the work of Gerhard Troost, Joseph Nicollet, and David Dale Owen, pp. 48–64, 106–15, and 122–25; and Tamara Miner Haygood, *Henry William Ravenel, 1814–1887: South Carolina Scientist in the Civil War Era*, History of American Science and Technology Series (Tuscaloosa: University of Alabama Press, 1987).
43. John G. Morris, "Address by Rev. John G. Morris, Chairman of the Subsection of Entomology," *Proc., AAAS* 30 (1881), 262.
44. Ibid.
45. Morris, "History of Entomology," 27.
46. LeBaron to Harris, April 21, 1841, MCZ Archives.
47. Haldeman wrote to Fitch, inquiring about his work, but the contact was not continued. Fitch to Haldeman, November 14, 1849, Haldeman Coll., ANSP; Jeffrey K. Barnes, *Asa Fitch and the Emergence of American Entomology: With an Entomological Bibliography and a Catalog of Taxonomic Names and Type Specimens*, New York State Museum Bulletin 461 (Albany: State Education Department, University of the State of New York, 1988), 49.
48. The geologists, who organized in 1840, had a similar restriction on membership. See Kohlstedt, *American Scientific Community*, 66–67.
49. Morris, "History of Entomology," 26–27.
50. Fitch apologized that the insects in the New York State Cabinet were not all correctly labeled. Haldeman may have interpreted this statement as a lack of systematic rigor on the part of Fitch and may thus have broken off contact with him. Fitch to Haldeman, November 14, 1849, Haldeman Coll., ANSP.
51. Haldeman to Harris, October 31, 1842, Harris Notebooks 14, MSB.
52. Melsheimer to Harris, November 24, 1842, Harris Notebooks 12, MSB.
53. Baird to Haldeman, November 24, 1849, Haldeman Coll., ANSP; William Healey Dall, *Spencer Fullerton Baird: A Biography* (Philadelphia: J. B. Lippincott, 1915), 81–82, 115–16; and E. F. Rivinus and E. M. Youssef, *Spencer Baird of the Smithsonian* (Washington, D.C.: Smithsonian Institution Press, 1992), 27–35.
54. LeConte to Haldeman, March 14, 1847, Haldeman Coll., ANSP.
55. Gerstner, "Academy of Natural Sciences," 184–87.

56. Haldeman's quarrel with Audubon on this matter is documented in his correspondence with Augustus A. Gould, who supported Haldeman. See especially Gould to Haldeman, December 14, 1844, Haldeman Coll., ANSP.
57. Baird's congenial relationship with Audubon is represented in Alice Ford, *John James Audubon* (Norman: University of Oklahoma Press, 1964), 200, 279, 371, 379, and 407; and Rivinus and Youssef, *Spencer Baird,* 30–31 and 42–43.
58. Harris to J. E. LeConte, February 19, 1830, LeConte Coll., APS. On this point see also John C. Greene, "American Science Comes of Age, 1780–1820," *Journal of American History* 55 (June 1968), 35.
59. Morris to Haldeman, October 30, 1843, Haldeman Coll., ANSP.
60. *Journal of the Boston Society of Natural History* 5 (1845), quoted in Weiss, *Pioneer Century,* 155–56. The phrase "America's Entomological Declaration of Independence" comes from R. P. Dow, "The Greatest Coleopterist," *JNYES* 22 (September 1914), 185; George Daniels notes the frequent calls for a declaration of scientific independence from various American scientists in the 1840s. See Daniels, *American Science in the Age of Jackson,* 46.
61. Melsheimer to Haldeman, March 27, 1844, Haldeman Coll., ANSP.
62. Ibid.
63. Ibid.
64. Melsheimer to Harris, November 24, 1842, Harris Notebooks 12, MSB.
65. Melsheimer to Haldeman, March 27, 1844, Haldeman Coll., ANSP.
66. Harris to Melsheimer, December 21, 1842, MCZ Archives; LeConte to Haldeman, March 14, 1848, Haldeman Coll., ANSP.
67. Haldeman to Harris, October 31, 1842, Harris Notebooks 14, MSB.
68. F. E. Melsheimer, Carl Ziegler, and John L. LeConte, "Descriptions of New Species of Coleoptera of the United States," *Proc., ANSP* 2 (October 1844), 98–118; (November 1844), 134–60; (March 1845), 213–23; (December 1845), 302–18; 3 (May 1846), 53–56; (February 1847), 158–81.
69. Haldeman to [?], March 25, 1844, Off. Corr., ANSP.
70. Schaum to LeConte, January 21, 1848, LeConte Coll., APS.
71. Ibid.
72. Melsheimer to Haldeman, September [1849?], Haldeman Coll., ANSP.
73. Joseph Henry to LeConte, August 2, 1852, LeConte Coll., APS.
74. Frederick Ernst Melsheimer, *Catalog of the Described Coleoptera of the United States,* revised by S. S. Haldeman and J. L. LeConte, Smithsonian Publication 62 (Washington, D.C.: Smithsonian Institution, 1853).
75. Schaum to Haldeman (referring to the plan for LeConte and Haldeman to edit the manuscript catalog), January 11, 1848, Haldeman Coll., ANSP.
76. Baird to Scudder, May 28, 1857, Samuel H. Scudder Correspondence, MSB; Weiss, *Pioneer Century,* 216. On the need for such catalogs, see Samuel Stehman Haldeman, "Report on the Progress of Entomology in the United States during 1849," *Proc., ANSP* 5 (1850–1851), 5. Later Smithsonian monographs include works by H. A. Hagen and P. R. Uhler on the Neuroptera, Baron Carl Robert Osten-Sacken on the Diptera, John G. Morris on the Lepidoptera, and John L. LeConte on the Coleoptera. Baird to Scudder, May 28, 1857, Scudder Corr., MSB.
77. Baatz, "Patronage, Science, and Ideology," iv, 65, and 74. With respect to lack of interest in publishing proceedings and in securing wealthy patrons the

members contrasted greatly with members of the Entomological Society of Philadelphia (discussed in Chapter 3), who sought both and who achieved permanence as a society.

78. Melsheimer to Haldeman, September [1849], Haldeman Coll., ANSP.
79. Haldeman to Baird, January 14, 1854, OIAS 26:323, SIA.
80. Gerstner, "Academy of Natural Sciences," 180–81; Morris delivered public lectures on natural history in 1839, but the society did not sponsor similar lectures. Carl Bode, *The American Lyceum: Town Meeting of the Mind* (1956; repr., Carbondale: Southern Illinois University Press, Arcturus Books, 1968), 72.
81. *PF* 7 (May 18, 1847), 146, reprinted from the *Quarterly Journal of Agriculture*.
82. Morris to Haldeman, Melsheimer, Ziegler, and Baird, [July 1846], Haldeman Coll., ANSP.
83. Melsheimer to Harris, September 23, 1838, MCZ Archives.
84. James M. Clarke, *James Hall of Albany, Geologist and Paleontologist* (Albany, N.Y.: E. E. Rankin, 1921), 237–38; Barnes, *Asa Fitch*, 40, 45.
85. LeConte to Haldeman, February 24, 1849, Haldeman Coll., ANSP. Haldeman insisted, however, that the entomological portion of the New York survey should continue. Haldeman to Erastus Vans and Harvey Coryell, April 23, 1849, [not sent?], Haldeman Coll., ANSP.
86. Gerstner, "Academy of Natural Sciences," 183.
87. Baatz, "Patronage, Science, and Ideology," 1–41 and 197–245.
88. Emmons was an example of what Alexander D. Bache called "modified charlatanism," whereby an authority in one field claimed competence in another. See Kohlstedt, *American Scientific Community*, 160.
89. LeConte to Haldeman, February 24, 1849, Haldeman Coll., ANSP. The tension between "amateurs" and "professionals" with respect to the issue of American scientific credibility in Europe is discussed in Gerstner, "Academy of Natural Sciences," 182–89; Kohlstedt, *American Scientific Community*, 79, 132, and 155–58; and John D. Holmfeld, "From Amateurs to Professionals in American Science: The Controversy over the Proceedings of an 1853 Scientific Meeting," *Proc., APS* 114 (February 1970), 35–36.
90. *PF* 13 (November 1853), 421.
91. LeConte to Haldeman, June 10, 1849, Haldeman Coll., ANSP.
92. Harris to Morris, September 17, 1839, MCZ Archives.
93. Charles E. Rosenberg, *No Other Gods: On Science and American Social Thought* (Baltimore, Md.: Johns Hopkins University Press, 1976), 136.
94. Morris to Stauffer, January 27, 1858, Jacob Stauffer Papers, APS.
95. See A. Hunter Dupree, "National Pattern of Learned Societies," in Oleson and Brown, eds., *The Pursuit of Knowledge*, 21.
96. Ibid., 24.
97. Alexandra Oleson, "Introduction," in Oleson and Brown, eds., *The Pursuit of Knowledge*, xvi.
98. Melsheimer to Haldeman, September 23, 1844, Haldeman Coll., ANSP.
99. LeConte to Haldeman, July 31, 1844, and May 12, 1846; Melsheimer to Haldeman, October 21, 1845, and November 9, 1845, in ibid.
100. LeConte to Haldeman, November 20, 1847, and Melsheimer to LeConte, November 23, 1847, in ibid.
101. LeConte to Melsheimer, December 1, 1848, LeConte Coll., APS.

102. S. S. Haldeman, "Materials Towards a History of Coleoptera Longicornia of the United States," *Trans., APS*, n.s. 10 (1853), 28. He thanked the LeContes, Melsheimer, and Hentz, whose specimens he examined.
103. LeConte to Haldeman, December 20, 1847, Haldeman Coll., ANSP.
104. G. H. Horn, "John Lawrence LeConte," *Proc., American Academy of Arts and Sciences* 19 (1884), 513.
105. LeConte (quoting Agassiz) to Haldeman, December 30, 1847, Haldeman Coll., ANSP.
106. Dupree, "National Pattern of Learned Societies," 25.
107. LeConte to Haldeman, May 14, 1844; Melsheimer to [Haldeman], June 24, 1842, Haldeman Coll., ANSP.
108. Morris to Harris, February 17, 1840, MCZ Archives.
109. Examples of such correspondence are LeConte to Haldeman, November 20, 1847, Haldeman Coll., ANSP; LeConte to Haldeman, December 15, 1846, in ibid.; Schaum to Morris, February 21, 1848, LeConte Coll., APS; and Melsheimer to J. E. LeConte, July 20, 1843, in ibid.
110. Haldeman to Harris, October 31, 1842, Harris Notebooks 14, MSB.
111. Hentz to Haldeman, October 22, 1842, Haldeman Coll., ANSP.
112. Schaum to LeConte, January 21, 1848, LeConte Coll., APS.
113. Kohlstedt, *American Scientific Community*, 139.
114. Morris, "History of Entomology," 27.
115. Haldeman to [?], March 25, 1844, Off. Corr., ANSP.
116. In 1851, the AAAS required prior submission of articles and abstracts and placed the final authority over publication in the *Proceedings* with the permanent secretary. See Kohlstedt, *American Scientific Community*, 139–40.
117. Dupree, "National Pattern of Learned Societies," 24.
118. LeConte to Haldeman, May 20, 1849, Haldeman Coll., ANSP.
119. See, for example, C. Zimmermann, "Synopsis of the Scolytidae of America North of Mexico, with notes . . . by J. L. LeConte," *Trans., AES* 2 (1868–1869), 141.
120. James Thompson to LeConte, October 2, 1848, LeConte Coll., APS.
121. For an evaluation of LeConte by the contemporary European coleopterist Hermann Schaum, see Horn, "John Lawrence LeConte," 511; for a recent evaluation, see Carl H. Lindroth, "Systematics Specializes," 130. The figures come from C. L. Marlatt, "A Brief History of Entomology," *Proc., ESW* 4 (February 1897), 116; Scudder, "John Lawrence LeConte," 280; and Dow, "The Greatest Coleopterist," 189.
122. Morris to Baird, February 15, 1859, OIAS 17:481, SIA.
123. Morris, "Address," 261–66.
124. See Chapter 3.

3. Of Cabinets and Collections

1. Morris to Haldeman, Ziegler, Melsheimer, Baird [July 1846], Haldeman Coll., ANSP; Ursula Göllner-Scheidung, "Zur Geschichte der Entomologie in Berlin mit besonderer Berücksichtigung des 19. Jahrhunderts," *Wissenschaftliche Zeitschrift der Humboldt-Universität zu Berlin, Mathematische-Naturwissenschaftliche Reihe* 34:3 (1985), 311.

2. Edwards to Lintner, January 6, 1877, Joseph A. Lintner Correspondence, State Geologist's Correspondence Files (series BO561), NYSA.
3. LeConte to Haldeman, February 6, 1849, and March 25, 1849, Haldeman Coll., ANSP.
4. LeConte to Haldeman, January 2, 1849, Haldeman Coll., ANSP.
5. LeConte to Baird, May 18, 1865, OIAS 30:231, SIA.
6. Gustav Wilhelm Belfrage Papers, SIA.
7. Obituaries of Morrison appeared in *Psyche* 4 (June 1885), 287 and *EA* 1 (August 1885), 100; another prominent collector was J. Böll, from Bremgarten, Switzerland, who also collected in Texas. *MCZ Reports* (1870), 22–24.
8. Packard to Belfrage, November 5, 1868, Belfrage Papers, SIA; LeConte also paid George Engelmann five cents per specimen. LeConte to Haldeman, December 18, 1847, Haldeman Coll., ANSP.
9. LeConte to Haldeman, July 21, 1851, Haldeman Coll., ANSP.
10. John A. Moore, "Zoology of the Pacific Railroad Surveys," *American Zoologist* 26 (1986), 335.
11. A. S. Packard, Jr., "Introduction," *Record of American Entomology for the Year 1868* (Salem, Mass.: Naturalist's Book Agency, Essex Institute Press, 1869). The *Record of American Entomology for the Year 1868* (1869, 1870, etc.), edited by A. S. Packard, appeared annually for the years of 1868–1873, with the year of publication always one year later ("for 1868" published in 1869, etc.). Packard counted forty-five publishing authors in 1868, fifty-two in 1869, and thirty-five in 1870, for an average of forty-four for these three years.
12. In 1878, the editor of *Psyche* estimated the total number of American entomologists to be 762. A year later, the editor of the *Canadian Entomologist* thought this estimate should be revised upward to 835. *CE* 11 (September 1879), 164. See the quantitative analysis of American entomologists in about 1870 in Chapter 9.
13. *AE* 2 (May 1870), 199.
14. Herman A. Hagen, "The History of the Origin and Development of Museums," *AN* 10 (February 1876), 86.
15. See Paul L. Farber, "The Type Concept in Zoology During the First Half of the Nineteenth Century," *JHB* 9 (Spring 1976), 93–119. Farber identifies three distinct ways in which the type concept was used: the classification type concept, the collection type concept, and the morphological type concept. Here the collection type concept is most important. See especially pp. 96–100.
16. Weiss, *Pioneer Century*, 277.
17. Osten-Sacken to Scudder, January 17, 1859, Scudder Corr., MSB.
18. Herman A. Hagen, "The Melsheimer Family and the Melsheimer Collection," *CE* 16 (1884), 195–96.
19. Uhler to Edwards, January 31, 1863, and Bickmore to Edwards, February 20, 1872, William Henry Edwards Correspondence in William Henry Edwards Collection, West Virginia and Regional History Collection, WVUL (microfilm from originals in WVSA).
20. Edwards to Lintner, December 18, 1861, and February 7, 1862, Lintner Corr., NYSA.

21. Paul L. Farber, "The Development of Taxidermy and the History of Ornithology," *Isis* 68 (December 1977), 550–56; Farber, "Type Concept," 97–98.
22. Morris to Haldeman, Melsheimer, LeConte, Baird [July 1846], Haldeman Coll., ANSP.
23. Alpheus Spring Packard, Jr., *Guide to the Study of Insects, and a Treatise on those injurious and beneficial to crops; for the use of Colleges, Farm Schools, and Agriculturalists* (Salem, Mass.: Naturalist's Book Agency, 1869), 90–92; James T. Bell, "How to Destroy Cabinet Pests," *CE* 9 (1877), 139–40.
24. Packard, *Guide*, 90.
25. John L. LeConte, "On the Preservation of Entomological Cabinets," *AN* 3 (August 1869), 307–309.
26. LeConte to Alexander Agassiz, April 28, 1875, quoted in seventeenth *MCZ Report* (1875), 36.
27. Packard, *Guide*, 92.
28. The history of the collection is based on Fitch to Haldeman, November 14, 1849, Haldeman Coll., ANSP; Mallis, *American Entomologists*, 40; *Trans., NYSAS* 12 (1852), 13; Clarke, *James Hall*, 382–86; George P. Merrill, ed., *Contributions to a History of American State Geological and Natural History Surveys,* U.S. National Museum Bulletin 109 (Washington, D.C.: Smithsonian Institution, 1920), 331–35, 355–60; Lintner to Scudder, April 19 and May 10, 1869, Scudder Corr., MSB; and Barnes, *Asa Fitch,* 41–42.
29. Joseph A. Lintner and F. G. Sanborn, "An Account of the Collections which Illustrate the Labors of Dr. Asa Fitch," *Psyche* 2 (September–December 1879), 274–75.
30. See Chicago Academy of Sciences to Secretary, Smithsonian Institution, October 30, 1871, OIS 2:715, SIA.
31. Rivinus and Youssef, *Spencer Baird,* 61–67, 81–97; William A. Deiss, "Spencer F. Baird and His Collectors," *JSBNH* 9:4 (1980), 635 and 638; Dall, *Baird,* especially 248–49, 255, and 313; William H. Goetzmann, *Exploration and Empire: The Explorer and the Scientist in the Winning of the American West* (New York: Alfred A. Knopf, 1971), 323 and passim.
32. LeConte to Haldeman, March 25, 1849, Haldeman Coll., ANSP.
33. Baird to Haldeman, March 10 and March 20, 1857, in ibid.
34. John L. LeConte, "Report Upon the Insects Collected on the Survey," in I. I. Stevens, *Narrative and Final Report of Explorations . . . for a Pacific Railroad near the Forty-Seventh and Forty-Ninth Parallels,* in U.S. Corps of Topographical Engineers, *Reports of Explorations and Surveys to Ascertain the Most Practicable and Economical Route for a Railroad from the Mississippi River to the Pacific Ocean,* vol. 12, pt. 3 (Washington, D.C.: U.S. Government Printing Office, 1860), 1–72. See also Moore, "Zoology of the Pacific Railroad Surveys," 331–36.
35. LeConte to Baird, May 10, 1859, OIAS 17:309, SIA.
36. LeConte to Haldeman, March 25, 1849, Haldeman Coll., ANSP.
37. Baird to Haldeman, February 3, 1850, in ibid.
38. John L. LeConte, "Descriptions of New Coleoptera from Texas, Chiefly Collected by the U.S. Boundary Commission," *Proc., ANSP* 6 (December 27, 1853), 439.
39. LeConte to Haldeman, February 6, 1849, Haldeman Coll., ANSP.

40. See H. W. Howgate to Baird, May 1, 1874, LeConte Coll., APS; and Le-Conte to George Davidson, April 4, 1859, George Davidson Collection (C-B 940), Box 15, Bancroft Library, University of California, Berkeley.
41. LeConte to Sergeant Feliner, March 26, 1859, OIAS 17:306, SIA.
42. LeConte to Baird, July 11, 1872, OIAS 6:523, SIA.
43. Goetzmann, *Exploration and Empire*, 474–75, 495–96, 497, 502, 527–28.
44. Mike Foster, "Ferdinand Vandiveer Hayden as Naturalist," *American Zoologist* 26 (1986), 343.
45. Powell to Carl Schurz, August 1, 1878, NA, RG 48, Records of the Secretary of the Interior, Letters Received Relating to the U.S. Entomological Commission, 1877–1880.
46. LeConte to Hayden, February 14, 1879, and March 3, 1879, in NA, RG 57, Records of the Geological Survey, Hayden Survey Records, Letters Received (1871–1879) and Personal Letters Received (1872–1879), microfilm, University of California, Berkeley, library. For other entomologists who supported Hayden, see Packard to Hayden, December 8, 1878, Personal Letters, December 1, 1873, Letters Received, and Thomas to Hayden, June 6, 1873, Letters Received, in ibid.
47. This thesis is developed in Joel J. Orosz, "Disloyalty, Dismissal, and a Deal: The Development of the National Museum at the Smithsonian Institution, 1846–1855," *Museum Studies Journal* 2:2 (1986), 22–33. See also Deiss, "Spencer F. Baird," 635 and 637–38; Dall, *Baird*, 248, 257–58, 313; and Nathan Reingold, ed., *Science in Nineteenth Century America: A Documentary History*, American Century Series (New York: Hill and Wang, 1964), 153.
48. LeConte to Baird, November 7, 1859, OIAS 17:318, SIA. On the family relationship, see Deiss, "Spencer F. Baird," 636, and LeConte to Baird, March 6, 1860, OIAS 22:276, and November 21, 1860, 22:285, SIA, in which LeConte asks Baird for information about the LeConte family.
49. LeConte to Baird, November 7, 1859, OIAS 17:318, SIA.
50. LeConte to Baird, April 2, 1861, OIAS 22:291, SIA. See also LeConte to Baird, March 12, 1859, OIAS 17:304, SIA; Cresson to Baird, July 16, 1860, OIAS 20:238; and Cresson to Baird, October 21, 1863, OIAS 25:387, in SIA.
51. LeConte to Baird, March 12, 1859, OIAS 17:304, SIA; Cresson to Baird, July 16, 1860, OIAS, 20:238; and Cresson to Baird, October 21, 1863, OIAS 25:387, SIA.
52. Cresson to Baird, March 7, 1859, OIAS 16:431, SIA; Ezra Townsend Cresson, *A History of the American Entomological Society, Philadelphia, 1859–1909* (Philadelphia: [American Entomological Society, 1911]), 13.
53. Constitution of the Entomological Society of Philadelphia, ESP Collection, Library, ANSP.
54. Cresson, *American Entomological Society*, 8
55. Sam Bass Warner, Jr., *The Private City: Philadelphia in Three Periods of Its Growth* (Philadelphia: University of Pennsylvania Press, 1968), 61–67, 82–84.
56. M. J. James, *The New Aurelians: A Centenary History of the British Entomological and Natural History Society, 1873–1973* (London: British Entomological and Natural History Society, 1973), 3–5.
57. Mallis, *American Entomologists*, 343–48.

58. Cresson to Baird, October 12, 1858, OIAS 10:270, and October 14, 1858, 10:271, SIA.
59. Cresson to Baird, March 9, 1857, OIAS 10:265, and April 8, 1857, 10:266, SIA.
60. Cresson to Entomological Society, May 9, 1859, ESP Coll., ANSP.
61. Cresson to Baird, December 13, 1859, OIAS 16:433, SIA.
62. Cresson to Baird, March 7, 1859, OIAS 16:431, SIA.
63. Minutes, AES, June 27, 1859, AES Collection, Library, ANSP.
64. Cresson to Baird, April 11, 1860, OIAS 20:235, SIA.
65. Ibid.; Minutes, AES, June 27, 1859, AES Collection, ANSP. LeConte's estimation of his own social standing may be inferred from the indignant response of his father, Major John Eatton LeConte, to Baird's suggestion that his son enlist as a sailor to go to South America: "Who has the arrangement of this business? That he would . . . propose to a person of good education and polished manners, to go as a common sailor? The proposition I consider as insulting to myself. It might as well be proposed [that I] enter as cook!" John E. LeConte to Baird, January 10, 1853, OIAS 12:269, SIA.
66. Cresson, *American Entomological Society,* 22–23.
67. Ezra Townsend Cresson, "Memoir of Thomas B. Wilson," *Proc., ESP* 5 (1865), 35–36.
68. Gerstner, "Academy of Natural Sciences," 188–89.
69. LeConte to Baird, January 10, 1857, OIAS 12:276, SIA.
70. LeConte to Baird, January 25, 1857, OIAS 12:277, SIA.
71. Wilson to Leidy, January 31, 1859, Thomas B. Wilson Collection, Library, ANSP.
72. Wilson to B. Howard Rand, January 1, 1864, and June 28, 1864, Wilson Coll., ANSP.
73. Cresson to Scudder, February 4, 1863, Scudder Corr., MSB.
74. Cresson to Baird, April 11, 1860, OIAS 20:235, SIA.
75. LeConte to Baird, December 8, 1859, OIAS 17:324, SIA.
76. John Pearsall to Entomological Society [1862], ESP Coll., ANSP.
77. LeConte to Baird, January 16, 1860, OIAS 22:270, SIA.
78. Cresson to Entomological Society, February 13, 1860, ESP Coll., ANSP.
79. Cresson to Baird, April 11, 1860, OIAS 20:235, SIA.
80. Ibid.
81. Charges of theft of specimens recurred among the entomologists. LeConte had earlier suspected someone of stealing from him. Other charges of theft appear in Augustus Radcliffe Grote to Scudder, December 6, 1875, Scudder Corr., MSB, and Walsh to Scudder, August 25, 1865, in ibid.
82. Cresson to Baird, April 11, 1860, OIAS 20:235, SIA.
83. Ibid.
84. Henry to Cresson, April 17, 1860, ESP Coll., ANSP.
85. Cresson to Baird, July 5, 1860, OIAS 20:236, SIA.
86. LeConte to [Knight?], November 26, 1860, ESP Coll., ANSP.
87. "Directions for Collecting and Preserving Insects," 1861, OIAS 20:243, SIA.
88. LeConte to Baird, July 10, 1861, OIAS 22:298, SIA.
89. LeConte to Belfrage, March 31, 1869, Belfrage Papers, SIA.
90. LeConte to Edwards, March 23, 1862, Edwards Corr., WVUL.

91. Quoted in Minutes, AES, January 13, 1862, AES Coll., ANSP.
92. Ibid.
93. LeConte to Baird, January 20, 1862, OIAS 22:300, SIA.
94. Cresson to Baird, February 29, 1872, OIS 109:348, SIA. The squabble among the entomologists bore some similarity to the competition among the paleontologists in the 1870s when Othaniel Charles Marsh and Edward Drinker Cope sought exclusive rights to paleontological collecting sites and specimens. See Nathan Reingold, "Cope and Marsh: The Battle of the Bones," in Reingold, ed., *Science in Nineteenth Century America*, 236–50.
95. Cresson to Baird, February 10, 1864, OIAS 25:387, and April 1, 1864, OIAS 25:393, SIA.
96. Uhler to Edwards, January 31, 1863, Edwards Corr., WVUL; *Proc., ESP* 2 (December 1863), 314, 3 (November 1864), 437, 440, and 4 (April 1865), 242; Minutes, ESP, April 9, 1866, printed in *Proc., ESP* 6 (1866–1867), v.
97. Cresson, *American Entomological Society*, 23.
98. American Entomological Society to Academy of Natural Sciences of Philadelphia, Petition to Be Made a Section, March 13, 1876, AES Collection, ANSP.
99. Cresson to Scudder, May 22, 1865, Scudder Corr., MSB.
100. Edward Lurie, *Louis Agassiz: A Life in Science* (Chicago: University of Chicago Press, 1960), chap. 1.
101. Ibid., 127 and passim.
102. Lurie, *Agassiz*, passim; *An Account of the Organization and Progress of the Museum of Comparative Zoology at Harvard College, in Cambridge, Massachusetts* (Cambridge, Mass.: Welch, Bigelow, 1871), p. 5 (hereafter cited as *Org. of MCZ*); Agassiz's fund raising is discussed in Howard S. Miller, *Dollars for Research: Science and Its Patrons in Nineteenth Century America* (Seattle: University of Washington Press, 1970), chap. 4, "The Personal Equation."
103. *Org. of MCZ*, 18.
104. Director's Report for 1859, *MCZ Report* (1860), 33.
105. Director's Report for 1860, *MCZ Report* (1861), 5.
106. Ibid.
107. *MCZ Report* (1864), 9, and (1866), 24.
108. Considering Agassiz's announced intention not to compete with other American museums in departments in which those museums were already strong (e.g., the Smithsonian and the Academy of Natural Sciences in mammals and birds), it is an interesting question whether he would have pursued entomology equally avidly if the Philadelphians' plans had materialized; *MCZ Report* (1866), 10. Riley later indicated that LeConte decided against giving his collection to the Entomological Society because the society was not prepared to give it proper care; Charles V. Riley, "Tribute to the Memory of John Lawrence LeConte," *Psyche* 4 (November–December 1883), 109–10; *MCZ Report* (1867), 9.
109. *MCZ Report* (1868), 10–11, 28–31.
110. Ibid. (1864), 30, (1870), 21, (1871), 26–27, (1875), 8.
111. Ibid. (1869), 22, (1870), 23, (1871), 26–27, and (1872), 25–26.
112. Carl Robert Osten-Sacken, *Record of My Life Work in Entomology*, with an appreciation and introductory preface by K. G. V. Smith (1903–1904; repr., Faringdon, England: E. W. Classey, 1977), [2–4], 4, 29, 77.

113. Mallis, *American Entomologists,* 185–91; *Psyche* 2 (January–February 1878), 97.
114. *Psyche* 1 (May 1874), 1.
115. J. R. Matthews, "History of the Cambridge Entomological Club," *Psyche* 81 (March 1974), 17–21, 31–33.
116. Baird to Scudder, April 7, 1859, Scudder Corr., MSB.
117. Baird to Scudder, January 11, 1863, in ibid.
118. Townend Glover wrote to Henry, in 1867, noting that the Smithsonian had no regular entomological collection and requesting therefore that some European insects just received be sent to the Department of Agriculture. Glover to Henry, July 23, 1867, OIS 69:508, SIA.
119. Remington Kellogg, "A Century of Progress in Smithsonian Biology," *Science* 104 (August 9, 1946), 132–41.
120. Baird to LeConte, December 26, 1881, John L. LeConte Coll., Library, ANSP.
121. The account of Glover is based on Charles Richards Dodge, "The Life and Entomological Work of the Late Townend Glover, first Entomologist of the U.S. Department of Agriculture," U.S. Department of Agriculture, Division of Entomology, *Bulletin* 18 (Washington, D.C.: U.S. Government Printing Office, 1888), quotation on p. 14.
122. Ibid., 14.
123. *Psyche* 1 (October 1876), 201–203.
124. Riley's career is discussed more fully in Chapter 5.
125. Riley to Baird, October 23, 1885, USNM, Accession 10, SIA.
126. F. W. True, "An Account of the U.S. National Museum," in George Brown Goode, ed., *The Smithsonian Institution, 1846–1896: The History of its First Half Century* (Washington, D.C.: Smithsonian Institution, 1897), 330. John Henry Comstock, during his brief tenure as federal entomologist, 1879–1881, carried forward the plan for establishing a national collection of insects; *CE* 11 (November 1879), 202–203.
127. Baird to LeConte, December 26, 1881, John L. LeConte Coll., ANSP.
128. Riley to Baird, October 23, 1885, USNM, Accession 10, SIA.
129. Ibid.
130. Ibid.; *Psyche* 3 (February 1882), 315.
131. The latter included Belfrage's American collection; *EA* 4 (October 1888), 130–31.

4. Agricultural Entomologists and Institutions

1. On the shift in scientific personnel, see Everett Mendelsohn, "The Emergence of Science as a Profession in Nineteenth-Century Europe," in Karl B. Hill, ed., *The Management of Scientists* (Boston: Beacon Press, 1964), 3–5, and Maurice Crosland, "The Development of a Professional Career in Science in France," *Minerva* 13:3 (1975), 38. American developments in agricultural entomology are discussed in Leland O. Howard, "The Rise of Applied Entomology in the United States," *AgH* 3 (July 1929), 131–39, and in early sections of two more recent works that focus primarily on aspects of insect control rather than on professionalization: John H. Perkins, *Insects, Experts, and the Insecticide Crisis: The Quest for New Pest Management Strate-*

gies (New York: Plenum Press, 1982), and James Whorton, *Before Silent Spring: Pesticides and Public Health in Pre-DDT America* (Princeton, N.J.: Princeton University Press, 1974). For the comparison with European developments, see George Ordish, *John Curtis and the Pioneering of Pest Control* (Reading, England: Osprey Press, 1974), and Vincent Köllar, *A Treatise on Insects Injurious to Gardeners, Foresters, and Farmers,* trans. Jane Loudon and Mary Loudon (London: William Smith, 1840).

2. Thomas L. Haskell, "Power to the Experts," *New York Review of Books* (October 13, 1977), 32–33.

3. Unless otherwise noted, biographical information is based on Mallis, *American Entomologists.* On the background of American scientists, see Donald Beaver, *The American Scientific Community: A Historical-Statistical Study* (New York: Arno Press, 1980), 155–56, and 203–10.

4. Beaver, *American Scientific Community,* 160–68 and 173.

5. Clark A. Elliott, "Sketch of Thaddeus William Harris" (Harvard University, Cambridge, Mass., 1981); Harris, *Report.*

6. LeBaron to Harris, April 21, 1841; Glover to Harris, December 7, 1852; Fitch to Harris, February 23, 1852, MCZ Archives.

7. On the duties of a state entomologist see Charles V. Riley, *Fifth Annual Report on the Noxious, Beneficial, and Other Insects of the State of Missouri* (Jefferson City, Mo.: State Printer, 1873), 27–29, and Walsh in *AE* 2 (May 1870), 197–99.

8. Fitch to Harris, February 26, 1855, MCZ Archives.

9. Arnold Mallis, ed., "The Diaries of Asa Fitch, M.D.," *Bull., ESA* 9 (1963), 264.

10. Ibid., 263.

11. Rosenberg, *No Other Gods,* 137–40.

12. Riley to Hagen, May 17, 1873, MCZ Archives.

13. Walsh to Hagen, September 11, 1868, in ibid.

14. Harris wrote, for example, that flies "subserve [a] highly important purpose, for which an all-wise Providence has designed them, namely . . . furnishing food to . . . other animals" (Harris, *Report,* 16).

15. Walsh to Hagen, September 11, 1868, MCZ Archives.

16. Edward H. Smith, "The Entomological Society of America: The First Hundred Years, 1889–1989," *Bull., ESA* 35:3 (Fall 1989), 14–15.

17. A. S. Packard, "Injurious and Beneficial Insects," *AN* 7 (September 1873), 525.

18. For example, see Benjamin D. Walsh, *First Annual Report on the Noxious Insects of the State of Illinois* (from the appendix to the *Transactions of the Illinois State Horticultural Society for 1867,* vol. 1, 1868), 104. Also printed as a separate pamphlet (Chicago: Prairie Farmer, 1868).

19. *AE* 2 (June 1870), 232–33.

20. The principal works I draw upon are George Rogers Taylor, "The National Economy Before and After the Civil War," in David Gilchrist and W. David Lewis, eds., *Economic Change in the Civil War Era: Proceedings of a Conference on American Economic Institutional Change, 1850–1873, and the Impact of the Civil War, [held] March 12–14, 1964* (Greenville, Del.: Eleutherian Mills-Hagley Foundation, 1965), 2–3; Clarence H. Danhof, *Agricultural Change: The Northern United States, 1820–1870* (Cambridge, Mass.: Harvard University Press,

1969), chaps. 1 and 2; Percy Wells Bidwell and John I. Falconer, *Agriculture in the Northern United States, 1620–1860* (1925; repr., Magnolia, Mass.: Peter Smith, 1941), 7; Paul Wallace Gates, *The Farmer's Age: Agriculture, 1815–1860* (New York: Holt, Rinehart and Winston, 1960), 100–102; Morton Rothstein, "Antebellum Wheat and Cotton Exports: A Contrast in Marketing Organization and Economic Development," *AgH* 40 (April 1966), 94; and Wayne Rasmussen, "The Civil War: A Catalyst of Agricultural Revolution," *AgH* 39 (October 1965), 187–95.

21. Bidwell and Falconer, *Agriculture in the Northern United States*, 263–64, 321–30; Paul Wallace Gates, *Agriculture and the Civil War* (New York: Alfred A. Knopf, 1965), 243–44.

22. Allen Bogue, *From Prairie to Corn Belt: Farming on the Illinois and Iowa Prairies in the Nineteenth Century* (Chicago: University of Chicago Press, 1963), 239–40; Gates, *Farmer's Age*, 261–62.

23. V. G. Dethier, *Man's Plague? Insects and Agriculture* (Princeton, N.J.: Darwin Press, 1976), 33–35.

24. Gates, *Farmer's Age*, 168; Bogue, *From Prairie to Corn Belt*, 125–27.

25. Köllar, *Treatise on Insects*, 196; Cyrus Thomas, "Insects Injurious to Vegetation in Illinois," *Trans., ISAS* 5 (1861–1864), 461.

26. Perkins, *Insects, Experts, and the Insecticide Crisis*, 209–12, 215–23.

27. Dethier, *Man's Plague?* 100.

28. Bogue, *From Prairie to Corn Belt*, 215; Bidwell and Falconer, *Agriculture in the Northern United States*, 452.

29. Margaret W. Rossiter, "The Organization of Agricultural Improvement in the United States," in Oleson and Brown, eds., *Pursuit of Knowledge*, 285–91; George Francis Lemmer, "Early Agricultural Editors and Their Farm Philosophies," *AgH* 31 (October 1957), 3.

30. Ulysses Prentiss Hedrick, *History of Horticulture in America to 1860* (New York: Oxford University Press, 1950), 492–510; Hamilton Traub, "Tendencies in the Development of American Horticultural Associations, 1800–1850," *National Horticultural Magazine* 9 (July 1930), 134–40.

31. Walsh to LeConte, February 14, 1869, LeConte Coll., APS.

32. In the 1860s, beef and dairy farmers and wool growers organized effective national lobbies. Their animals represented the same long-term financial commitment as did orchards for the horticulturists (and Walsh considered beef and dairy farmers to be as intelligent and progressive as the fruit growers). Ibid. and Gates, *Agriculture and the Civil War*, 139–41.

33. Walsh to Hagen, October 12, 1868, MCZ Archives; Walsh to Baird, May 27, 1860, USNM, Accession 10, SIA.

34. Fitch to Harris, February 26, 1855, MCZ Archives.

35. Mallis, *American Entomologists*, 45–46; Walsh to Hagen, April 13, 1862, MCZ Archives.

36. Charles V. Riley, *Third Annual Report on the Noxious, Beneficial, and Other Insects of the State of Missouri* (Jefferson City, Mo.: State Printer, 1871), 4.

37. *PE* 1 (August 27, 1866), 10.

38. Fitch to Harris, August 11, 1852, MCZ Archives.

39. *CE* 1 (September 1868), 49. Other estimates appear in *Trans., ISHS* (1865), in *Trans., ISAS* 6 (1865–1866), 333–34; *Trans., ISHS* (1868), in *Trans., ISAS* 7

(1867–1868), 547; *Third Annual Report, MSBA* (1867), vii; and Charles V. Riley, *First Annual Report on the Noxious, Beneficial, and Other Insects of the State of Missouri* (Jefferson City, Mo.: State Printer, 1869).

40. James H. Cassedy, *American Medicine and Statistical Thinking, 1800–1860* (Cambridge, Mass.: Harvard University Press, 1984), 217.

41. James X. Corgan, "Early American Geological Surveys," in Corgan, ed., *Geological Sciences in the Antebellum South*, 39–40.

42. The emphasis on minerals in the state surveys is discussed in Mary C. Rabbitt, *Minerals, Lands, and Geology for the Common Defense and General Welfare, Vol. 1, Before 1879: A History of Public Land, Federal Science and Mapping Policy, and Development of Mineral Resources in the United States* (Washington, D.C.: U.S. Government Printing Office, 1979), 4, 40, 49, and 53.

43. Richard M. Foose and John Lancaster, "Edward Hitchcock: New England Geologist, Minister, and Educator," in William M. Jordan, ed., "History of Geology in the Northeast," *Northeastern Geology* 3:1 (1981), 13–16; Massachusetts, *Reports on the Fishes, Reptiles, and Birds of Massachusetts, By the Commissioners on the Zoological and Botanical Survey of the State* (Boston: Dutton and Wentworth, 1839), iii.

44. Harris's *Report* (1841) may be compared to Köllar, *Treatise on Insects*, published in 1840, and John Curtis, *Farm Insects: Being the Natural History and Economy of the Insects Injurious to the Field Crops of Great Britain and Ireland, and also those which infest Barns and Granaries, With Suggestions for their Destruction* (Glasgow: Blakie and Son, 1860).

45. Massachusetts, *Reports* (1839), 6; Harris to Haldeman, November 15, 1842, in Haldeman Coll., ANSP.

46. Michele Alexis LaClergue Aldrich, "New York Natural History Survey, 1836–1845" (Ph.D. diss., University of Texas, Austin, 1974), 68, 81, 94, 132, 262, and 332.

47. Clarke, *James Hall*, 135; Merrill, *Geological and Natural History Surveys*, 331–40.

48. *Trans., NYSAS* 14 (1854), 701.

49. Barnes, *Asa Fitch*, 47; Haldeman wrote to Jacob Stauffer, a fellow entomologist, that Emmons's 1854 volume was "probably the most erroneous one ever put forth under the name of Entomology" (Haldeman to Stauffer, June 29, 1856, Stauffer Papers, APS). In 1873, when LeConte was lobbying for an entomological division in the federal government, he singled out Emmons's report as "a monument . . . of presumption and ignorance . . . , a permanent example of misplaced confidence and liberality." John L. LeConte, "Hints for the Promotion of Economic Entomology," *Proc., AAAS* 22 (1873), 15–17 (also published in *AN* 7 (November 1873), 710–22. The 1854 volume was eventually revised by a later New York state entomologist, J. A. Lintner, who corrected the names.

50. Barnes, *Asa Fitch*, 7–8, 18–21, and 32–33.

51. Fitch to Harris, February 26, 1855, MCZ Archives.

52. Barnes, *Asa Fitch*, 49.

53. *Trans., NYSAS* 10 (1850), 522, and 11 (1851), 748–50.

54. Asa Fitch, *First Report on the Noxious, Beneficial, and Other Insects of the State of New York* (Albany: New York Agricultural Society, 1855), 701–702.

55. *Trans., NYSAS* 12 (1852), 179–82; 13 (1853), 182; and 14 (1854), iii; Fitch to Harris, July 18, 1854, MCZ Archives.
56. Barnes, *Asa Fitch,* iii, 56–57. Townend Glover (discussed below) was also appointed in 1854, but his duties at first included other areas besides entomology.
57. Fitch, *First Report* (1855), 705–706; Barnes, *Asa Fitch,* 60. See, for example, Walsh, *First Annual Report* (1868).
58. Barnes, *Asa Fitch,* 61.
59. Glover to Harris, December 7, 1852, and December 27, 1852, MCZ Archives; *Trans., NYSAS* 11 (1851), 178; 12 (1852), 11; Mallis, *American Entomologists,* 63–64; Dupree, *Science in the Federal Government,* 112.
60. Thomas A. Dickinson, "Francis Gregory Sanborn," *Proceedings of the Worcester Society of Antiquities* (1884), 157–58.
61. *Trans., ISHS,* in *Trans., ISAS* 5 (1862), 740.
62. *Trans., ISHS* (1868), in *Trans., ISAS* 7 (1867–1868), 547.
63. Charles V. Riley, "In Memoriam," *AE* 2 (December–January 1869–1870), 65–68.
64. Walsh to Baird, September 17, 1859, OIAS 18:402, SIA.
65. For example, the Illinois State Agricultural Society gave awards to Walsh and Cyrus Thomas for their essays on entomology. *Trans., ISAS* 5 (1861–1864), 67.
66. *Trans., ISHS* (1866), in *Trans., ISAS* 6 (1865–1866), 365, and *Trans., ISAS* 6 (1865–1866), 18.
67. *Trans., ISHS* (1868), in *Trans., ISAS* 7 (1867–1868), 548; Walsh, *First Annual Report* (1867), 11–12.
68. *Third Annual Report, MSBA* (1867), vii.
69. Quoted in Riley, *First Annual Report* (1869), 4.
70. *Proc., MSHS,* in the *Third Annual Report, MSBA* (1867), 415–17.
71. *Fourth Annual Report, MSBA* (1868), iii; George Francis Lemmer, *Norman J. Colman and Colman's Rural World: A Study in Agricultural Leadership* (Columbia: University of Missouri Press, 1953), 44; Riley, *Fifth Annual Report* (1873), 22. The first Missouri geological survey under G. C. Swallow (1853–1862) had been charged with reporting on agricultural and manufacturing resources as time allowed, but at the time the survey ended, no botanical or zoological work had been done. A second geological survey, under Albert D. Hager, was organized in 1870, after the creation of an independent office of state entomologist, and the work of the two bureaus was kept separate. Merrill, *Geological and Natural History Surveys,* 275–86.
72. G. B. Goode, "A Memorial Appreciation of Charles Valentine Riley," *Science,* n.s. 3 (February 14, 1896), 219.
73. Leland O. Howard, "Progress in Economic Entomology in the United States," in *U.S. Department of Agriculture, Yearbook of Agriculture, 1899* (Washington, D.C.: U.S. Government Printing Office, 1900), 140.
74. *Proc., MSHS,* in *Third Annual Report, MSBA* (1867), 415–17; *Fourth Annual Report, MSBA* (1868), iii; William Edward Ogilvie, *Pioneer Agricultural Journalists: Brief Biographical Sketches of Some of the Early Editors in the Field of Agricultural Journalism* (Chicago: Arthur G. Leonard, 1927), 41.
75. Mallis, *American Entomologists,* 107–10; *CE* 1 (August 1, 1868), 1.

76. *CE* 1 (September 15, 1868), 9; G. P. Holland, "Entomology in Canada: Address for the Opening at the Royal Ontario Museum of the Special Exhibit Marking the Centennial of Entomology in Canada, 1863," in Glenn B. Wiggins, ed., *Centennial of Entomology in Canada, 1863–1963: A Tribute to Edmund M. Walker,* Royal Ontario Museum, Life Sciences Contribution 69 (Toronto: University of Toronto Press, 1966), 7–8.

77. *CE* 1 (August 1, 1868), 1.

78. *CE* 2 (October 1, 1869), 10, 13–16, and *CE* 3 (July 1870), 111–12.

79. *CE* 1 (September 1868), 1, and *CE* 2 (May 15, 1869), 91.

80. Holland, "Entomology in Canada," 9. Bethune recounts the history of the society in *CE* 3 (September 1871), 121–23. See also *CE* 2 (January 1870), 41–42, and (December 1870), 177–78; *CE* 3 (July 1871), 41–48; *CE* 4 (April 1872), 77, and (October 1872), 189–90; *CE* 5 (October 1873), 182; *CE* 9 (October 1877), 188–89.

81. Merrill, *Geological and Natural History Surveys,* 428–35; Anne Millbrooke, "Henry Darwin Rogers and the First State Geological Survey of Pennsylvania," *Northeastern Geology* 3:1 (1981), 73; enclosure in John H. Brennan to G. W. Brewer, February 3, 1858, and J. Jay Smith to Stauffer, January 9, 1858, Stauffer Papers, APS.

82. Merrill, *Geological and Natural History Surveys,* 158–63, 175.

83. Ibid., 203–31.

84. Ibid., 131–35.

85. A. E. Verrill, "Sidney Irving Smith," *Science* 64 (1926), 58; Margaret W. Rossiter, *The Emergence of Agricultural Science: Justus Liebig and the Americans, 1840–1880,* Yale Studies in the History of Science and Medicine 9 (New Haven: Yale University Press, 1975), 160.

86. *Psyche* 1 (February 1876), 140.

87. *Psyche* 1 (March–April 1877), 32.

88. Merrill, *Geological and Natural History Surveys,* 368–75.

89. On this topic, see Laurence R. Veysey, *The Emergence of the American University* (Chicago: University of Chicago Press, 1965).

90. William Morton Wheeler, "Entomology at Harvard University," in *Notes Concerning the History and Contents of the Museum of Comparative Zoology,* by Members of the Staff (Cambridge, Mass.: Harvard University, MCZ, 1936), 22–26, 31.

91. Wheeler, "Entomology at Harvard," 27–28.

92. Agassiz to Hagen, June 3, 1868, MCZ Archives; Wheeler, "Entomology at Harvard," 30–31.

93. Dickinson, "Francis Gregory Sanborn," 158–59; Harris to Haldeman, November 15, 1842, Haldeman Coll., ANSP.

94. Earle D. Ross, *Democracy's College: The Land Grant Movement in the Formative Stage* (Ames: Iowa State College Press, 1942), 39–42 and 65–67; Lyman Carrier, "The U.S. Agricultural Society, 1852–1860: Its Relation to the Origin of the United States Department of Agriculture and the Land Grant Colleges," *AgH* 11 (October 1937), 279–81; Lemmer, "Early Agricultural Editors," 18–20.

95. Fitch to Harris, February 23, April 6, and August 11, 1852, MCZ Archives; Robert Silverman and Mark Beach, "A National University for Upstate New York," *American Quarterly* 22:3 (Fall 1970), 701–13; Herbert Osborn,

Fragments of Entomological History: Some Personal Recollections of Men and Events (Columbus, Ohio: Author, 1937), 15.

96. Wallace Eugene Houk, "A Study of Some Events in the Development of Entomology and Its Application in Michigan" (Ph.D. diss., Michigan State University, Lansing, 1954), 8–26 and 293–96; *Psyche* 1 (July 1876), 179.

97. Lintner to Scudder, July 10, 1867, Scudder Corr., MSB.

98. Anna Botsford Comstock, *The Comstocks of Cornell: John Henry Comstock and Anna Botsford Comstock,* edited by Glenn W. Herrick and Ruby Green Smith (New York: Comstock Publishing Associates, 1953), 1–96; Edward H. Smith, "The Comstocks and Cornell: In the People's Service," *AR* 21 (1976), 1–25; Pamela M. Henson, "Evolution and Taxonomy: J. H. Comstock's Research School in Evolutionary Entomology at Cornell University, 1874–1930" (Ph.D. diss., University of Maryland, 1990), 105–24.

99. Mallis, *American Entomologists,* 126–38; Henson, "Evolution and Taxonomy," 323–93. For example, L. O. Howard, Comstock's student, served as Riley's assistant and later succeeded Riley as chief of the Division of Entomology in the Department of Agriculture.

100. Merrill, *Geological and Natural History Surveys,* 242–49; *Psyche* 2 (May–August 1878), 166.

101. Osborn, *Fragments,* 15, 99–109; *Psyche* 1 (March 1875), 59, quoting Charles L. Flint in the *Report of the Commissioner of Agriculture* for 1872; *Report of the Commissioner of Agriculture* (1873), 323–52.

102. Monte A. Calvert, *The Mechanical Engineer in America, 1830–1910* (Baltimore, Md.: Johns Hopkins University Press, 1967).

103. Osborn, *Fragments,* 105.

104. Estimate based on ten "full-time equivalent" positions at an estimated average of $2,000 per year ($1,500 salary plus $500 in other support). The estimate of support per person is based on Fitch, who was supported at $1,000 per year in the 1850s and Riley, who was supported at $3,000 per year in the 1870s. Overall, entomologists seem to have earned about $1,200 in salary, about the amount paid a second-rank engineer. See Mark Aldrich, "Earnings of American Civil Engineers, 1820–1859," *Journal of Economic History* 31 (1971), 416.

105. George Ordish, "Scientific Pest Control and the Influence of John Curtis," *Journal of the Royal Society of Arts* 116 (1968), 298–309.

106. Lintner to Scudder, December 7, 1861, Scudder Corr., MSB; Edwards to Lintner, April 7, 1869, Lintner Corr., NYSA.

107. Lintner to Scudder, July 19, 1867, Scudder Corr., MSB.

108. Lintner to Scudder, April 6, 1874, in ibid.

109. Scudder refers to Lintner as Fitch's "successor," as quoted in a letter from Lintner to Scudder, April 6, 1874, in ibid. Lintner's successor, Ephriam Porter Felt, states that Lintner began his duties as state entomologist in 1874. See Ephriam Porter Felt, *Memorial of Life and Entomological Work of Joseph Albert Lintner, Ph.D.,* Bulletin of the New York State Museum 5:24 (Albany: University of the State of New York, 1899), title page.

110. Felt, *Joseph Albert Lintner,* 303–304; John Henry Comstock to Lintner, March 30, 1880. Lintner Corr., NYSA; *Psyche* 3 (June 1880), 83; Mallis, *American Entomologists,* 52–54. In 1883, the offices of the New York state entomologist and the state paleontologist were made departments of the

New York State Museum of Natural History, and in 1889, the museum was incorporated as a part of the state university system. Merrill, *Geological and Natural History Surveys*, 353–60.

111. F. W. Golding, "Biographical Sketch of William LeBaron, *EA* 2 (October 1885), 122–25; Mallis, *American Entomologists*, 49–50.

112. *Trans., ISAS* 4 (1859–1860), 649; E. O. Essig, *A History of Entomology* (New York: Macmillan, 1931), 771.

113. Mallis, *American Entomologists*, 50–52.

114. Ibid., 55–60.

115. The main sources of this sketch are Mallis, *American Entomologists*, 297–301; Ralph W. Dexter, "The Development of A. S. Packard, Jr., as a Naturalist and an Entomologist," Pts. 1, 2, *Bull., BES* 52 (June, October 1957), 57–66, 101–12; T. D. A. Cockerell, "Alpheus Spring Packard, Jr.," in National Academy of Sciences, *Biographical Memoirs*, vol. 9 (Washington, D.C.: National Academy of Sciences, 1920), 181–236, and Weiss, *Pioneer Century*, 241–43.

116. Packard to Scudder, April 17, 1865, Scudder Corr., MSB.

117. Dexter, "Packard," 101.

118. Howard, "Progress in Economic Entomology," 141; A. S. Packard, Jr., "First Annual Report on Injurious and Beneficial Insects of Massachusetts," *AN* 5 (August 1871), 423–27, and "Second Annual Report on Injurious and Beneficial Insects of Massachusetts," *AN* 7 (April 1873), 241–44.

119. A. S. Packard, Jr., "Report on the Rocky Mountain Locust and other Insects . . . in the Western States and Territories," in F. V. Hayden, ed., *Ninth Annual Report of the U.S. Geological and Geographical Survey of the Territories, for 1875* (Washington, D.C.: U.S. Government Printing Office, 1877), 589–810.

120. Sanborn to Edwards, May 18, 1872, Edwards Corr., WVUL.

121. LeConte, "Promotion of Economic Entomology," 10–22.

122. The history of the Entomological Club is based on *CE* 4 (October 1872), 182–85; 5 (September 1873), 165–66; 6 (September 1874), 162–63; 7 (September 1875), 177; *Proc., AAAS* 23 (1874), 153–54; 24 (1875), 337; 25 (1876), 343; see also Smith, "Entomological Society of America," 13.

5. The Balance of Nature

1. *PE* 1 (August 27, 1866), 110.

2. *PE* 1 (August 27, 1866), 108.

3. Quoted in Mallis, *American Entomologists*, 44.

4. Quoted in Frank N. Egerton III, "Changing Concepts of the Balance of Nature," *QRB* 48 (1973), 333.

5. Ibid., 333–36.

6. William Kirby and William Spence, *An Introduction to Entomology; or, Elements of the Natural History of Insects*, 4 vols. (London: Longman, Hurst, Rees, Orme, and Brown, 1816–1826), vol. 1, p. 250.

7. Ibid., 79–80.

8. Harris to Hentz, May 16, 1825, in Harris, *Entomological Correspondence*, 5.

9. Joseph A. Lintner, "Entomological Contributions," *New York State Museum of Natural History, Twenty-fourth Annual Report* (Albany: Angus, 1872), 154–67.

10. Egerton, "Concepts of Balance of Nature," 324–38.

11. Ibid., 334.

12. Riley, *First Annual Report* (1869), 109.

13. *Trans., ISAS* 4 (1859–1860), 338–39.

14. Charles V. Riley, "Great Truths in Applied Entomology; Address to Georgia State Agricultural Society, February 12, 1884," in *Report of the Commissioner of Agriculture* (1884), 324. For similar statements, see Thomas, "Insects Injurious to Vegetation in Illinois," 407–408; Charles V. Riley, *Second Annual Report on the Noxious, Beneficial, and Other Insects of the State of Missouri* (Jefferson City, Mo.: State Printer, 1870), 8; and Packard, *Guide,* 76.

15. Riley, *Second Annual Report* (1870), 11–12.

16. *PE* 1 (September 29, 1866), 117–18; see also Riley, *Second Annual Report* (1870), 11–12.

17. *Trans., ISAS* 4 (1859–1860), 340.

18. *PE* 1 (September 29, 1866), 119.

19. See, for example, Dr. Hull, "A Few of Our Insect Enemies and Friends," *Report, MSHS* in *First Annual Report, MSBA* (Jefferson City, Mo.: State Printer, 1865), 453–54.

20. Egerton, "Ecological Studies and Observations Before 1900," in Frank N. Egerton, ed., *History of Ecology* (New York: Arno Press, 1977), 337–38; Edward H. Smith, "The Grape Phylloxera: A Celebration of Its Own," *AE* (Winter 1992), 215. Riley, however, was spectacularly successful with the introduction of American grapevine stock to France, which saved the French wine industry from the ravages of the phylloxera. Ibid., 216–19.

21. Egerton, "Ecological Studies," 338; Stephen A. Forbes Correspondence, University of Illinois Archives, Urbana, Illinois.

22. Packard, *Guide,* 72.

23. Thomas, "Insects Injurious to Vegetation in Illinois," 407–408.

24. *First Annual Report, MSBA* (1865), 455–56. See also pp. 342–45, where the horticultural society proposed legislation to protect insectivorous birds.

25. E. A. Clifford, "Birds Injurious and Beneficial to the Horticulturist," *Trans., ISHS* (1864) in *Trans., ISAS* 5 (1861–1864), 925–27.

26. Gurdon Evans, "Agricultural Survey of Madison County," *Trans., NYSAS* 11 (1851), 744–45.

27. Theodore Whaley Cart, "The Struggle for Wildlife Protection in the United States, 1870–1900: Attitudes and Events Leading to the Lacey Act" (Ph.D. diss., University of North Carolina, Chapel Hill, 1971), 57–58.

28. *Trans., ISHS* (1862), in *Trans., ISAS* 5 (1861–1864), 24.

29. A. S. Packard, Jr., "Nature's Means of Limiting the Numbers of Insects," *AN* 8 (May 1874), 273–74.

30. Quoted in Dr. George Gasey, "The Study of Natural History," *First Annual Report, MSBA* (1865), 183.

31. *Psyche* 3 (August 1882), 379–80.

32. Charles V. Riley, "The Rocky Mountain Locust," *AN* 11 (November 1877), 665–66.

33. On the sparrow episode, and particularly the role of Elliott Coues, see Paul Russell Cutright and Michael J. Brodhead, *Elliott Coues: Naturalist and Frontier Historian* (Chicago: University of Chicago Press, 1981).

34. *PE* 2 (January 1867), 44–46.

35. Walsh to Edwards, March 27, 1867, Edwards Corr., WVUL.

36. Cart, "Wildlife Protection," 56.

37. *AE* 2 (April 1870), 168.

38. Packard, "Limiting the Numbers of Insects," 276.

39. Cart, "Wildlife Protection," 57–59.

40. Ibid., 56.

41. Stephen A. Forbes, "The Food of Birds," in *Trans., ISHS* for 1878, n.s. 12 (1879), 140–45, and Stephen A. Forbes, "The Food of Birds," in *Bulletin of the Illinois State Laboratory of Natural History* 1:3 (1880), 80–148; see also U.S. Entomological Commission, *First Annual Report for the Year 1877 relating to the Rocky Mountain Locust and the best means of preventing its injuries and of guarding against its invasions* (Washington, D.C.: U.S. Government Printing Office, 1878), 27, which gives the results from the examination of birds' stomachs.

42. Cassedy, *American Medicine and Statistical Thinking,* 4, 20, 62, 67, 230; Keir B. Sterling, *Last of the Naturalists: The Career of C. Hart Merriam,* rev. ed., Natural Sciences in America Series (New York: Arno Press, 1977), 91–92, 97–112, 122–23, and 134; Cart, "Wildlife Protection," 59–60.

43. *PF* 13 (June 1853), 230.

44. *PF* 6 (March 1846), 2–3.

45. *PF* 12 (December 1852), 564.

46. Evans, "Agricultural Survey of Madison County," 749.

47. *First Annual Report, MSBA* (1865), 454–55.

48. Henry Shimer, "Insects Injurious to the Potato," *AN* 3 (April 1869), 94–99.

49. William LeBaron, *Second Annual Report on the Noxious Insects of the State of Illinois* (Springfield, Ill.: State Printer, 1872), 7.

50. Examples of this function of entomological knowledge may be found in John Delafield, "A General View and Agricultural Survey of the County of Seneca," *Trans., NYSAS* 10 (1850), 522, 525–26, and Evans, "Agricultural Survey of Madison County," 749–50.

51. Rossiter, *Emergence of Agricultural Science,* 45–46, 122–23.

52. B. P. [John Pitkin?] Norton, in *Trans., NYSAS* 11 (1851), 167.

53. Ibid., 165. See also p. 171 on the importance of controlled experiments.

54. Forbes, "Food of Birds," 143–44.

55. Weiss, *Pioneer Century,* 264.

6. A Weevil, a Fly, a Bug, and a Beetle

1. Ulysses Prentiss Hedrick, *A History of Agriculture in the State of New York* (Albany: J. B. Lyon for the New York State Agriculture Society, 1933), 386; *Trans., NYSAS* 13 (1853), 11; Gates, *Agriculture and the Civil War,* 141.

2. Hedrick, *Agriculture in New York,* 399.

3. *PF* 10 (1850), 275.

4. *PF* 13 (December 1853), 452–53.

5. Hedrick, *Agriculture in New York,* 400.

6. *PF* 8 (December 1848), 373.

7. *PF* 11 (February 1851), 60, and *PF* 13 (December 1853), 452–53.

8. *Trans., NYSAS* 14 (1854), 701; Walsh, *First Annual Report* (1868), 11–12.

9. Hedrick, *Agriculture in New York*, 399–401.

10. Herbert Osborn, *The Hessian Fly in the United States*, U.S. Department of Agriculture, Bureau of Entomology, Bulletin 16 (Washington, D.C.: U.S. Government Printing Office, 1898), 7.

11. Ibid., 14.

12. F. G. W. Jones and Margaret G. Jones, *Pests of Field Crops*, 2d ed. (New York: St. Martin's Press, 1974), 179–80.

13. The question whether the Hessian fly was introduced from Europe or was native to America became the subject of debate between Riley and Hagen. Modern authorities support Fitch's original hypothesis that it was introduced. Osborn, *The Hessian Fly*, 8; Jones and Jones, *Pests of Field Crops*, 2.

14. Asa Fitch, "The Hessian Fly: Its History, Character, Transformation, and Habits," *Trans., NYSAS* 6 (1846), 316–33.

15. Ibid., 331.

16. The group to which it belongs has since been revised, and the present name is *Mayetiola (-Phytophaga) destructor* (Say). Jones and Jones, *Pests of Field Crops*, 179.

17. Fitch, "The Hessian Fly," 108; Osborn, *The Hessian Fly*, 10.

18. *PF* 4 (April 1844), 93, *PF* 7 (May 1847), 146, and *PF* 11 (August 1851), 335–36.

19. E. C. Herrick, "Observations Communicated at the Request of the Honorable H. L. Ellsworth: The Hessian Fly," *Report of the Commissioner of Patents, 1844* (Washington, D.C.: U.S. Government Printing Office, 1845), 161–67, reprinted in *PF* 5 (July 1845), 169; Asa Fitch, "The Hessian Fly," *American Journal of Agriculture and Science* (1846), revised in the *Trans., NYSAS* 6 (1846), 316–72.

20. *PF* 1 (May 1841), 33, and *PF* 4 (April 1844), 93; Margaretta Morris, "Observations of the development of the Hessian Fly," *Proc., ANSP* 1 (1841–1843), 66–68; Margaretta Morris, "On the *Cecidomyia destructor*, or Hessian Fly," *Trans., APS*, n.s. 8 (1843), 49–52; see *PF* 11 (December 1851), 578–79 (where Fitch distinguished between the Hessian fly, the joint worm, and Morris's wheat midge), and Packard, *Guide*, 375–76 (on the similarity of the wheat midge and the Hessian fly).

21. *PF* 7 (September 1847), 270–71, and (October 1847), 302–303.

22. Osborn, *The Hessian Fly*, 19.

23. Harlow B. Mills, "Weather and Climate," in *Insects: The Yearbook of Agriculture 1952* (Washington, D.C.: U.S. Government Printing Office, 1952), 424 (hereafter cited as *USDA, Insects*); G. J. Haeussler and R. W. Leiby, "Surveys of Insect Pests," in ibid., 444–46; Osborn, *The Hessian Fly*, 25.

24. A. S. Packard, Jr., *The Hessian Fly*, U.S. Entomological Commission Bulletin 4 (Washington, D.C.: U.S. Government Printing Office, 1880), 3.

25. Osborn, *The Hessian Fly*, 18.

26. Jones and Jones, *Pests of Field Crops*, 180.

27. Ibid.; C. M. Packard and John H. Martin, "Resistant Crops, the Ideal Way," in *USDA, Insects*, 429–33.

28. Packard, *The Hessian Fly*, 21–31.

29. Ibid., 31; Osborn, *The Hessian Fly,* 42; Jones and Jones, *Pests of Field Crops,* 180.
30. Osborn, *The Hessian Fly,* 41.
31. Ibid., 46; Packard and Martin, "Resistant Crops," 429.
32. *PF* 7 (November 1847), 345; Packard, *The Hessian Fly,* 31; Osborn, *The Hessian Fly,* 42.
33. Hedrick, *Agriculture in New York,* 332.
34. *PF* 3 (April 1843), 68.
35. *PF* 5 (January 1845), 31; Osborn, *The Hessian Fly,* 44–45.
36. Packard, *Guide,* 375; Osborn, *The Hessian Fly,* 38.
37. Jones and Jones, *Pests of Field Crops,* 180.
38. W. A. Baker and O. R. Mathews, "Good Farming Helps Control Insects," in *USDA, Insects,* 437–40; C. M. Packard, "Cereal and Forage Insects," in *USDA, Insects,* 581–84; Jones and Jones, *Pests of Field Crops,* 3.
39. Claude Wakeland, "The Chinch Bug," in *USDA, Insects,* 611.
40. Ibid., 612.
41. Edwin Way Teale, "The Life of the Chinch Bug," in Teale, *The Strange Lives of Familiar Insects* (New York: Dodd, Mead, 1962), 121–29.
42. *PF* 10 (August 1850), 245; *PF* 15 (September 1855), 296.
43. *PF* 5 (October 1845), 254.
44. *PF* 10 (September 1850), 281.
45. Teale, "Life of the Chinch Bug," 122; Leland O. Howard, *The Chinch Bug,* U.S. Department of Agriculture, Division of Entomology, Bulletin 17 (Washington, D.C.: U.S. Government Printing Office, 1888), 8; "Chinch Bug Notes," in *Report of the Commissioner of Agriculture* (1881–1882), 87–88.
46. *PF* 10 (August 1850), 245.
47. *PF* 5 (December 1845), 287–88.
48. *PF* 10 (September 1850), 281.
49. *PF* 5 (December 1845), 287–88.
50. Howard, *Chinch Bug,* 44–46.
51. *PF* 10 (September 1850), 281.
52. Howard, *Chinch Bug,* 23 and 44; Teale, "Life of the Chinch Bug," 128.
53. Howard, *Chinch Bug,* 25 and 45; Teale, "Life of the Chinch Bug," 122.
54. Howard, *Chinch Bug,* 32–38, 46; Riley, *Second Annual Report* (1870), 29.
55. Howard, *Chinch Bug,* 38; Teale, "Life of the Chinch Bug," 126; Wakeland, "Chinch Bug," 611.
56. "Chinch Bug Notes," 87; *PF* 10 (September 1850), 281.
57. Riley, *Second Annual Report* (1870), 24; Howard, *Chinch Bug,* 44–47.
58. Cyrus Thomas, *The Chinch Bug: Its History, Characters, and Habits, and the means of Destroying it or Counteracting its Injuries,* Department of the Interior, U.S. Entomological Commission, Bulletin 5 (Washington, D.C.: U.S. Government Printing Office, 1879), 39; Cyrus Thomas, "Temperature and Rainfall as affecting the Chinch Bug," *AE,* n.s. 1 (October 1880), 240–42; Cyrus Thomas, "The Relation of Meteorological Conditions to Insect Development," *Tenth Annual Report of the State Entomologist of Illinois* (Springfield, Ill.: State Printer, 1881), 47–59.
59. Weather data for the years 1873 to 1880 were furnished by the U.S. Signal Corps. Earlier records came from observers in the U.S. Army Medical Department and the Smithsonian meteorological project. Benjamin D.

Whitaker (a volunteer weather observer for the Smithsonian and the agricultural department of the U.S. Patent Office) to LeConte, August 13, 1856, LeConte Coll., APS. See James Roger Fleming, *Meteorology in America, 1800–1870* (Baltimore, Md.: Johns Hopkins University Press, 1990), 68–70, 75–76, and 125–28.

60. Thomas, "Meteorological Conditions," 49–51.
61. Ibid., 54–55.
62. *AN* 14 (October 1881), 820; "Chinch Bug Notes," 87; Howard, *Chinch Bug,* 8 and 46.
63. Mills, "Weather and Climate," 424; Haeussler and Leiby, "Surveys of Insect Pests," 444–46; Wakeland, "Chinch Bug," 612; W. C. Allee, Orlando Park, Alfred Emerson, and Thomas Park, *Principles of Animal Ecology* (Philadelphia: W. B. Saunders, 1949), 27.
64. Cyrus Thomas in *AN* 14 (July 1880), 511; Howard, *Chinch Bug, 5.*
65. Hedrick, *Agriculture in New York,* 339–41; S. Fraser, "Potato," in L. H. Bailey, ed., *Cyclopedia of American Agriculture,* vol. 2 (New York: Macmillan, 1907), 519–27.
66. Gates, *Farmer's Age,* 263–66; Hedrick, *Agriculture in New York,* 339–41.
67. Benjamin D. Walsh, "The New Potato Bug, and its Natural History," *PE* 1 (October 1865), 1–4, also reprinted in *First Annual Report, MSBA* (1865), 200–206; R. A. Casagrande, "The Colorado Potato Beetle: 125 Years of Mismanagement," *Bull., ESA* 33:3 (1987), 142–43.
68. Walsh, "New Potato Bug," 1–4; Walsh, *First Annual Report* (1868), 138.
69. Walsh, "New Potato Bug," 3; Riley, *First Annual Report* (1869), 104.
70. Walsh, "New Potato Bug," 1; Walsh in *AE* 1 (November 1868), 42; *CE* 2 (July 1, 1870), 115; Charles V. Riley, *Seventh Annual Report on the Noxious, Beneficial, and Other Insects of the State of Missouri* (Jefferson City, Mo.: State Printer, 1875), 1–3; Jones and Jones, *Pests of Field Crops,* 131.
71. Townend Glover, in *Report of the Commissioner of Agriculture* (1866), 27; Charles V. Riley, *Fourth Annual Report on the Noxious, Beneficial, and Other Insects of the State of Missouri* (Jefferson City, Mo.: State Printer, 1872), 5.
72. *CE* 3 (July 1871), 41–42; Riley, *Third Annual Report* (1871), 97–98.
73. *EMM* 12 (July 1875), 45; Riley, *Seventh Annual Report* (1875), 3.
74. Walsh, "New Potato Bug," 3; Riley, *First Annual Report* (1869), 107; Jones and Jones, *Pests of Field Crops,* 132; Packard, *Guide,* 508.
75. Riley, *Fourth Annual Report* (1872), 16–17; LeBaron, *Second Annual Report* (1869), 63.
76. Shimer, "Insects Injurious to the Potato," 92–96.
77. Walsh, "New Potato Bug," 3; LeBaron, *Second Annual Report* (1869), 68; Glover, in *Report of the Commissioner of Agriculture* (1866), 27; Benjamin D. Walsh, "The New Potato Bug," *PE* 2 (November 1866), 15.
78. Walsh, "New Potato Bug," 4.
79. Walsh, in *AN* 1 (July 1869), 219; LeBaron, *Second Annual Report* (1869), 68; Glover in *Report of the Commissioner of Agriculture* (1870), 75.
80. Whorton, *Before Silent Spring,* 15–20.
81. C. H. Fernald, "Presidential Address, 1896," in AAEE, "Annual Meetings and Presidential Addresses: Annual Meetings 1–24 (1889–1911)" (University of California Library, Davis; typescript), 4–5.
82. Riley, *Third Annual Report* (1871), 99–100; Glover, [experiments on poison],

Report of the Commissioner of Agriculture (1873), 165–67; Whorton, *Before Silent Spring*, 33.

83. Riley, *Seventh Annual Report* (1875), 8–13; Charles V. Riley, *Eighth Annual Report on the Noxious, Beneficial, and Other Insects of the State of Missouri* (Jefferson City, Mo.: State Printer, 1876), 5–7.

84. Whorton, *Before Silent Spring*, 34–64.

85. J. C. Headley, "Economics of Agricultural Pest Control," *AR* 17 (1972), 273; Fernald, "Presidential Address," 5; Whorton, *Before Silent Spring*, 85; Thomas R. Dunlap, "The Triumph of Chemical Pesticides," *ER* (May 1978), 39–47. R. A. Casagrande has concluded that, because of increasing insect resistance, chemical insecticides have lost their usefulness as single agents in the control of the Colorado potato beetle. He recommends that more attention be given to the farm culture practices initially advocated by Shimer, Riley, Walsh, and other early entomologists. Casagrande, "Colorado Potato Beetle," 146 and 148–49.

7. The Rocky Mountain Locust Plague

1. Representatives of professional entomologists' writings are Leland O. Howard, *A History of Applied Entomology (Somewhat Anecdotal)*, Smithsonian Miscellaneous Collections 84 (Washington, D.C.: Smithsonian Institution, 1930), and Sir Boris Uvarov, *Grasshoppers and Locusts: A Handbook of General Acridology*, 2 vols. (Cambridge, England: Cambridge University Press, 1966). Agricultural historians writing about locusts include John T. Schlebecker, "Grasshoppers in American Agricultural History," *AgH* 27 (July 1953), 85–93; Gilbert C. Fite, *The Farmers' Frontier, 1865–1900* (New York: Holt, Rinehart and Winston, 1966); and Annette Atkins, *Harvest of Grief: Grasshopper Plagues and Public Assistance in Minnesota, 1873–78* (St. Paul: Minnesota Historical Society Press, 1984). Historians of science are represented by Dupree, *Science in the Federal Government*, 148–62, and Ralph W. Dexter, "The Organization and Work of the U.S. Entomological Commission (1877–1882)," *Melsheimer Entomological Series* 26 (1979), 28–32. Atkins, *Harvest of Grief*, notes the locust plague in works of fiction. See pp. 17, 19, 21, 30, and 32.

2. This definition and the discussion of locusts is based on C. P. Friedlander, *The Biology of Insects* (New York: Pica Press, 1977), 162; John Stoddard Kennedy, "Continuous Polymorphism in Locusts," in John Stoddard Kennedy, ed., *Insect Polymorphism* (London: Royal Entomological Society, 1961), 80–90, and Uvarov, *Grasshoppers and Locusts*, vol. 1, pp. 379–89, and vol. 2, pp. 321–23, 368–70, 522–23.

3. This density-dependent type of polymorphism is known in only a few insect species. Kennedy, "Continuous Polymorphism in Locusts," 80.

4. Schlebecker, "Grasshoppers," 85.

5. Benjamin D. Walsh in *PE* 2 (October 1866), 1–5; Riley, *Seventh Annual Report* (1875), 121–58; Riley, *Eighth Annual Report* (1876), 57–96; Charles V. Riley, *Destructive Locusts: A Popular Consideration of a few of the more injurious Locusts (or Grasshoppers) in the United States, together with the best means of Destroying them*, U.S. Department of Agriculture, Division of Entomology, Bulletin 25 (Washington, D.C.: U.S. Government Printing Office, 1891), 10; Fite, *Farm-*

ers' Frontier, 58–73; Paul W. Riegert, *From Arsenic to DDT: A History of Entomology in Western Canada* (Toronto: University of Toronto Press, 1980), 24–25, 29–31, and 35.

6. Charles V. Riley, *The Locust Plague in the United States: Being More Particularly a Treatise on the Rocky Mountain Locust or so-called Grasshopper, as it occurs East of the Rocky Mountains, with Practical Recommendations for its Destruction* (Chicago: Rand McNally, 1877), 85–87.

7. Schlebecker, "Grasshoppers," 86.

8. Charles V. Riley, "The Locust Plague; How to Avert it," *Proc., AAAS* 24 (1875), 215–16; Gilbert C. Fite, "Daydreams and Nightmares: The Late Nineteenth Century Frontier," *AgH* 40 (October 1966), 289–91.

9. Fite, "Daydreams and Nightmares," 289–91; Atkins, *Harvest of Grief,* 58–83.

10. Atkins, *Harvest of Grief,* 121.

11. Riley, "The Locust Plague," 215; U.S. Entomological Commission, *First Annual Report* (1878), 1.

12. LeConte, "Promotion of Economic Entomology," 10–22.

13. See Dupree, *Science in the Federal Government,* 155–56.

14. Riley to LeConte, December 3, 1873, LeConte Coll., APS.

15. Earle D. Ross, "The U.S. Department of Agriculture in the Commissionership," *AgH* 20 (July 1946), 134–36.

16. LeConte "Promotion of Economic Entomology," 20.

17. Earle D. Ross, a historian of the Department of Agriculture, is critical of the scientists' position. He notes that in the disputes between scientists and the commissioners the scientists' side was usually the only one represented to the public and that the scientists were not opposed to the spoils system when they could select their own candidates. Ross, "Department of Agriculture," 138–39.

18. Riley to LeConte, December 3, 1873, LeConte Coll., APS.

19. Ibid., and Riley to LeConte, December 23, 1874, in ibid.

20. Riley to LeConte, December 23, 1874, in ibid.

21. LeConte, "Promotion of Economic Entomology," 20. Some entomologists evaluated Glover's work more favorably. William LeBaron encouraged Glover to publish his illustrations and notes, and these publications were reviewed favorably in the *Canadian Entomologist.*

22. Riley, "The Locust Plague," 216; see also Riley, *Seventh Annual Report* (1875), v; Fite, *Farmers' Frontier,* 70.

23. *Report of the Commissioner of Agriculture* (1876), 17.

24. Thomas to Hayden, March 18, 1876, Hayden Survey, Letters Received.

25. Thomas to Scudder, February 18, 1874, Scudder Corr., MSB.

26. Ibid.

27. Conference of Governors, *The Rocky Mountain Locust, or Grasshopper, Being the Report of Proceedings of a Conference of the Governors of Several Western States and Territories . . . held at Omaha, Nebraska, on the Twenty-fifth and Twenty-sixth Days of October, 1876, to consider the Locust Problem; also a Summary of the Best Means now known for Counteracting the Evil* (Saint Louis, Mo.: R. R. Studley Company, 1876), 24 (hereafter cited as *Conference of Governors*).

28. John L. LeConte, "Methods of Subduing Insects Injurious to Agriculture," *CE* 7 (September 1875), 168–69, 171.

29. Riley to LeConte, January 23, 1875, LeConte Coll., APS.

30. Riley, *Seventh Annual Report* (1875), v.
31. Senator H. B. Anthony to LeConte, February 7, 1875, LeConte Coll., APS. A bill that incorporated their proposal was introduced at Riley's request in the *Congressional Record,* 44th Cong., 1st sess., Senate, vol. 4, pt. 2 (March 7, 1876), 1502.
32. John L. LeConte, "On the Method of Subduing Insects Injurious to Agriculture," *Proc., AAAS* 24 (1875), 203 [same address published with minor changes in *CE* 7 (1875)]; Riley, "The Locust Plague," 221.
33. Riley to LeConte , February 1, 1875, and June 18, 1875, LeConte Coll., APS.
34. *Congressional Record* 44th Cong., 1st sess., Senate, vol. 4, pt. 2 (March 7, 1876), 1502 and 1510.
35. Ibid., 1502–1504.
36. Ibid., 1502–1503.
37. Thomas to Hayden, March 18, 1876, Hayden Survey, Letters Received.
38. Ibid.
39. *Congressional Record,* 44th Cong, 1st. sess., Senate, vol. 4, pt. 2 (March 7, 1876), 1502–1503.
40. Ibid., 1503 and 1508.
41. Ibid., 1503.
42. Ibid., 1504–1505.
43. *Nation* 22 (March 16, 1876), 169.
44. *Nation* 22 (March 30, 1876), 208.
45. *Congressional Record,* 44th Cong, 1st. sess., Senate, vol. 4, pt. 2 (March 7, 1876), 1504–1505, 1510–11, 1541, and 1558. See LeConte's account of the bill's failure in *CE* 8 (September 1876), 177–78.
46. Thomas to Hayden, October 12, 1876, Hayden Survey, Letters Received.
47. Ibid.
48. Thomas to Hayden, October 28, 1876, in ibid.
49. Dexter, "Organization and Work of the U.S. Entomological Commission," 28.
50. Thomas to Hayden, October 28, 1876, Hayden Survey, Letters Received.
51. Thomas to Hayden, January 6, 1877, and January 23, 1877, in ibid.; Congressman Erastus Wells to Riley, January 29, 1877, in ibid.
52. Riley to Hayden, January 3, 1877, January 13, 1877, and January 18, 1877, and Thomas to Hayden, January 19, 1877, in ibid.
53. *U.S. Statutes at Large* 19 (1876), 357.
54. Aldrich, "Earnings of American Civil Engineers," 416.
55. Thomas to Hayden, October 28, 1876, and March 10, 1877, Hayden Survey, Letters Received.
56. Riley to [Hayden] [January 1877], in ibid.
57. Thomas to Hayden, January 19, 1877, in ibid.
58. LeBaron to LeConte, December 5, 1873, LeConte Coll., APS.
59. LeConte to Hilgard, March 3, 1877, OIS 164:384, SIA.
60. LeConte to Baird, May 14, 1860, OIAS 22:294, SIA; LeConte to Aubrey [H. Smith], March 15, 1870, LeConte Coll., APS; LeConte to Scudder, April 29, 1879, Scudder Corr., MSB.
61. Hilgard to LeConte, March 10, 1877; F. A. P. Barnard to LeConte, March 14, 1877; Henry to Barnard, March 12, 1877, LeConte Coll., APS.
62. LeConte to Hilgard, April 22, 1877, Hilgard Family Papers (C-B 972), Bancroft Library, University of California, Berkeley.

63. Riley to LeConte and to the President, March 12, 1877, LeConte Coll., APS; Flagg to E. W. Hilgard, December [?], 1876, and July 14, 1877, Hilgard Papers, Bancroft Library; Riley's obituary of Flagg in *Rural World*, April 3, 1878, in Scrapbook 19, p. 199, Charles V. Riley Papers, SIA.

64. See copies of "Letters to the President of the United States recommending J. L. LeConte for Commissioner of Agriculture," LeConte Coll., APS.

65. "Abstract of Statement of Expenditures of Department of Agriculture, from July 1, 1862, to June 30, 1876, with Explanatory Notes," in ibid.

66. LeConte to Morton McMichael, March 20, 1877, in ibid.

67. LeConte to Scudder [not sent?], March 24, 1877, in ibid.

68. Hilgard to LeConte, June 15, 1877, and A. Lowden Snowden to LeConte, June 29, 1877, in ibid.; LeConte to Hilgard, April 22, 1877, Hilgard Papers, Bancroft Library; Lester D. Stephens, "The Appointment of the Commissioner of Agriculture in 1877: A Case Study in Political Ambition and Patronage," *Southern Quarterly* 15:4 (1977), 383–84.

69. Mallis, *American Entomologists,* 248.

70. On the expansion of the activities of the Entomological Commission, see Carl Schurz, [Note on Report to President], in NA, RG 48, *Records of the Secretary of the Interior, Letters Relating to the U.S. Entomological Commission.* On expansion of the Division of Entomology, see Howard, *History of Applied Entomology,* 86–89, and *Report of the Commissioner of Agriculture* (1879), 14. In 1880, the Entomological Commission was transferred from the Department of the Interior to the Department of Agriculture, but it still had a separate budget and operated independently from the department. See Gustavus A. Weber, *The Bureau of Entomology: Its History, Activities, and Organization,* Institute for Government Research, Science Monographs of the U.S. Government 60 (Washington, D.C.: Institute for Government Research, 1930), 7.

71. Dupree, *Science in the Federal Government,* 151–62.

72. Ibid., 164–69.

73. Stirling, *Last of the Naturalists,* 89, 91–92, 98–99, 104–105, 110, 111; Jenks Cameron, *The Bureau of Biological Survey: Its History, Activities, and Organization,* Institute for Government Research, Service Monographs of the U.S. Government 54 (Baltimore, Md.: Johns Hopkins University Press, 1929), 15–23.

74. *PE* 2 (October 1866), 1–5; Walsh to Scudder, February 21, 1869, Scudder Corr., MSB.

75. Thomas published reports on the Orthoptera for Hayden in 1872, 1873, and 1875. In 1875, Packard and Uhler made a study of locusts in the West that was published in Hayden's ninth report. See Dexter, "Organization and Work of the U.S. Entomological Commission," 28.

76. The other significant nineteenth-century improvement in grasshopper control (besides the "hopperdozer") was the use of poison bran, which was introduced after the period of this study. The first success of poison bran was reported in the San Joaquin Valley in 1885; by the 1890s, poison bran had become the primary control measure against grasshoppers in the United States and Canada. Schlebecker, "Grasshoppers," 87–89; Riegert, *From Arsenic to DDT,* 39, 68–69, 197.

77. Howard, *History of Applied Entomology,* 83.

78. Riley, *Eighth Annual Report,* 96, 148; C. H. Fernald, "Evolution of Economic Entomology," *Science,* n.s. 4 (October 16, 1896), 545.

79. Riley, "The Locust Plague," 221.
80. Friedlander, *The Biology of Insects,* 164–66.
81. Summary from Riley, *Seventh Annual Report* (1875), 161–62; Alexander S. Taylor, "An Account of the Grasshoppers and Locusts of America, Condensed from an Article written and furnished by Alexander S. Taylor, Esq., of Monterey, California," in *Annual Report of the Smithsonian Institution for 1858* (Washington, D.C.: U.S. Government Printing Office, 1859), 200–13; *Zool. Rec.,* 459, 461; Motschulsky quoted in LeConte, "Methods of Subduing Insects," 204.
82. Friedlander, *The Biology of Insects,* 164–66.
83. Uvarov, *Grasshoppers and Locusts,* vol. 2, p. 338.
84. Charles V. Riley, "The Philosophy of the Movement of the Rocky Mountain Locust," *Proc., AAAS* 27 (August 1878), 271–77; Uvarov, *Grasshoppers and Locusts,* vol. 2, p. 338.
85. Riley, *Locust Plague,* 26, 62–63; Riley, *Seventh Annual Report* (1875), 169–71.
86. C. B. Williams, "Insect Migration," *AR* 2 (1957), 165; W. J. Brown, "Taxonomic Problems with Closely Related Species," *AR* 4 (1959), 78; Ashley B. Gurney and A. R. Brooks, "Grasshoppers of the Mexicanus Group, Genus *Melanoplus* (Orthoptera: Acrididae)," *Proc., USNM* 110: 3416 (Washington, D.C.: Smithsonian Institution, 1959), 1–2.
87. Friedlander, *The Biology of Insects,* 164; D. L. Gunn, "The Biological Background of Locust Control," *AR* 5 (1960), 281–82. Riley considered the Rocky Mountain locust a special form of a species normally confined to its permanent area in the West, which under special conditions developed a migratory instinct. The similarity between this and Uvarov's later development of the phase theory is apparent, but Riley and his contemporaries did not develop this aspect into a full theory of locust development and behavior. See Riley, *The Locust Plague,* 199.
88. Williams, "Insect Migration," 167.
89. For a comparison with European meteorology, see Fleming, *Meteorology in America,* 46–54 and 164–70.
90. Ibid., 70, 143–44, 150–59.
91. A. S. Packard, Jr., "Migrations of the Destructive Locust of the West," *AN* 11 (January 1877), 22. On the prominent place given meteorological investigations in the planning for the commission, see Riley to Hayden, January 14, 1876, *Hayden Survey, Letters Received;* Riley to Hayden [February 12, 1877?] NA, RG 48, *Records of the Secretary of the Interior, Letters Received Relating to the Entomological Commission;* Hayden to Schurz (transmitting report from Packard), April 23, 1877, in ibid.; Riley, *Eighth Annual Report* (1876), 105.
92. Riley, *The Locust Plague,* 50–51, 57, 103, 216–17; Fedora Petrovich Köppen, who investigated locusts in southern Russia in the 1860s, also concluded that the locust migrations were directed by a combination of instinct and wind movement. *Zool. Rec.* (1867), 459, 461.
93. Packard, "Migrations of the Destructive Locust," 22.
94. U.S. Entomological Commission, *First Annual Report* (1878), 2.
95. Ibid., 423–25.
96. Uvarov, *Grasshoppers and Locusts,* vol. 2, pp. 136–38; Uvarov states that the Packard-Thomas explanation of wind-directed movement of swarms was

correct but that this finding was lost to later investigators, who considered their statements to be opinions and not scientific conclusions based on substantial evidence. Their explanation of wind influence was rediscovered in the twentieth century, and it was subsequently recognized that the Packard-Thomas evidence and conclusions were scientifically valid. The continuity or discontinuity in locust theory over the past century is a topic beyond the bounds of this study, but on the basis of evidence from the nineteenth century, Uvarov's account would seem to underestimate the continuing influence of the Entomological Commission and the Division of Entomology in the nineteenth and twentieth centuries. See below and also Williams, "Insect Migration," 176; and Gunn, "Locust Control," 283.

97. Riley, *Destructive Locusts,* 10; Riegert, *From Arsenic to DDT,* 35, 68.
98. Riegert, *From Arsenic to DDT,* 214. *Entomopthora grull.* Fres.
99. Ibid., 22, 214.
100. Williams, "Insect Migration," 165.
101. Uvarov, *Grasshoppers and Locusts,* vol. 2, pp. 528–29.
102. Riegert, *From Arsenic to DDT,* 197; Vernon M. Stern, "Economic Thresholds," *AR* 18 (1973), 261.
103. U.S. Entomological Commission, *First Annual Report* (1878), 417–20.
104. *Insect Life* 1 (December 1888), 194–95; John Sterling Kingsley, ed., *The Riverside Natural History,* 6 vols. (New York: Houghton Mifflin, 1886), vol. 2, pp. 194–95; C. F. G. Cumming, "Locusts and Farmers of America," *Nineteenth Century* 17 (January 1885), 134–52; "An Insect Plague," *All the Year Round* (March 28, 1885), 30–32; Howard, *History of Applied Entomology,* 83–103.
105. Howard, *History of Applied Entomology,* 82; Riley, *Destructive Locusts,* 7.
106. Quoted in Dexter, "Organization and Work of the U.S. Entomological Commission," 31.

8. Profile of the American Entomological Community About 1870

1. The survey was so comprehensive, and the editing so capable, that one historian of science has cited the *Record of American Entomology* as the forerunner of the *Biological Abstracts* and the *Annual Review of Entomology,* begun almost a century later. See Ralph W. Dexter, "A. S. Packard's Annual Record of American Entomology, 1871–1873," *Bull., BES* 59–60 (1964–1965), 36; *RAE* (1868–1873).
2. See Appendix 1 for the list of American entomological authors.
3. Pamela Gilbert, *A Compendium of the Biographical Literature on Deceased Entomologists* (London: British Museum [Natural History], 1977). Other standard references include *DAB; CDAB;* the *Cyclopaedia of American Biography; Who Was Who in America: A Component Volume of Who's Who in American History* (Chicago: Marquis Who's Who, 1968); *The New Century Cyclopedia of Names* (New York: Appleton Century Crofts, 1954); Charles Van Doren, ed., *Webster's American Biographies* (Springfield, Mass.: G. & C. Merriam Company, 1974); and *The Canadian Encyclopedia* (Edmonton, Alberta: Hurtig Publishers, 1988).
4. Robert V. Bruce, "A Statistical Profile of American Scientists, 1846–1876,"

in George H. Daniels, ed., *Nineteenth-Century American Science: A Reappraisal* (Evanston, Ill.: Northwestern University Press, 1972), 63–94.

5. Kohlstedt, "Profile of a Voluntary Scientific Community," chap. 8, *American Scientific Community*, 190–223.

6. Clark A. Elliott, "The American Scientist, 1800–1863; His Origins, Career, and Interests" (Ph.D. diss., Case Western Reserve University, 1970). Elliott has published his results in "The American Scientist in Antebellum Society: A Quantitative View," *Social Studies of Science* 5 (1975), 93–108; "Models of the American Scientist: A Look at Collective Biography," *Isis* 73 (March 1982), 77–93; and "The *Royal Society Catalogue* as an Index to Nineteenth Century American Science," *Journal of the American Society for Information Science* 21 (November–December 1970), 396–401.

7. Bruce, "Statistical Profile," 74.

8. Kohlstedt, *American Scientific Community*, 223.

9. Elliott, "American Scientist," table 4, 49.

10. Beaver, *American Scientific Community*, 205–208, 118–19, passim.

11. The figure for Europe includes one person born in the European colony of Brazil.

12. Bruce, "Statistical Profile," table 6, p. 76; Kohlstedt, *American Scientific Community*, table 13, p. 209; and Elliott, "American Scientist," table 8, pp. 59–60, and table 10, p. 68.

13. Bruce, "Statistical Profile," table 6, p. 76; Kohlstedt, *American Scientific Community*, table 13, p. 209; and Elliott, "American Scientist," table 8, pp. 59–60.

14. David Elliston Allen, *The Naturalist in Britain: A Social History* (Worcester: Allen Lane, 1976), 103 (quotation), 50, 74–99, and 101–105; Harriett Ritvo, "Animal Pleasures: Popular Zoology in Eighteenth and Nineteenth-Century England," *Harvard Library Bulletin* 33 (1985), 240–41; Adrian Desmond, "The Making of Institutional Zoology in London, 1822–1836, Part 1," *History of Science* 23 (1985), 156–58, 167–68, 176–77; and Adrian Desmond, "The Making of Institutional Zoology in London, 1822–1836, Part 2," *History of Science* 23 (1985), 234, 240–43.

15. Elliott, "American Scientist," 87 and 233.

16. Bruce, "Statistical Profile," table 6, p. 76; Kohlstedt, *American Scientific Community*, table 13, p. 209; Elliott, "American Scientist," table 8, pp. 59–60. Note that the percentages are not always strictly comparable, because each author defines regions like "South" in different ways. I have followed the same scheme as Elliott in defining regions.

17. Ronald L. Numbers and Janet S. Numbers, "Science in the Old South: A Reappraisal," *Journal of Southern History* 48 (May 1982), 166, 168–69, 170, and table 3 on p. 167.

18. Ibid., table 3, p. 167. My midwest is north central in Numbers's terminology.

19. Based on Bruce, "Statistical Profile," table 11, p. 82.

20. Elliott did not include this category in his breakdown, presumably because it was not significant enough to single out.

21. Edward L. Graef, "Some Early Brooklyn Entomologists," *Bull., BES* 9 (June 1914), 47–51.

22. Zimmermann to LeConte, September 6, 1847, and Schaum to LeConte, June 21, 1854, LeConte Coll., APS.

23. Quoted in Muriel Louise Blaisdell, "Darwinism and Its Data: The Adaptive Coloration of Animals" (Ph.D. diss., Harvard University, 1976), 23.

24. On the pervasiveness of natural theology in the 1830s and 1840s, see Daniels, *American Science in the Age of Jackson,* 51–57 and 174–82.
25. For an account of the Boston society, see Kohlstedt, "Boston Society of Natural History," 178–82.
26. *CE* 41 (1909), 220; *RAE* (1869), 60; *The Naturalists' Directory, Part 2, North America and the West Indies,* edited by F. W. Putnam (Salem, Mass.: Essex Institute, 1866); *The Naturalists' Directory for 1878, Containing the Names of the Naturalists of America North of Mexico,* edited by Samuel E. Cassino (Salem, Mass.: Naturalists' Agency, 1878).
27. *Entomological News* 31 (1920), 119–20; *Naturalists' Directory* (1878); *RAE* (1873), 79; *CE* 6 (September 1874), 163.
28. *Entomological News* 5 (June 1894), 161–63; *Naturalists' Directory* (1866 and 1878).
29. *RAE; Naturalists' Directory;* Cresson, *American Entomological Society.*
30. Incidentally, fifty-nine of those ninety-eight (or 60 percent) are on the list of entomological authors cited in the *RAE,* which figure confirms that these individuals were indeed the core of the entomological community.
31. See Chapter 3.
32. $212 + 129 (.61 \times 212) = 341$.
33. $341 - 63 = 278$.
34. The proportion of published to unpublished scientists, 1 : 9, compares roughly to Reingold's estimate that there were between 5 and 10 nonpublishing scientists for each one who published. See Nathan Reingold, "Definitions and Speculations: The Professionalization of Science in America in the Nineteenth Century," in Oleson and Brown, eds., *The Pursuit of Knowledge,* 62.
35. Carl Robert Osten-Sacken, "Diptera," *RAE* (1869), 30.
36. Lintner to Scudder, June 15, 1870, Scudder Corr., MSB.
37. William Henry Edwards, *The Butterflies of North America,* 3 vols. (vol. 1: New York: Hurd and Houghton, 1874; vols. 2 and 3: Boston: Houghton Mifflin, 1884 and 1897), vol. 1, *Paphia* and *Colias III.* The three-volume work is hereafter cited as Edwards, *BNA.*
38. This subject is explored at more length in my paper "American Women Entomologists, 1840–1880," presented at the Columbia History of Science Society, Friday Harbor, Washington, April 21, 1990.
39. On this point, and for a wider discussion of women scientists in this period, see Margaret W. Rossiter, *Women Scientists in America: Struggles and Strategies to 1940* (Baltimore, Md.: Johns Hopkins University Press, 1982), xv; Sally Gregory Kohlstedt, "In from the Periphery: American Women in Science, 1830–1880," *Signs: A Journal of Women in Culture and Society* 4 (Autumn 1978), 81; Joan Hoff Wilson, "Dancing Dogs of the Colonial Period: Women Scientists," *Early American Literature* 7 (Winter 1973), 227; Barbara Welter, "The Cult of True Womanhood, 1820–1860," in Jane E. Friedman and William G. Shade, eds., *Our American Sisters: Women in American Life and Thought* (Boston: Allyn and Bacon, 1973), 96–112; and Deborah Jean Warner, "Science Education for Women in Antebellum America," *Isis* 69 (1978), 58–67.
40. "Margaretta Hare Morris," in Elliott, ed., *Biographical Dictionary of American Science,* 185–86; Kohlstedt, "In from the Periphery," 85; Rossiter, *Women Scientists,* 76.

41. The exchange is reported in *RAE* (1870), 12.
42. *CE* 5 (June 1873), 106.
43. Mary Treat, "Controlling Sex in Butterflies," *AN* 7 (March 1873), 129–32.
44. *CE* 5 (June 1873), 106.
45. Charles V. Riley, "Controlling Sex in Butterflies," *AN* 7 (September 1873), 518–19. On other occasions, Riley cited observations made by Treat, and he considered her a reliable and valued correspondent. Riley, *Third Annual Report* (1871), 164, note "t."
46. Riley to Scudder, June 22, 1870, Scudder Corr., MSB.
47. Gladys L. Baker, "Women in the U.S. Department of Agriculture," *AgH* 50 (January 1976), 192.
48. Rossiter, *Women Scientists,* 328.
49. *CE* 10 (September 1878), 176–77.
50. Ibid.
51. *CE* 11 (September 1879), 172; Mallis, *American Entomologists,* 52; Smith to LeConte, March 25, 1880, LeConte Coll., APS.
52. Kohlstedt, "In from the Periphery," 85 n. 14; *DAB* (1959), s.v. "Charlotte deBernier Taylor (1806–1861)."
53. Michele Aldrich discusses women as paleontological illustrators in "Women in Paleontology in the United States, 1840–1960," *Earth Sciences* 1 (1982–1983), 14–22.
54. Mallis, *American Entomologists,* 22–25.
55. Ibid., 290.
56. Ibid.
57. *Proc., ESW* 23 (April 1921), 91.
58. Emphasis added. *Trans., NYSAS* 12 (1852), 252.
59. Edwards to Lintner, February 23, 1863, Lintner Corr., NYSA.
60. Morris was presumably *not* the Margaretta Morris noted above. Apparently she had one or more friends who wanted to join the session. E. O. Kendall to LeConte, February 5, 1855, LeConte Coll., APS.
61. Fred[erick] W. Grant to LeConte, June 3, 1856, LeConte Coll., APS.
62. All quotations above from ibid.
63. Walsh to LeConte, August 28, 1868, LeConte Coll., APS.

9. Acceptance and Implications of Evolution

1. Charles V. Riley, "Darwin's Work in Entomology," *Proceedings of the Biological Society of Washington, D.C.* 1 (1882), 71–72.
2. Jeanne Remington and Charles L. Remington, "Darwin's Contributions to Entomology," *AR* 6 (1961), 1–6. Eighty-five entomologists are cited by name in *The Descent of Man;* Blaisdell, "Darwinism and Its Data," 269.
3. Charles V. Riley, "Mimicry as Illustrated by these two Butterflies, with some Remarks on the Theory of Natural Selection," in *Third Annual Report* (1871), 173. The special relevance of Darwin to botany is discussed in Mae Allan, *Darwin and his Flowers: The Key to Natural Selection* (New York: Taplinger Publishing, 1977).
4. Riley, "Darwin's Work in Entomology," 70–72.
5. Edward J. Pfeifer, "United States," in Thomas F. Glick, ed., *The Comparative Reception of Darwinism* (Austin: University of Texas Press, 1974), 184–85, and Edward Justin Pfeifer, "The Reception of Darwinism in the United States,

1859–1880" (Ph.D. diss., Brown University, 1957), 50–54, 87, 97–98, 160–74. Pfeifer notes the general lack of overt references to Darwinian theory in American scientific writing. See also Rosenberg, *No Other Gods*, 3, who notes that the most remarkable aspect of Darwinism in the United States was the lack of conflict it inspired. Bert James Lowenberg notes the initially cautious attitude of leading American scientists toward Darwinism and their general conversion to evolution by the mid-1870s: "The Reaction of American Scientists to Darwinism," *American Historical Review* 38 (July 1933), 687–701.

6. Pfeifer, "Reception of Darwinism," 54; Benjamin D. Walsh, "On Certain Entomological Speculations of the New England School of Naturalists," *Proc., ESP* 3 (August 1864), 207–11, 223–29, 236–41.
7. Walsh to Scudder, November 14, 1863, Scudder Corr., MSB.
8. Walsh's articles are noted in Pfeifer, "United States," 184–85.
9. Darwin to Walsh, October 21, 1864, Charles Darwin–Benjamin D. Walsh Correspondence, microfilm of eighteen letters, APS, from photostats in the Library of the Field Museum of Natural History, Chicago, Illinois.
10. Darwin to Walsh, December 4, 1864, in ibid.
11. Darwin to Walsh, April 20, 1866, in ibid.
12. Benjamin D. Walsh, "On the Insects, Coleopterous, Hymenopterous, and Dipterous, inhabiting the Galls of Certain Species of Willow, Part 1, Diptera," *Proc., ESP* 3 (December 1864), 635. See pp. 634–40 for Walsh's explanation of the willow galls in relation to Darwinism. See also his article "On Phytophagic Varieties and Phytophagic Species," *Proc., ESP* 3 (November 1864), 403–30, reviewed in *Zool. Rec.* 1 (1864), 332, where Walsh's interpretation of known facts in terms of evolutionary theory is summarized. Riley, in 1871, cited similar examples of variation in moth larvae due to differences in food plants as an indication of evolutionary influences. See Riley, *Third Annual Report* (1871), 141. For related aspects of Walsh's work in evolutionary theory, see Brown, "Taxonomic Problems," 79.
13. James D. Dana, "On Cephalization, No. IV: Explanations drawn out by the Statements of an Objector," *AJS*, n.s. 41 (March 1866), 163–74; A. S. Packard, Jr., "On Certain Entomological Speculations: A Review," *Proc., ESP* 6 (November 1866), 209–18; Benjamin D. Walsh, "Prof. Dana and his Entomological Speculations," *Proc., ESP* 6 (June 1866–1867), 116–21.
14. Ernst Mayr, "The Nature of the Darwinian Revolution," *Science* 176 (1972), 983–88.
15. Ernst Mayr, *Population, Species, and Evolution: An Abridgment of Animal Species and Evolution* (Cambridge, Mass.: Harvard University Press, 1970), 4–5.
16. William R. Coleman, *Georges Cuvier, Zoologist: A Study in the History of Evolution Theory* (Cambridge, Mass.: Harvard University Press, 1964), 186.
17. Ibid., 67–68, 74, 147–48.
18. Ibid., 3; E. S. Russell, *Form and Function: A Contribution to the History of Animal Morphology* (London: John Murray, 1916), 31–33.
19. The evolutionists would later claim that these advances were possible because the relationships Cuvier discovered were based on common ancestry. It was some time, however, before comparative anatomists developed their evidence in this fashion. See Russell, *Form and Function*, 31–40, 246–47, 260, 303; Coleman, *Cuvier* (1964), xiii.
20. Russell, *Form and Function*, 3, 182. See also Mary P. Winsor, who maintains

that the basis of Agassiz's opposition to Darwinism was primarily psychological: "Louis Agassiz and the Species Question," *Studies in the History of Biology* 3 (1979), 89–90, 112.

21. Mayr, *Population, Species, and Evolution*, 4–5.
22. Ibid., 4–5. On essentialist and related contemporary objections to Darwinism, see also David L. Hull, *Darwin and His Critics: The Reception of Darwin's Theory of Evolution by the Scientific Community* (Cambridge, Mass.: Harvard University Press, 1973), 55, 67–73.
23. Packard, "On Certain Entomological Speculations: A Review," 218 (emphasis added).
24. *Trans., ISAS* 4 (1859–1860), 631–32 (emphasis in original).
25. "G" [Augustus R. Grote] in *PE* 1 (April 30, 1866), 59.
26. Walsh to Scudder, December 17, 1864, Scudder Corr., MSB; see also Walsh to LeConte, August 15, 1868, LeConte Coll., APS.
27. Walsh to Scudder, April 6, 1865, Scudder Corr., MSB.
28. Mary Alice Evans, "Mimicry and the Darwinian Heritage," *JHI* 26 (1965), 213; Barbara Beddall, ed., *Wallace and Bates in the Tropics: An Introduction to the Theory of Natural Selection. Based on the Writings of Alfred Russel Wallace and Henry Walter Bates* (London: Macmillan, 1969), 104–105.
29. From Bates's paper in *Transactions of the Linnaean Society of London* 23 (1862), quoted in Beddall, ed., *Wallace and Bates*, 214.
30. Darwin to Bates, November 20 [1862], in Charles Darwin, *The Life and Letters of Charles Darwin, Including an Autobiographical Chapter*, edited by his son, Francis Darwin, 3 vols. (1888; repr., New York: Johnson Reprint, 1969), vol. 2, p. 183.
31. Darwin to Walsh, March 27, 1865, Darwin-Walsh Corr., APS; Remington and Remington, "Darwin's Contributions to Entomology," 8.
32. Evans, "Mimicry," 213; Blaisdell, "Darwinism and Its Data," 142; Riley, "Mimicry as Illustrated by these two Butterflies," 161; Benjamin D. Walsh and Charles V. Riley, "Imitative Butterflies," *AE* 1 (June 1869), 189–93.
33. Walsh and Riley, "Imitative Butterflies," 191.
34. Ibid., 192.
35. Riley, "Mimicry as Illustrated by these two Butterflies," 165.
36. Ibid., 160.
37. Ibid., 167; Samuel H. Scudder, "Is Mimicry Advantageous?" *Nature* 3 (December 22, 1870), 147.
38. Riley, "Mimicry as Illustrated by these two Butterflies," 170.
39. Grote to Scudder, October 30, 1869, Scudder Corr., MSB.
40. Grote to Scudder, May 18, 1872, in ibid.
41. Notice of forthcoming article by Samuel H. Scudder, "Mimicry in Butterflies Explained by Natural Selection," to be published in *Psyche*, as noted in *CE* 8 (October 1876), 181 [apparently the article did not appear]; Samuel H. Scudder, "Presidential Address, Subsection of Entomology," *Proc., AAAS* 29 (1880), 612–13.
42. Edward J. Pfeifer, "The Genesis of American Neo-Lamarckism," *Isis* 56 (1965), 157–58; see also Theodore John Greenfield, "Variation, Heredity, and Scientific Explanation in the Evolutionary Theories of Four American Neo-Lamarckians, 1867–1897" (Ph.D. diss., University of Wisconsin, Madison, 1986).

43. Peter J. Bowler, "Edward Drinker Cope and Evolution," *Isis* 68 (June 1977), 259.
44. Ibid., 250–51; Ralph W. Dexter, "The Impact of Evolutionary Theories on the Salem Group of Agassiz Zoologists (Morse, Hyatt, Packard, Putnam)," *Historical Collections of the Essex Institute* 115 (1979), 165–67; A. S. Packard, Jr., *Lamarck, the Founder of Evolution: His Life and Work* (New York: Longmans, Green, 1901), 390–98.
45. Greenfield, "American Neo-Lamarckians," 64–69.
46. Greenfield, "American Neo-Lamarckians," 69–74; Edward S. Morse, "Address of Professor Edward S. Morse, Vice President, Section B (What American Zoologists Have Done for Evolution)," *Proc., AAAS* 25 (1876), 158–59; Packard, *Lamarck,* 390; Stephen Bocking, "Alpheus Spring Packard in the Evolution Debate," *JHB* 21:3 (Fall 1988), 439–48.
47. Packard, *Lamarck,* 402.
48. Dexter, "Impact of Evolutionary Theories," 163–64.
49. Pfeifer, "Neo-Lamarckism," 159; Greenfield, "American Neo-Lamarckians," 10–14, 55–57, 66.
50. John L. LeConte, "Presidential Address," *Proc., AAAS* (1875), 8, 12, 15.
51. Charles V. Riley, "On the Causes of Variation in Organic Forms," *Proc., AAAS* 37 (1888), 265, 271.
52. Pfeifer, "Neo-Lamarckism," 162–63.
53. Packard, *Lamarck,* 400.
54. Riley, "Causes of Variation," 234–40, 255–56.
55. Paul F. Boller, Jr., *American Thought in Transition: The Impact of Evolutionary Naturalism, 1865–1900* (Chicago: Rand McNally, 1969), 3; Loren Eiseley, *Darwin's Century: Evolution and the Men Who Discovered It* (1958; repr. New York: Anchor, 1961), 217, 240–46; Greenfield, "American Neo-Lamarckians," 2–8, 27, and 118–22.
56. *Psyche* 2 (January–February 1878), 97.
57. See *Psyche* 3 (January 1880), 7–8; 4 (January–February 1883), 12.
58. *Psyche* 2 (January–February 1877), 1.
59. *Psyche* 1 (May 1875), 70; Dimmock to Edwards, September 11, 1878, Edwards Coll., WVUL.
60. Coleman, *Cuvier* (1964), 182, 186.
61. Ibid., 2, 14, 29, 55, 171; Willi Hennig, "Phylogenetic Systematics," *AR* 10 (1965), 100–101.
62. *CE* 12 (September 1880), 163.
63. *Psyche* 2 (March 1879), 217.
64. Ibid.
65. Pfeifer, "United States," 184–85, 192–94.
66. Riley, "John Lawrence LeConte," 109.
67. *Proc., AAAS* (1875), 5–6.
68. *CE* 4 (June 1872), 102–105.
69. Ibid.
70. *CE* 4 (October 1872), 184–86. See also Walker's communication in *CE* 3 (October 1871), 141. Other references to evolution include those by Riley, *CE* 3 (September 1871), 119; Hagen, "On Genera," *CE* 8 (October 1876), 196; Grote, *CE* 8 (March 1876), 56–58; and Grote, *CE* 11 (February 1879), 40.

71. *Bull., BES* 1 (May 1878), 1.
72. *EA* 1 (April 1885), 1, 20.
73. The discussion of the London society is based on Remington and Remington, "Darwin's Contributions to Entomology," 9–10, and Blaisdell, "Darwinism and Its Data," 242–89.
74. Opinion on the climate of debate varies. Remington and Remington describe the society as a source of determined opposition to Darwin (p. 9), whereas Blaisdell detects an atmosphere of friendly coexistence (pp. 264 and 281–82).
75. Blaisdell, "Darwinism and Its Data," 268; Remington and Remington, "Darwin's Contributions to Entomology," 9.
76. Blaisdell, "Darwinism and Its Data," 288–89.
77. Blaisdell notes that field naturalists like Bates and Wallace were especially receptive to Darwinism, whereas "closet" naturalists like Westwood were more generally opposed. The London society had a strong contingent of closet taxonomists. See Blaisdell, "Darwinism and Its Data," 63, 243, 252, and 376–77.
78. Cassedy, *American Medicine and Statistical Thinking,* traces nineteenth-century Americans' uses of statistics in medicine and public health. Just as America, with its great variety of races, offered excellent opportunities to study phrenology (p. 149), America's geography also offered a great natural laboratory for the study of natural selection. See also pp. 4, 20, 62, 69, 158, 173–77, and 191.
79. Boller, *American Thought in Transition,* 11–12; Jon H. Roberts, *Darwinism and the Divine in America: Protestant Intellectuals and Organic Evolution, 1859–1900* (Madison: University of Wisconsin Press, 1988), ix–x, 15, 40, 84.
80. Bert James Lowenberg, "The Controversy over Evolution in New England," *New England Quarterly* 8 (June 1935), 233–56.
81. Fitch to Harris, August 11, 1852, MCZ Archives; on the role of natural theology, see also Roberts, *Darwinism and the Divine,* 4–8.
82. Lowenberg, "Evolution in New England," 251.
83. Riley, *Third Annual Report* (1871), 173.
84. Walsh and Riley, "Imitative Butterflies," 190.
85. Riley, "Mimicry as Illustrated by these two Butterflies," 174–75.
86. Walter B. Hendrickson, "An Illinois Scientist Defends Darwinism: A Case Study in the Diffusion of Scientific Theory," *Transactions of the Illinois Academy of Sciences* 65 [314] (1972), 25–29; Benjamin D. Walsh, "Imported Insects: The Gooseberry Sawfly," *PE* 1 (September 1866), 118.
87. Morse, "Address of Professor Edward S. Morse," 138.
88. Ibid.
89. Pfeifer, "United States," 197–98.
90. F. M. Webster, "Presidential Address, 1897," p. 4, in AAEE, "Annual Meetings and Presidential Addresses."

10. William Henry Edwards and Polymorphism in Butterflies

1. Alfred Russel Wallace, "Mimicry, and Other Protective Resemblances Among Animals," in *Contributions to the Theory of Natural Selection: A Series of Essays* (1870; repr., New York: Arno Press, 1973), 45–47.
2. Polymorphism occurs in many organisms and in a wide variety of characters.

For a modern definition and exposition, see Julian Huxley, "Morphism and Evolution," *Heredity* 9, pt. 1 (April 1955), 1–52, especially 3–5 and 42–43. For the treatment of polymorphism and mimicry in the pre-Darwinian era, see Blaisdell, "Darwinism and Its Data," chap. 1, "Natural Theology and Nature's Disguises," especially the treatment of analogies in the insect world by Kirby (p. 19) and John O. Westwood (p. 29). The contrast between Darwinians and natural theologians is discussed in ibid., 77–81, 276, and 373–75.

3. [William Edwards], *Memoirs of Colonel William Edwards . . . with notes and additions by his son, William W. Edwards, and his Grandson, William Henry Edwards* (Washington, D.C.: Press of W. F. Roberts, 1897), 12–52; C. J. S. Bethune, "William Henry Edwards," *CE* 41 (August 1909), 245–46.

4. [Edwards], *Memoirs,* 104–105.

5. William Stanton, *The Great United States Exploring Expedition of 1838–1842* (Berkeley: University of California Press, 1975), 294–304; [Ebenezer] Emmons to whom it may Concern (recommending Edwards), in Albert Hopkins to [Edwards], June 30, 1842, Edwards Corr., WVUL; William Henry Edwards, "The Entomological Reminiscences of William Henry Edwards," edited by Cyril F. Dos Passos, *JNYES* 59 (September 1951), 134.

6. William H. Edwards, *A Voyage up the River Amazon, including a residence at Pará* (London: John Murray, 1847); Prince Adalbert of Prussia preceded Edwards in 1842, but he did not publish until 1849. See George Woodcock, *Henry Walter Bates: Naturalist of the Amazons* (New York: Barnes and Noble, 1969), 10–12.

7. Woodcock, *Bates,* 28–32; Beddall, *Wallace and Bates,* 20–21; H. Lewis McKinney, *Wallace and Natural Selection* (New Haven: Yale University Press, 1972), 8–9.

8. Edwards, "Reminiscences," 136–37; Bethune, "William Henry Edwards," 248.

9. Bethune, "William Henry Edwards," 245–48; Alfred Goldsborough Mayor, "Samuel Hubbard Scudder, 1837–1911," *Biographical Memoirs,* vol. 27 (Washington, D.C.: National Academy of Sciences, 1924), 83, 86; Edwards, "Reminiscences," 139–40, 149.

10. Edwards, "Reminiscences," 142–43, 150, 157.

11. Bethune, "William Henry Edwards," 245–48; *Proceedings of the Entomological Society of London* (1909), ixxxix; Edwards, "Reminiscences," 144–45.

12. Wallace, "Mimicry," 75–76; William H. Edwards, "An Abstract of Dr. Aug. Weismann's Paper on 'The Seasonal-Dimorphism of Butterflies.' [Leipzig 1875.] To which is Appended a Statement of Some Experiments made upon Papilio Ajax," *CE* 7 (December 1875), 228. In this article, the genus *Arashnia* is translated as *Vanessa.*

13. *EMM* (February 1878), 211–12; Edwards, *BNA* 2 (1884), *P. turnus* and *glaucus.* For another nineteenth-century discovery of sexual dimorphism, see Alfred Russel Wallace, "The Malayan Papilionidae," in Wallace, *Contributions to the Theory of Natural Selection,* 147–48.

14. See Wallace to Edwards, January 7, 1864, Edwards Corr., WVUL (sending articles). Edwards was in contact with Bates in 1868 and probably with Darwin as well. See Darwin to Bates, April 15 (1868), in Robert M. Stecher, ed., "The Darwin-Bates Letters: Correspondence between Two Nineteenth-

Century Travellers and Naturalists, Part 2," *Annals of Science* 25 (June 1969), 111–12.

15. Wallace, "The Malayan Papilionidae," 148–49.

16. Ibid., 145; Edwards cited Wallace's summary (as reprinted in 1870) in *BNA*, vol. 1, s.v. *P. ajax.*

17. Quoted in Wallace, "The Malayan Papilionidae," 130–32. Wallace called attention to the analogy between butterflies and orchids (one of Darwin's specialties) as subjects for the study of natural selection. Orchids exhibited the only known example of mimicry in the plant kingdom and the only case of conspicuous polymorphism among plants. Ibid., 185. The observation that butterflies' wings contain unique clues to evolution was confirmed in the 1880s and later through the taxonomic studies of John Henry Comstock. See Henson, "Evolution and Taxonomy."

18. Edwards, *BNA*, vol. 2, notes following *Grapta III.*

19. *Bull., BES* 1 (May 1878), 1; George P. Hulst, "Hints on the Rearing of Lepidoptera," 2 parts, *Bull., BES* 2 (September 1879–January 1880), 63–73, and 4 (June 1881), 13–14. See also *Psyche* 4 (May–June 1883), 53; L. Lowell Elliott to Edwards, March 28, 1882, Edwards Corr., WVUL.

20. See, for example, Riley to Scudder, June 15, 1870, Scudder Corr., MSB, and Packard, *Guide*, 95.

21. William Saunders in *CE* 1 (February 15, 1869), 53, and (April 15, 1869), 74–77.

22. *CE* 1 (April 15, 1869), 82; Lintner to Scudder, April 19, 1869, and May 25, 1869, Scudder Corr., MSB.

23. Edwards, *BNA*, vol. 1, s.v. *P. ajax;* Edwards, "Reminiscences," 145.

24. Edwards in *CE* 2 (July 1, 1870), 115.

25. Edwards in *CE* 2 (August 31, 1870), 133–44.

26. William H. Edwards, "Notes on the Early Stages of Some of our Butterflies," *CE* 5 (December 1873), 223–25; Thomas E. Bean reporting on *Pieris vernalis* and *Pieris protodice, CE* 9 (November 1877), 201–203.

27. P. C. Zeller to Edwards, August 3, 1879, Edwards Corr., WVUL.

28. Edwards, *BNA*, vol. 2, notes following *Grapta III.*

29. Ibid.

30. Edwards, *BNA*, vol. 1, s.v. *P. ajax.*

31. Ibid.; *CE* 2 (December 1870), 162, and *CE* 3 (August 1871), 70.

32. Though only two names, *ajax* and *marcellus*, were in use before Edwards changed the terminology, all three forms had been accurately described and illustrated by various authorities under the two names. The species is now known as *Eurythides marcellus*, or the zebra swallowtail butterfly; see Lee D. Miller and F. Martin Brown, *A Catalogue/Checklist of the Butterflies of America North of Mexico*, Memoir no. 2 (Lepidopterists' Society, 1981), 60. For the history of the nomenclature, see Edwards, *BNA*, vol. 1, s.v. *P. ajax;* Walter Rothschild and Karl Jordan, "A Revision of the American Papilios," *Novitates Zoologicae* 13 (August 1906), 690; and John R. Beattie, *The Rhopalocera Directory: A Comprehensive Index to Butterfly Names Found in the Systematic Index of the* Zoological Record *1864–1971 and in* Berichte über die wissenschaftlichen Leistungen im Gebiete der Entomologie *1834–1863* (Berkeley, Calif.: JB Indexes, 1976). I thank Kenelm W. Philip for these and other references on lepidopteran nomenclature.

33. Edwards, *BNA*, vol. 1, s.v. *Grapta V.*
34. William H. Edwards, "Rearing Butterflies from the Egg," *CE* 3 (August 1871), 70.
35. In vol. 1, Edwards had already named them varieties *(Grapta, interrogationis var. umbrosa* and var. *fabricii),* and then he apparently later added his opinion that they should be considered forms rather than varieties without changing the text. See Edwards, *BNA*, vol. 1, s.v. *Grapta IV* and *V.*
36. Lintner to Scudder, September 28, 1874, Scudder Corr., MSB.
37. Edwards, *BNA*, vol. 1, s.v. *Grapta V.*
38. William H. Edwards, "On the Identity of Grapta Dryas with Comma," *CE* 5 (October 1873), 184; Edwards, *BNA*, vol. 2, s.v. *Grapta I.*
39. William H. Edwards, "Notes on a remarkable Variety of Papilio turnus and descriptions of two species of Diurnal Lepidoptera," *Trans., AES* 2 (1868), 208–209.
40. Theodore L. Mead, "Notes Upon Some Butterfly Eggs and Larvae," *CE* 7 (September 1875), 161–62.
41. William H. Edwards, "History of Phyciodes Tharos, A Polymorphic Butterfly," *CE* 9 (March 1877), 51–55.
42. Edwards, *BNA*, vol. 1, s.v. *Grapta V.*
43. Bates to Edwards, September 14, 1871, Edwards Corr., WVUL.
44. Ibid.
45. Wallace to Edwards, February 15, 1873, in ibid. Wallace also suggested that Edwards try certain experiments reported by T. Wood in the London Entomological Society Proceedings in which the color of Lepidoptera pupae changed when they were placed in different surroundings. Wallace to Edwards, November 5, 1873, in ibid.
46. Weismann to Edwards, March 16, 1872, in ibid. It is not clear whether Weismann sent the original German monograph, *Über den Einfluss der Isolirung auf der Artbildung* (1872), or an English translation.
47. Gloria Robinson, "August Friedrich Leopold Weismann," *DSB* (1970), 232–39.
48. August Weismann, *Studies in the Theory of Descent; with Notes and Additions by the Author,* translated and edited by Raphael Meldola with a prefatory notice by Charles Darwin, 2 vols. (London: Sampson Low, Marston, Searle, and Rivington, 1882), vol. 1, pp. ix, 117; Frederick B. Churchill, "August Weismann and a Break from Tradition," *JHB* 1 (1968), 91; *Zool. Rec.* 14 (1877), 119 ins.
49. Weismann, *Studies*, vol. 1, pp. 9–10.
50. Ibid., 10–12, 16.
51. Ibid., 19–22.
52. Ibid., 13–14, 29, 38–42.
53. Ibid., 18, 24, 29–30.
54. Greenfield, "American Neo-Lamarckians," 33, 43, and 122.
55. Edwards, *BNA*, vol. 2, s.v. *Phyciodes tharos.*
56. Edwards, "Abstract," 237; William H. Edwards, "Experiments Upon the Effect of Cold Applied to Chrysalids of Butterflies," *Psyche* 3 (February 1880), 17–18; Weismann, *Studies*, vol. 1, p. 39.
57. Edwards, "Abstract," 234–38; Weismann, *Studies*, vol. 1, pp. 30–32.
58. Edwards, "Effects of Cold," 15.

59. Edwards, "Abstract," 228–40.
60. William H. Edwards, "History of Phyciodes Tharos, a Polymorphic Butterfly," *CE* 9 (January 1877), 5–8, same title, *CE* 9 (March 1877), 51–53; Edwards, *BNA*, vol. 2, s.v. *Phyciodes tharos;* Edwards's experiments on Phyciodes are reported in Weismann, *Studies,* vol. 1, pp. 140–48.
61. Edwards, *BNA*, vol. 1, s.v. *Grapta satyrus.*
62. A form of *Alope* named *olympus* was found in the range from the Rocky Mountains to the Pacific coast. See William J. Holland, *The Butterfly Book: A Popular Guide to a Knowledge of the Butterflies of North America* (New York: Doubleday, Page, 1913), 215.
63. William H. Edwards, "On Certain Species of Satyrus," *CE* 12 (February 1880), 21–31; Edwards, *BNA*, vol. 2, s.v. *Satyrus alope.*
64. William H. Edwards, "On Certain Species of Satyrus," pts. 1, 2, 3, 4, *CE* 12 (February, March, May, June 1880), 21–32, 51–54, 90–94, 109–14; Edwards, *BNA* vol. 2, s.v. *Satyrus alope.*
65. See Huxley, "Morphism and Evolution," 42, who states that in certain cases polymorphism "may give rise to secondarily monomorphic populations, which, if isolated, may then evolve into distinct species or subspecies." See also pp. 12–13. For a modern definition of speciation, see Michael T. Ghiselin, *The Triumph of the Darwinian Method* (Berkeley: University of California Press, 1969), 90: "A species is a population, a unit of evolution and of reproductive activity. . . . the development of a barrier to reproduction is called *speciation,* and the products resulting from it are . . . *biological species.*"
66. Bates wrote to Edwards, for example, that variation could be caused by climate or by natural selection but that in both cases previous selection predetermined a species toward variation in a certain direction. He advised Edwards to watch for this in his work. Bates to Edwards, April 19, 1877, Edwards, Corr., WVUL; see Mayr, *Population, Species, and Evolution,* 95, for the problem of blending characteristics in the thinking of early Darwinians; and Ghiselin, *Triumph of the Darwinian Method,* 179–80, for Darwin's concept of environmentally invoked variation.
67. William H. Edwards, *Synopsis of North American Butterflies* (New York: Hurd and Houghton, 1874), 52.
68. Edwards, *BNA*, vol. 2, preface; see Zeller to Edwards, August 3, 1879, Edwards, Corr., WVUL, for a corroborating opinion that Edwards was ahead of the Germans in the knowledge of preparatory stages.
69. Blaisdell, "Darwinism and Its Data," 102–103.
70. Ibid., 106–24; H. Lewis McKinney, "Henry Walter Bates," *DSB*, vol. 1, pp. 500–504.
71. Bates to Edwards, June 1, 1875, Edwards, Corr., WVUL. See also Bates to Edwards, April 19, 1877, in ibid.
72. Wallace to Edwards, February 15, 1873, April 4, 1887, and June 18, 1889, in ibid.
73. Weismann to Edwards, March 16, 1872, in ibid.; Weismann, *Studies* (1882), vol. 1, pp. xviii–xix, 126–40; M. Robert [Raphael] Meldola to Edwards, January 27, 1879 (expressing interest in incorporating Edwards's work in the new edition), Edwards, Corr., WVUL.
74. Weismann, *Studies,* vol. 1, pp. 18, 54–61, 111–13; *Zool. Rec.* 12 (1875), 405.
75. Churchill, "Weismann," 100–105.

76. Greenfield, "American Neo-Lamarckians," 118–21.
77. William R. Coleman, *Biology in the Nineteenth Century: Problems of Form, Function, and Transformation* (1971; repr., Cambridge: Cambridge University Press, 1977), 38–40; Gloria Robinson, *A Prelude to Genetics: Theories of a Material Substance of Heredity: Darwin to Weismann* (Lawrence, Kans.: Coronado Press, 1979), chap. 7, "Weismann's Concept of the Continuity of the Germ Plasm," especially pp. 133–35, 137, 141, 151, 154, and 165.
78. Churchill, "Weismann," 91, 100, 104–105.
79. Frederick B. Churchill, "Lepidopteran Research and the Rise of Classical Genetics, 1880–1920," abstract of paper, *Proceedings of the Sixteenth International Congress of the History of Science,* Bucharest, Romania, August 26–September 3, 1981; Frederick B. Churchill, preface to L. I. Blacher, *The Problem of the Inheritance of Acquired Characteristics: A History of a priori and Empirical Methods Used to Find a Solution,* English translation edited by F. B. Churchill (1971; repr., New Delhi, India: Amerind Publishing for the Smithsonian Institution Libraries and the National Science Foundation, 1982), vii–viii.
80. H. J. Müller, "Die Wirkung verschiedener Licht-Dunkel-Relationen auf die Saisonformenbildung von *Araschnia levana,*" *Naturwissenschaften* 43 (1956), 503–504; H. J. Müller, "Probleme der Insektendiapause," *Zoologischer Anzeiger,* Supplementband 29 (1965), 203–204, 206.
81. Edwards, "Reminiscences," 160–82.
82. Bethune, "William Henry Edwards," 247.
83. *CE* 10 (December 1878), 221.
84. See the analysis of Edwards's correspondence in Chapter 8.
85. Edwards to Lintner, December 9, 1878, Lintner Corr., NYSA.
86. Edwards, "History of Phyciodes Tharos" (March 1877), 51.
87. Bates to Edwards, April 19, 1877, and Zeller to Edwards, August 3, 1879, Edwards, Corr., WVUL.
88. Edwards to Lintner, October 7, 1878, Lintner Corr., NYSA.

11. The Yucca Moth

1. Testimonies to Darwin's influence appear in William Trelease, review of Hermann Müller, *The Mutual Relations between Flowers and the Insects which Serve to Cross Them, AN* 13 (July 1879), 451–52; and the section on insects in Kingsley, ed., *The Riverside Natural History,* vol. 2, p. 94.
2. The most important later works are *The Variation of Animals and Plants under Domestication* (1868), *The Effects of Cross and Self Fertilization on the Vegetable Kingdom* (1876), and *The Different Forms of Flowers on Plants of the same Species* (1877).
3. Ghiselin, *Triumph of the Darwinian Method,* 135–39.
4. Ibid., 136–38.
5. Quoted in ibid., 136.
6. Joseph Ewan, "George Engelmann," *DSB,* 130–31; *DAB,* 1958 ed., s.v. "George Engelmann"; Hendrickson, "An Illinois Scientist Defends Darwinism," 25–26.
7. George Engelmann, "The Flower of Yucca and its Fertilization," *Bulletin of the Torrey Botanical Club* 3 (July 1872), 33; George Engelmann, "Notes on the

Genus *Yucca*," *Trans., SLAS* 3 (April 1873), 17–54, 210–14; Charles V. Riley, "On a new Genus in the Lepidopterous Family *Tineidae*, with Remarks on the Fertilization of *Yucca: Pronuba*, Nov. Genus," *Trans., SLAS* 3 (1873), 59.

8. In addition to the orchid book, Riley cited Darwin's *The Variation of Animals and Plants Under Domestication* (1868), Asa Gray's *How Plants Behave*, and Fritz Müller (specific title not given), who gave examples of unusual dependencies of plants on specific insects for their pollination; see Riley, "Fertilization of *Yucca*," 57.

9. Ibid., 59–64.

10. Reported in *CE* 4 (October 1872), 182.

11. Quoted in Charles V. Riley, "Further Notes on the Pollination of Yucca and on Pronuba and Prodoxus," *Proc., AAAS* 29 (1880), 623.

12. Riley, "Fertilization of *Yucca*," 63–64.

13. Charles V. Riley, "On a New Genus in the Lepidopterous Family Tineidae: with Remarks on the Fertilization of Yucca," [same article as in the *Transactions of the St. Louis Academy of Science* with slightly altered title and additional notes and correspondence], in Riley, *Fifth Annual Report* (1873), 150–60; Charles V. Riley, "The Yucca Moth—*Pronuba yuccasella* Riley," in Charles V. Riley, *Sixth Annual Report on the Noxious, Beneficial, and other Insects of the State of Missouri* (Jefferson City, Mo.: State Printer, 1874), 131–35; Charles V. Riley, "Supplementary Notes on *Pronuba Yuccasella*," *Trans., SLAS* 4 (1874), 178–80; Charles V. Riley, "On the Ovaposition of the Yucca Moth," *AN* 7 (October 1873), 619–23.

14. Riley, "Further Notes on the Pollination of Yucca," 637–69. Riley continued to work on the group for many years, notably in collaboration with William Trelease. See Charles V. Riley, "Further Remarks on *Pronuba yuccasella* and on the Pollination of Yucca," *Trans., SLAS* 8 (1878), 568–73; *Insect Life* 1 (June 1889), 367–71 (where he defended his priority in the discovery of pronuba); Charles V. Riley, "Some Interrelations of Plants and Insects," *Proc., BSW* 7 (May 1892), 81–104; Charles V. Riley, "The Yucca Moth and Yucca Pollination," *Annual Report of the Missouri Botanical Garden* 3 (1892), 99–158; Charles V. Riley, "Further Notes on Yucca Insects and Yucca Pollination," *Proc., BSW* 8 (June 1893), 41–54; William Trelease, "Detail Illustrations of Yucca," *Annual Report of the Missouri Botanical Garden* 3 (1892), 159–66; William Trelease, "Further Studies on Yuccas and their Pollination," *Annual Report of the Missouri Botanical Garden* 4 (1893), 181–226; Charles V. Riley–William Trelease Correspondence, November 17, 1891, to June 7, 1892, William Trelease Papers, Cornell University Library, Ithaca, New York. For a modern account of the yucca pollination process, see Michael Proctor and Peter Yeo, *The Pollination of Flowers* (London: Collins, 1973), 316–18.

15. Charles V. Riley, "The True and the Bogus Yucca Moth; with Remarks on the Pollination of Yucca," *AE* 3 (1880), 142–45.

16. Thomas Meehan, "On the Fertilization of Yucca," *North American Entomologist* 1 (November 1879), 34.

17. Riley, "Supplementary Notes on Pronuba Yuccasella," 179–80; Riley, "Fertilization of Yucca," in Riley, *Fifth Annual Report* (1873), 158–59.

18. As quoted in *Vick's Illustrated Monthly Magazine* for March 1880, in Riley, "Further Notes on the Pollination of Yucca," 624.

19. Riley, "Further Notes on the Pollination of Yucca," 626–28.

12. The Debate over Entomological Nomenclature

1. *CE* 4 (November 1872), 212.
2. *CE* 4 (October 1872), 183.
3. *CE* 4 (November 1872), 212.
4. Morris to Edwards, November 30 [1873], Edwards, Corr., WVUL; P. R. Uhler, "Meeting of the American Association for 1873," *CE* 5 (September 1873), 165–66.
5. *CE* 5 (November 1873), 214.
6. William H. Edwards, "Some Remarks on Entomological Nomenclature," *CE* 5 (February 1873), 35–36.
7. *CE* 7 (October 1875), 183.
8. John L. LeConte, "On Some Changes in the Nomenclature of North American Coleoptera which have been Recently Proposed," *CE* 6 (October 1874), 186–87.
9. "Report of the Committee on Zoological Nomenclature," *Proc., AAAS* 26 (1877), 42–43; David Heppell, "The Evolution of the Code of Zoological Nomenclature," in Alwyne Wheeler and James H. Price, eds., *History in the Service of Systematics: Papers from the Conference to Celebrate the Centenary of the British Museum (Natural History), 13–16 April 1981* (London: Society for the Bibliography of Natural History, 1981), 136–37.
10. "Report of the Committee on Zoological Nomenclature," 47.
11. William H. Edwards, "Notes on Entomological Nomenclature, Part I," *CE* 8 (March 1876), 41–52.
12. Ibid., 46.
13. Essig, *History of Entomology,* 664–66.
14. *CE* 7 (July 1875), 121; see also Riley to Scudder, June 22, 1872, Scudder Corr., MSB.
15. Edwards, "Notes on Entomological Nomenclature, Part I," 48.
16. Edwards to Lintner, September 14, 1876, Lintner Corr., NYSA. See Edwards to Lintner, February 20, 1875, in ibid., for the same charge leveled at Scudder.
17. Samuel H. Scudder, "Canons of Systematic Nomenclature for the Higher Groups" (reprinted from the *American Journal of Science,* May 1872), *CE* 5 (March 1873), 55–59.
18. Augustus R. Grote in *CE* 4 (November 1872), 214–16; Augustus R. Grote, "On Mr. Scudder's Systematic Revision of Some of the American Butterflies," *CE* 5 (April 1873), 62–63; and Augustus R. Grote, "On Jacob Hübner and His Works on the Butterflies and Moths," *CE* 8 (July 1876), 131–35.
19. *CE* 4 (November 1872), 212.
20. *CE* 9 (May 1877), 91.
21. *CE* 8 (October 1876), 180.
22. *CE* 6 (January 1874), 18–19.
23. Edwards, "Some Remarks on Entomological Nomenclature," 35–36.
24. H. K. Morrison, "Specific Nomenclature," *CE* 5 (April 1873), 70–71.
25. John Henry Comstock, "Evolution and Taxonomy: An Essay on the Application of the Theory of Natural Selection in the Classification of Animals and Plants, Illustrated by a Study of the Evolution of the Wings of Insects, and by a Contribution to the Classification of the Lepidoptera," in *The Wilder*

Quarter Century Book: A Collection of Original Papers dedicated to Professor Burt Green Wilder at the close of his Twenty-fifth Year of Service in Cornell University (1868–1893) by some of his former Students (Ithaca, N.Y.: Comstock Publishing, 1893), 37; H. F. Wilson and M. H. Doner, *The Historical Development of Insect Classification* (St. Louis, Mo.: John S. Swift, 1937), 10–25; Pamela Henson traces the development of evolutionary taxonomy in entomology through the career of John Henry Comstock. His proposals in the 1880s and 1890s had received wide acceptance only by the 1920s. See Henson, "Evolution and Taxonomy," 96–98, 125–26, 138–41, 144–54, and 174–84.

26. Charles Darwin, *On the Various Contrivances by which British and Foreign Orchids are Fertilized by Insects, and on the Good Effects of Intercrossing* (London: J. Murray, 1862), 159–60, 330–31.
27. Ghiselin, *Triumph of the Darwinian Method,* 82–83 and 93.
28. Walsh to Edwards, March 27, 1867, Edwards Corr., WVUL. See also Walsh to Scudder, August 10, 1863, and Riley to Scudder, February 25, 1875, Scudder Corr., MSB; and Riley in *CE* 3 (September 1871), 119.
29. Riley to Scudder, June 22, 1872, Scudder Corr., MSB. Emphasis in the original.
30. Ibid. Scudder's friend Lintner had also objected to what he considered overly fine distinctions, such as Scudder's microscopic investigation of genital armor, though Lintner did not tie his objection to Darwinism. Lintner to Scudder, June 14, 1871, in ibid.
31. William H. Edwards, "Argynnis Myrina and its alleged Abnormal Peculiarities," *CE* 7 (October 1875), 193. Emphasis in original.
32. Edwards to Lintner, May 3, 1875, Lintner Corr., NYSA.
33. Edwards to Lintner, February 6, 1873, in ibid. On the need to consider all stages in classification, see also Edwards to Lintner, July 30, 1871, in ibid.
34. Edwards to Lintner, March 3, 1875, and July 21, 1876, in ibid.
35. Edwards to Lintner, July 21, 1876, and September 2, 1876 (referring to articles in the *Canadian Entomologist* and the *Transactions of the American Entomological Society*), in ibid.
36. *CE* 8 (September 1876), 179–80.
37. *CE* 8 (October 1876), 183–84.
38. Edwards to Lintner, November 22, 1876, Lintner Corr., NYSA.
39. S. H. Peabody to Edwards, October 7, 1876, Edwards Corr., WVUL.
40. Edwards to Lintner, September 7, 1876, Lintner Corr., NYSA.
41. Edwards to Lintner, October 1, 1876, in ibid.
42. Edwards to Lintner, February 20, 1875, and November 26, 1876, in ibid.; Zeller to Edwards, August 3, 1879, Edwards Corr., WVUL.
43. Edwards to Lintner, March 12, 1877, Lintner Corr., NYSA.
44. Edwards to Lintner, February 10, 1877, in ibid.
45. William Henry Edwards, "Introduction" to *Catalogue of the Lepidoptera of America North of Mexico* (Philadelphia: American Entomological Society, 1877), 1–4.
46. Edwards to Lintner, April 15, 1877, Lintner Corr., NYSA.
47. Edwards to Lintner, April 6, 1877, in ibid.
48. *CE* 9 (May 1877), 97–98.
49. Dall is quoted in Edwards to Lintner, April 22, 1877, Lintner Corr., NYSA.
50. Edwards to Lintner, May 2, 1877, in ibid.

51. "Report of the Committee on Zoological Nomenclature," 26–27.
52. Ibid., 44.
53. Ibid., 31.
54. LeConte, "On Some Changes in Nomenclature," 226.
55. "Report of the Committee on Zoological Nomenclature," 41–47. See also Edwards to Lintner, April 22 and May 2, 1877 (reporting his presentation of the lepidopterists' case to Dall), Lintner Corr., NYSA; *CE* 9 (September 1877), 172. Edwards had suggested that the year 1842 (when the British Association first adopted formal rules of nomenclature) be established as the cutoff date for the consideration of changes in the standard nomenclature. *CE* 8 (June 1876), 117.
56. Edwards to Lintner, April 3, 1878, Lintner Corr., NYSA. Edwards also objected to Hermann Strecker's publication of a checklist of American butterflies that differed from his, primarily in its non-Darwinian concept of varieties and species. Edwards to Lintner, January 1, 1878, and October 16, 1878, in ibid.

Conclusion

1. *CE* 10 (September 1878), 170–76.
2. Ibid.; *CE* 3 (July 1871), 591; *CE* 11 (September 1879), 164.
3. Kohlstedt, *American Scientific Community*, 193.
4. *CE* 10 (September 1878), 171–72.
5. Ibid., 173.
6. *CE* 9 (October 1877), 183.
7. Morse, "Address of Professor Edward S. Morse," 154.
8. Ibid., 141, 148–49, 152, 154, and 158.
9. S. A. Forbes, "Presidential Address, 1908," pp. 3–6, in AAEE, "Annual Meetings and Presidential Addresses."
10. This theme is explored at length in Whorton, *Before Silent Spring*, and Perkins, *Insects, Experts, and the Insecticide Crisis.*
11. *AE* 1 (July 1869), 209.
12. Rossiter, *Women Scientists*, chap. 3, "Women's Work in Science," 51–72.
13. Zimmermann to LeConte, August 5, 1846, LeConte Coll., APS.

Bibliography

Primary Sources

Manuscripts

ALBANY, NEW YORK

New York State Archives, University of the State of New York, State Education Department
  State Geologist's Correspondence Files, Series B0561
    Joseph A. Lintner Correspondence
New York State Library, University of the State of New York, State Education Department
  Personal Papers of James Hall, accession number PG16478
  Personal Papers of Joseph Lintner, accession number KI13168

BERKELEY, CALIFORNIA

The Bancroft Library, University of California, Berkeley
  Hilgard Family Papers (C-B 972)
  George Davidson Collection (C-B 940)

BOSTON, MASSACHUSETTS

Museum of Science, Boston
  Boston Society of Natural History Records
    Samuel H. Scudder Correspondence
    Thaddeus William Harris Notebooks

CAMBRIDGE, MASSACHUSETTS

Museum of Comparative Zoology Archives, Harvard University
Correspondence of entomologists

ITHACA, NEW YORK

Cornell University Library
William Trelease Papers

MORGANTOWN, WEST VIRGINIA

West Virginia University Libraries
William Henry Edwards Correspondence in William Henry Edwards Collection, West Virginia and Regional History Collection, West Virginia University Libraries. Microfilm. (Originals in William Henry Edwards Collection, West Virginia State Archives, Charleston.)
William Henry Edwards Entomological Notebooks in William Henry Edwards Collection, West Virginia and Regional History Collection, West Virginia University Libraries. Microfilm. (Originals in West Virginia State Archives, Charleston.)

PHILADELPHIA, PENNSYLVANIA

Academy of Natural Sciences of Philadelphia, Library
Academy of Natural Sciences of Philadelphia, Official Correspondence
American Entomological Society Collection
Entomological Society of Philadelphia Collection
John L. LeConte Collection
Samuel S. Haldeman Collection
Thomas B. Wilson Collection
American Philosophical Society
Charles Darwin–Benjamin D. Walsh Correspondence. (Microfilm of eighteen letters. From photostats in the Field Museum of Natural History, Library, Chicago, Illinois.)
Jacob Stauffer Papers
John Eatton LeConte–Thaddeus William Harris Correspondence. APS film 580.1 and 554.1–2.
LeConte Collection. (Contains correspondence and family papers of John Eatton and John L. LeConte.)

URBANA, ILLINOIS

University of Illinois Archives
Stephen A. Forbes Correspondence

WASHINGTON, D.C.

Smithsonian Institution Archives
Charles V. Riley Papers
Gustav Wilhelm Belfrage Papers

Official Incoming Assistant Secretary, Correspondence
Official Incoming Secretary, Correspondence
U.S. National Museum. Accession Records
U.S. National Archives
Record Group 48. Records of the Secretary of the Interior, Letters Received
Relating to the U.S. Entomological Commission, 1877–1880
Record Group 57. Records of the Geological Survey
Hayden Survey Records
Letters Received, 1871–1879 (Microfilm.)
Personal Letters Received, 1872–1879 (Microfilm.)

Nineteenth-Century Periodicals and Works in Series

American Entomologist: An Illustrated Magazine of Popular and Practical Entomology. Vols. 1–2 (1868–1870), vol. 3 (1880).
Bulletin of the Brooklyn Entomological Society. Vols. 1–3 (1878– 1880).
Canadian Entomologist. Vols. 1–13 (1868–1880).
Congressional Record. 1876.
Entomologica Americana. Brooklyn, N.Y. Vols. 1–6 (1885–1890).
Entomological News. Entomological Section of the Academy of Natural Sciences of Philadelphia. 1890–1895.
Entomologist's Monthly Magazine. London. Vols. 1–17 (1864–1880).
Fitch, Asa. *First [-Fourteenth] Report on the Noxious, Beneficial, and other Insects of the State of New York.* Albany: State Agricultural Society, 1855–1872. Title varies. See Barnes, *Asa Fitch,* Appendix A, "Entomological Publications by Dr. Asa Fitch," 76–82, for exact titles and the publication history of Fitch's *Reports.*
Illinois State Agricultural Society. *Transactions.* Vols. 4–8 (1859–1870). Springfield, Ill.: State Printer, 1861–1871.
Illinois State Horticulture Society. *Transactions* (1861–1870). Published in Illinois State Agricultural Society. *Transactions.* Vols. 4–8 (1859–1870). Springfield, Ill.: State Printer, 1861– 1871.
Insect Life: Devoted to the Economy and Life Habits of Insects, Especially in their Relations to Agriculture. Vols. 1–7. Washington, D.C.: U.S. Department of Agriculture, 1888–1895.
LeBaron, William. *First [-Fourth] Annual Report on the Noxious Insects of the State of Illinois.* Springfield, Ill.: State Printer, 1871–1874.
Lintner, Joseph A. *First Annual Report on the Injurious and other Insects of the State of New York.* Albany: Weed, Parsons, 1882.
———. *Second [Annual] Report on the Injurious and other Insects of the State of New York.* Albany: Weed, Parsons, 1885.
Meisel, Max. *Bibliography of American Natural History: The Pioneer Century, 1769–1865* . . . 3 vols. Brooklyn, N.Y.: Premier Publishing, 1924–1929.
Missouri State Board of Agriculture. *First [-Sixth] Annual Report.* Jefferson City, Mo.: State Printer, 1865–1870.
Missouri State Horticultural Society. *Proceedings* (1865–1870). In Missouri State Board of Agriculture, *Annual Reports* (1865–1870). Jefferson City, Mo.: State Printer, 1866–1871.

The Naturalists' Directory for 1878, Containing the Names of the Naturalists of America North of Mexico. Edited by Samuel E. Cassino. Salem, Mass.: Naturalists' Agency, 1878.

The Naturalists' Directory, Part 2, North America and the West Indies. Edited by F. W. Putnam. Salem, Mass.: Essex Institute, 1866.

New York State Agricultural Society. *Transactions.* Vols. 5–15. Albany, N.Y.: State Printer, 1845–1855.

Papilio: Organ of the New York Entomological Society, Devoted Exclusively to Lepidoptera. Vols. 1–4 (1881–1884).

Practical Entomologist. Published by the Entomological Society of Philadelphia. Vols. 1–2 (1865–1867).

Prairie Farmer: Devoted to Western Agriculture, Mechanics, and Education. Chicago, Ill. Vols. 1–15 (1841–1855).

Proceedings of the American Association for the Advancement of Science. Vols. 19–30 (1870–1881).

Proceedings of the Entomological Society of Philadelphia. Vols. 1–6 (1863–1867).

Psyche: Organ of the Cambridge Entomological Club. Cambridge, Mass. Vols. 1–3 (1874–1882).

Record of American Entomology for the Year 1868 [–1873]. Edited by A. S. Packard., Jr. Salem, Mass.: Naturalist's Book Agency, Essex Institute Press, 1869–1874.

Riley, Charles V. *First [–Ninth] Annual Report on the Noxious, Beneficial, and Other Insects of the State of Missouri.* Jefferson City, Mo.: State Printer, 1869–1877.

The Riverside Natural History. Edited by John Sterling Kingsley, 6 vols. New York: Houghton Mifflin, 1886.

U.S. Department of Agriculture. *Report of the Commissioner of Agriculture.* Washington, D.C.: U.S. Government Printing Office, 1862–1880.

U.S. Entomological Commission. *First Annual Report for the year 1877 relating to the Rocky Mountain Locust and the best means of preventing its injuries and of guarding against its invasions.* Washington, D.C.: U.S. Government Printing Office, 1878.

———. *Second Annual Report for the year 1878 . . .* Washington, D.C.: U.S. Government Printing Office, 1880.

Walsh, Benjamin D. *First Annual Report on the Noxious Insects of the State of Illinois.* From the Appendix to the *Transactions of the Illinois State Horticultural Society for 1867,* vol. 1 (1868). (Separate pamphlet edition printed by the Prairie Farmer Company, Chicago, Ill., 1868.)

Zoological Record. Vols. 1–12 (1864–1880). London: J. Van Voorst, 1864–1880.

Books and Articles

An Account of the Organization and Progress of the Museum of Comparative Zoology at Harvard College, in Cambridge, Massachusetts. Cambridge, Mass.: Welch, Bigelow, 1871.

Bell, James T. "How to Destroy Cabinet Pests." *Canadian Entomologist* 9 (1877), 139–40.

"Chinch Bug Notes." *Report of the Commissioner of Agriculture* (1881–1882), 87–88.

Clifford, E. A. "Birds Injurious and Beneficial to the Horticulturist." *Transactions of the Illinois State Horticultural Society,* 1864. In *Transactions of the Illinois State Agricultural Society* 5 (1861–1864), 925–27.

Conference of Governors. *The Rocky Mountain Locust, or Grasshopper, Being the Report of Proceedings of a Conference of the Governors of Several Western States and Territories . . . held at Omaha, Nebraska, on the 25th and 26th Days of October, 1876, to consider the Locust Problem; also a Summary of the Best Means now known for Counteracting the Evil.* Saint Louis, Mo.: R. R. Studley, 1876.

Cumming, C. F. G. "Locusts and Farmers of America." *Nineteenth Century* 17 (January 1885), 134–52.

Curtis, John. *Farm Insects: Being the Natural History and Economy of the Insects Injurious to the Field Crops of Great Britain and Ireland, and also those which infest Barns and Granaries, With Suggestions for their Destruction.* Glasgow: Blakie and Son, 1860.

Dana, James D. "On Cephalization; No. IV: Explanations drawn out by the Statements of an Objector." *American Journal of Science,* n.s. 41 (March 1866), 163–74.

"The Darwin-Bates Letters: Correspondence Between Two Nineteenth-Century Travellers and Naturalists, Part 2." Edited by Robert M. Stecher. *Annals of Science* 25 (June 1969), 95–125.

Darwin, Charles. *The Descent of Man and Selection in Relation to Sex.* Introduction by John Tyler Bonner and Robert M. May. 1871. Repr. Princeton, N.J.: Princeton University Press, 1981.

———. *The Life and Letters of Charles Darwin, Including an Autobiographical Chapter.* Edited by his son, Francis Darwin, 3 vols. 1888. Repr. New York: Johnson Reprint, 1969.

———. *On the Origin of Species by Means of Natural Selection; or, The Preservation of Favoured Races in the Struggle for Life.* Facsimile of the first edition with an introduction by Ernst Mayr. 1859. Repr. Cambridge, Mass.: Harvard University Press, 1964.

———. *On the Various Contrivances by which British and Foreign Orchids are Fertilized by Insects, and on the Good Effects of Intercrossing.* London: J. Murray, 1862.

Delafield, John. "A General View and Agricultural Survey of the County of Seneca." *Transactions of the New York State Agricultural Society* 10 (1850), 356–616.

Dodge, Charles Richards. *The Life and Entomological Work of the Late Townend Glover, first Entomologist of the U.S. Department of Agriculture.* U.S. Department of Agriculture, Division of Entomology, Bulletin 18. Washington, D.C.: U.S. Government Printing Office, 1888.

[Edwards, William]. *Memoirs of Colonel William Edwards . . . with notes and additions by his son, William W. Edwards, and his Grandson, William Henry Edwards.* Washington, D.C.: Press of W. F. Roberts, 1897.

Edwards, William Henry. "An Abstract of Dr. Aug. Weismann's Paper on 'The Seasonal Dimorphism of Butterflies.' [Leipzig 1875.] To which is Appended a Statement of Some Experiments made upon Papilio Ajax." *Canadian Entomologist* 7 (December 1875), 228–41.

———. "Argynnis Myrina and its alleged Abnormal Peculiarities." *Canadian Entomologist* 7 (October 1875), 189–95.

———. *The Butterflies of North America,* 3 vols. Vol. 1: New York: Hurd and Houghton, 1874. Vols. 2 and 3: Boston: Houghton Mifflin, 1884 and 1897.

————. *Catalogue of the Lepidoptera of America North of Mexico.* Philadelphia: American Entomological Society, 1877.

————. "The Entomological Reminiscences of William Henry Edwards." Edited by Cyril F Dos Passos. *Journal of the New York Entomological Society* 59 (September 1951), 129–86.

————. "Experiments Upon the Effect of Cold Applied to Chrysalids of Butterflies." Pts. 1, 2, 3. *Psyche* 3 (1880): (January), 1–4; (February), 15–19; (June), 75–76.

————. "History of Phyciodes Tharos, A Polymorphic Butterfly." Pts. 1, 2. *Canadian Entomologist* 9 (1877): (January), 1–10; (March), 51–58.

————. *List of Species of the Diurnal Lepidoptera of America North of Mexico.* Boston: Houghton Mifflin, 1884.

————. "Notes on the Early Stages of Some of our Butterflies." *Canadian Entomologist* 5 (December 1873), 223–25.

————. "Notes on Entomological Nomenclature." Pts. 1, 2, 3. *Canadian Entomologist* 8 (1876): (March), 41–52; (May), 81–94; (June), 113–19.

————. "Notes on a remarkable Variety of Papilio Turnus, and descriptions of two species of Diurnal Lepidoptera." *Transactions of the American Entomological Society* 2 (1868), 207–10.

————. "On Certain Species of Satyrus." Pts. 1, 2, 3, 4. *Canadian Entomologist* 12 (1880): (February), 21–32; (March), 51–55; (May), 90–94; (June), 109–15.

————. "On the Identity of Grapta Dryas with Comma." *Canadian Entomologist* 5 (October 1873), 184.

————. "Rearing Butterflies from the Egg." *Canadian Entomologist* 3 (August 1871), 70.

————. "Some Remarks on Entomological Nomenclature." *Canadian Entomologist* 5 (February 1873), 21–36.

————. *Synopsis of North American Butterflies.* New York: Hurd and Houghton, 1874.

————. *A Voyage up the River Amazon, including a residence at Pará.* London: John Murray, 1847.

Engelmann, George. "The Flower of Yucca and its Fertilization." *Bulletin of the Torrey Botanical Club* 3 (July 1872), 33.

————. "Notes on the Genus *Yucca*." *Transactions of the St. Louis Academy of Science* 3 (April 1873), 17–54 and 210–14.

"Entomological Society of Pennsylvania." *American Journal of Science* 44 (1843), 199–200.

Evans, Gurdon. "Agricultural Survey of Madison County." *Transactions of the New York State Agricultural Society* 11 (1851), 658–777.

Fitch, Asa. "The Hessian Fly: Its History, Character, Transformation, and Habits." From the *American Journal of Agriculture and Science,* revised in *Transactions of the New York State Agricultural Society* 6 (1846), 316–72.

Forbes, Stephen A. "The Food of Birds." In *Bulletin of the Illinois State Laboratory of Natural History* 1:3 (1880), 80–148.

————. "The Food of Birds." In *Transactions of the Illinois State Horticultural Society for 1878,* n.s. 12 (1879), 140–45.

Grote, Augustus R. "On Jacob Hübner and His Works on the Butterflies and Moths." *Canadian Entomologist* 8 (July 1876), 131–35.

————. "On Mr. Scudder's Systematic Revision of Some of the American Butterflies." *Canadian Entomologist* 5 (April 1873), 62–63.

Haldeman, Samuel Stehman. "Materials Towards a History of the Coleoptera Longicornia of the United States." *Transactions of the American Philosophical Society,* n.s. 10 (1853), 27–66.

———. "Report on the Progress of Entomology in the United States during 1849." *Proceedings of the Academy of Natural Sciences of Philadelphia* 5 (1850–1851), 5–7.

Harris, Thaddeus William. *Entomological Correspondence of Thaddeus William Harris, M.D.* Edited by S. H. Scudder. Occasional Paper of the Boston Society of Natural History 1. Boston: Boston Society of Natural History, 1869.

———. *A Report on the Insects of Massachusetts, Injurious to Vegetation.* 1841. Repr. New York: Arno Press, American Environmental Studies, 1970.

Harvard Museum of Comparative Zoology Reports, 1864–1874. Cambridge, Mass.: Welsh, Bigelow, 1871–1873. (Later reports were published separately by Harvard University Press.)

Herrick, E. C. "Observations Communicated at the Request of the Honorable H. L. Ellsworth: The Hessian Fly." In *Report of the Commissioner of Patents, 1844,* 161–67. Washington, D.C.: U.S. Government Printing Office, 1845. (Reprinted in *Prairie Farmer* 5 [July 1845], 169.)

Howard, Leland O. *The Chinch Bug.* U.S. Department of Agriculture, Division of Entomology, Bulletin 17. Washington D.C.: U.S. Government Printing Office, 1888.

Hull, Dr. "A Few of Our Insect Enemies and Friends." In Report of the Missouri State Horticultural Society in the *First Annual Report of the Missouri State Board of Agriculture,* 451–55. Jefferson City, Mo.: State Printer, 1865.

Hulst, George P. "Hints on the Rearing of Lepidoptera." Pts. 1, 2. *Bulletin of the Brooklyn Entomological Society* 2 (September 1879–January 1880), 63–73; and 4 (June 1881), 13–14.

"An Insect Plague." *All the Year Round* (March 28, 1885), 30–32.

Kirby, William, and William Spence. *An Introduction to Entomology; or, Elements of the Natural History of Insects.* 4 vols. London: Longman, Hurst, Rees, Orme, and Brown, 1816–1826.

Köllar, Vincent. *A Treatise on Insects Injurious to Gardeners, Foresters, and Farmers.* Translated by Jane Loudon and Mary Loudon. London: William Smith, 1840.

LeConte, John L. "Descriptions of New Coleoptera from Texas, Chiefly Collected by the U.S. Boundary Commission." *Proceedings of the Academy of Natural Sciences of Philadelphia* 6 (December 27, 1853), 439–48.

———. "Hints for the Promotion of Economic Entomology." *Proceedings of the American Association for the Advancement of Science* 22 (1873), 10–22. (Also published with extended title in *American Naturalist* 7 [November 1873], 710–22.)

———. "On the Method of Subduing Insects Injurious to Agriculture." *Proceedings of the American Association for the Advancement of Science* 24 (1875), 202–207. (Also published with minor changes in title and text in the *Canadian Entomologist* 7 [September 1875], 167–72.)

———. "On the Preservation of Entomological Cabinets." *American Naturalist* 3 (August 1869), 307–309.

———. "On Some Changes in the Nomenclature of North American Coleoptera which have been Recently Proposed." *Canadian Entomologist* 6 (October 1874), 186–97.

———. "Presidential Address." *Proceedings of the American Association for the Advancement of Science* 24 (1875), 1–18.

———. "Report Upon the Insects Collected on the Survey." In I. I. Stevens, *Narrative and Final Report of Explorations . . . for a Pacific Railroad near the Forty-Seventh and Forty-Ninth Parallels.* U.S. Corps of Topographical Engineers, *Reports of Explorations and Surveys to Ascertain the Most Practicable and Economical Route for a Railroad from the Mississippi River to the Pacific Ocean,* vol. 12, pt. 3, pp. 1–72. Washington, D.C.: U.S. Government Printing Office, 1860.

Lintner, Joseph A. "Entomological Contributions." In *New York State Museum of Natural History, Twenty-fourth Annual Report,* 154–67. Albany: Angus, 1872.

Lintner, Joseph A., and F. G. Sanborn. "An Account of the Collections which Illustrate the Labors of Dr. Asa Fitch." *Psyche* 2 (September– December 1879), 273–76.

"Locusts and Farmers of America." *Nineteenth Century* 17 (January 1885), 134–52.

Massachusetts. *Reports on the Fishes, Reptiles, and Birds of Massachusetts, By the Commissioners on the Zoological and Botanical Survey of the State.* Boston: Dutton and Wentworth, 1839.

Mead, Theodore L. "Notes Upon Some Butterfly Eggs and Larvae." *Canadian Entomologist* 7 (September 1875), 161–62.

Meehan, Thomas. "On the Fertilization of Yucca." *North American Entomologist* 1 (November 1879), 33–36.

Melsheimer, Frederick Ernst. *Catalogue of the Described* Coleoptera *of the United States.* Revised by S. S. Haldeman and J. L. LeConte. Smithsonian Institution Publication 62. Washington, D.C.: Smithsonian Institution, 1853.

Melsheimer, F. E., Carl Ziegler, and John L. LeConte. "Descriptions of New Species of Coleoptera of the United States." Pts. 1, 2, 3, 4, 5, 6. [Communicated by the Entomological Society of Pennsylvania], *Proceedings of the Academy of Natural Sciences of Philadelphia* 2 (October 1844), 98–118; (November 1844), 134–60; (March 1845), 213–23; (December 1845), 302–18; 3 (May 1846), 53–66; (February 1847), 158–81.

Melsheimer, Frederick Valentine. "A Catalogue of the Insects of Pennsylvania." 1806. In Robert Snetsinger, *Frederick Valentine Melsheimer, Parent of American Entomology.* University Park: Entomological Society of Pennsylvania, 1973.

Morris, John G. "Address by Rev. John G. Morris, Chairman of the Subsection of Entomology." *Proceedings of the American Association for the Advancement of Science* 30 (1881), 261–66.

———. "Contributions Toward a History of Entomology in the United States." *American Journal of Science* 51 [n.s. 1] (January 1846), 17–27.

Morris, Margaretta. "Observations of the development of the Hessian Fly." *Proceedings of the Academy of Natural Sciences of Philadelphia* 1 (1841–1843), 66–68.

———. "On the *Cecidomyia destructor,* or Hessian Fly." *Transactions of the American Philosophical Society,* n.s. 8 (1843), 49–52.

Morrison, H. K. "Specific Nomenclature." *Canadian Entomologist* 5 (April 1873), 70–71.

Morse, Edward S. "Address of Professor Edward S. Morse, Vice President, Section B (What American Zoologists Have Done for Evolution)." *Proceedings of the American Association for the Advancement of Science* 25 (1876), 137–76.

Packard, Alpheus Spring, Jr. "First Annual Report on Injurious and Beneficial Insects of Massachusetts." *American Naturalist* 5 (August 1871), 423–27.

———. *Guide to the Study of Insects, and a Treatise on those injurious and beneficial to crops; for the use of Colleges, Farm Schools, and Agriculturalists.* Salem, Mass.: Naturalist's Book Agency, 1869.

———. *The Hessian Fly.* U.S. Entomological Commission Bulletin 4. Washington, D.C.: U.S. Government Printing Office, 1880.

———. "Injurious and Beneficial Insects." *American Naturalist* 7 (September 1873), 524–48.

———. *Lamarck, the Founder of Evolution: His Life and Work.* New York: Longmans, Green, 1901.

———. "Migrations of the Destructive Locust of the West." *American Naturalist* 11 (January 1877), 22–29.

———. "Nature's Means of Limiting the Numbers of Insects." *American Naturalist* 8 (May 1874), 270–82.

———. "On Certain Entomological Speculations—A Review." *Proceedings of the Entomological Society of Philadelphia* 6 (November 1866), 209–18.

———. "Report on the Rocky Mountain Locust and other Insects . . . in the Western States and Territories." In *Ninth Annual Report of the U.S. Geological and Geographical Survey of the Territories . . . for 1875,* edited by F. V. Hayden, 589–810. Washington, D.C.: U.S. Government Printing Office, 1877.

———. "Second Annual Report on Injurious and Beneficial Insects of Massachusetts." *American Naturalist* 7 (April 1873), 241–44.

"Report of the Committee on Zoological Nomenclature." *Proceedings of the American Association for the Advancement of Science* 26 (1877), 23–56.

Riley, Charles V. "Controlling Sex in Butterflies." *American Naturalist* 7 (September 1873), 513–21.

———. "Darwin's Work in Entomology." *Proceedings of the Biological Society of Washington, D.C.* 1 (1882), 70–80.

———. *Destructive Locusts: A Popular Consideration of a few of the more injurious Locusts (or 'Grasshoppers') in the United States, together with the best means of Destroying them.* U.S. Department of Agriculture, Division of Entomology, Bulletin 25. Washington, D.C.: U.S. Government Printing Office, 1891.

———. "Further Notes on the Pollination of Yucca and on Pronuba and Prodoxus." *Proceedings of the American Association for the Advancement of Science* 29 (1880), 617–39.

———. "Further Notes on Yucca Insects and Yucca Pollination." *Proceedings of the Biological Society of Washington, D.C.* (June 1893), 41–54.

———. "Further Remarks on *Pronuba yuccasella* and on the Pollination of Yucca." *Transactions of the Academy of Science of St. Louis* 8 (1878), 568–73.

———. "Great Truths in Applied Entomology; Address to the Georgia State Agricultural Society, February 12, 1884." In *Report of the Commissioner of Agriculture,* 323–29. Washington, D.C.: U.S. Government Printing Office, 1884.

———. "In Memoriam [B. D. Walsh]." *American Entomologist* 2 (December–January 1869–1870), 65–68.

———. "The Locust Plague; How to Avert it." *Proceedings of the American Association for the Advancement of Science* 24 (1875), 215–21.

———. *The Locust Plague in the United States: Being More Particularly a Treatise on the Rocky Mountain Locust or so-called Grasshopper, as it occurs East of the Rocky Mountains, with Practical Recommendations for its Destruction.* Chicago: Rand McNally, 1877.

———. "Mimicry as Illustrated by these two Butterflies, with some Remarks on the Theory of Natural Selection." In *Third Annual Report on the Noxious, Beneficial, and other Insects of the State of Missouri* by Charles V. Riley, 159–75. Jefferson City, Mo.: State Printer, 1871

———. "On the Causes of Variation in Organic Forms." *Proceedings of the American Association for the Advancement of Science* 37 (1888), 225–73.

———. "On a new Genus in the Lepidopterous Family *Tineidae*, with Remarks on the Fertilization of *Yucca: Pronuba*, Nov. Genus." *Transactions of the St. Louis Academy of Science* 3 (1873), 55–64.

———. "On a New Genus in the Lepidopterous Family Tineidae: with Remarks on the Fertilization of Yucca." [Same article as in *Transactions of the St. Louis Academy of Science*, with slightly altered title and additional notes and correspondence]. In *Fifth Annual Report on the Noxious, Beneficial, and other Insects of the State of Missouri* by Charles V. Riley, 150–60. Jefferson City, Mo.: State Printer, 1873.

———. "On the Ovaposition of the Yucca Moth." *American Naturalist* 7 (October 1873), 619–23.

———. "The Philosophy of the Movement of the Rocky Mountain Locust." *Proceedings of the American Association for the Advancement of Science* 27 (August 1878), 271–77.

———. "The Rocky Mountain Locust." *American Naturalist* 11 (November 1877), 663–73.

———. "Some Interrelations of Plants and Insects." *Proceedings of the Biological Society of Washington, D.C.* 7 (May 1892), 81–104.

———. "Supplementary Notes on Pronuba Yuccasella." *Transactions of the St. Louis Academy of Science* 4 (1874), 178–80.

———. "Tribute to the Memory of John Lawrence LeConte." *Psyche* 4 (November–December 1883), 107–10.

———. "The True and the Bogus Yucca Moth; with Remarks on the Pollination of Yucca," *American Entomologist* 3 (1880), 141–45.

———. "The Yucca Moth—*Pronuba yuccasella* Riley." In *Sixth Annual Report on the Noxious, Beneficial, and other Insects of the State of Missouri* by Charles V. Riley, 131–35. Jefferson City, Mo.: State Printer, 1874.

———. "The Yucca Moth and Yucca Pollination." *Annual Report of the Missouri Botanical Garden* 3 (1892), 99–158.

Say, Thomas. *American Entomology: A Description of the Insects of North America.* Edited by John L. LeConte. 1824–1828. 2 vols. Repr. New York: J. W. Bouthon, 1869.

———. *American Entomology; or, Descriptions of the Insects of North America.* 3 vols. Philadelphia: S. A. Mitchell, 1824, 1825, 1828.

———. "Letters from Thomas Say to John F Melsheimer, 1816–1825." Edited by William J. Fox. Pts. 1, 2, 3, 4, 5, 6, 7, 8, 9. *Entomological News* 12 (1901): (April), 110–13; (May), 138–41; (June), 173–77; (September), 203–205; (October), 233–36; (November), 281–83; (December), 314–16; 13 (January 1902), 9–11; (February 1902) [A Note on the Insect Collection of Thomas Say], 38–40.

Scudder, Samuel H. "Canons of Systematic Nomenclature for the Higher Groups." *Canadian Entomologist* 5 (March 1873), 55–59. (Reprinted from the *American Journal of Science*, May 1872.)

———. "Is Mimicry Advantageous?" *Nature* 3 (December 22, 1870), 147.

————. "John Lawrence LeConte, 1825–1883." In *Biographical Memoirs of the National Academy of Sciences,* vol. 2, pp. 261–93. Washington, D.C.: National Academy of Sciences, 1886.

————. "Presidential Address, Subsection of Entomology." *Proceedings of the American Association for the Advancement of Science* 29 (1880), 609–15.

————, ed. "Some Old Correspondence Between Harris, Say, and Pickering." Pts. 1, 2, 3, 4, 5. *Psyche* 6 (1891): (March), 57–60; (August), 121–24; (September), 137–41; (November), 169–72; (December), 185–87.

Shaler, N. S. "A State Survey for Massachusetts." *American Naturalist* 9 (March 1875), 156–59.

Shimer, Henry. "Insects Injurious to the Potato." *American Naturalist* 3 (April 1869), 91–99.

Taylor, Alexander S. "An Account of the Grasshoppers and Locusts of America, Condensed from an Article written and furnished by Alexander S. Taylor, Esq., of Monterey, California." In *Annual Report of the Smithsonian Institution for 1858,* 200–13. Washington, D.C.: U.S. Government Printing Office, 1859.

Thomas, Cyrus. *The Chinch Bug: Its History, Characters, and Habits, and the means of Destroying it or Counteracting its Injuries.* U.S. Department of the Interior, U.S. Entomological Commission, Bulletin 5. Washington, D.C.: U.S. Government Printing Office, 1879.

————. "Insects Injurious to Vegetation in Illinois." In *Transactions of the Illinois State Agricultural Society* 5 (1861–1864), 401–68.

————. "The Relation of Meteorological Conditions to Insect Development." In *Tenth Annual Report of the State Entomologist of Illinois,* 47–59. Springfield, Ill.: State Printer, 1881.

————. "Temperature and Rainfall as affecting the Chinch Bug." *American Entomologist,* n.s. 1 (October 1880), 240–42.

Treat, Mary. "Controlling Sex in Butterflies." *American Naturalist* 7 (March 1873), 129–32.

Trelease, William. "Detail Illustrations of Yucca." *Annual Report of the Missouri Botanical Garden* 3 (1892), 159–66.

————. "Further Studies on Yuccas and their Pollination." *Annual Report of the Missouri Botanical Garden* 4 (1893), 181–226.

————. Review of *The Mutual Relations between Flowers and the Insects which serve to Cross them* by Hermann Müller. *American Naturalist* 13 (July 1879), 451–52.

Uhler, P. R. "Meeting of the American Association for 1873." *Canadian Entomologist* 5 (September 1873), 165–66.

Wallace, Alfred Russel. *Contributions to the Theory of Natural Selection: A Series of Essays.* 1870. Repr. New York: Arno Press, 1973.

Walsh, Benjamin D. "Imported Insects: The Gooseberry Sawfly." *Practical Entomologist* 1 (September 1866), 117–25.

————. "The New Potato Bug." *Practical Entomologist* 2 (November 1866), 13–16.

————. "The New Potato Bug, and its Natural History." *Practical Entomologist* 1 (October 1865), 1–4. (Also in *First Annual Report of the Missouri State Board of Agriculture* [1865], 200–206.)

————. "On Certain Entomological Speculations of the New England School of Naturalists." *Proceedings of the Entomological Society of Philadelphia* 3 (August 1864), 207–49.

————. "On the Insects, Coleopterous, Hymenopterous, and Dipterous inhabiting the Galls of certain Species of Willow: Part 1, Diptera." *Proceedings of the Entomological Society of Philadelphia* 3 (December 1864), 543–644.

————. "On Phytophagic Varieties and Phytophagic Species." *Proceedings of the Entomological Society of Philadelphia* 3 (November 1864), 403–30.

————. "Prof. Dana and his Entomological Speculations." *Proceedings of the Entomological Society of Philadelphia* 6 (June 1866), 116–21.

Walsh, Benjamin D., and Charles V. Riley. "Imitative Butterflies." *American Entomologist* 1 (June 1869), 189–93.

Weismann, August. *Studies in the Theory of Descent; with Notes and Additions by the Author.* Translated and edited by Raphael Meldola with a prefatory notice by Charles Darwin. 2 vols. London: Sampson, Low, Marston, Searle, and Rivington, 1882.

Ziegler, D. "Descriptions of New North American Coleoptera." *Proceedings of the Academy of Natural Sciences of Philadelphia* 2 (1844–1845), 43–47, 266–72.

Zimmermann, C. "Synopsis of the Scolytidae of America North of Mexico, with notes . . . by J. L. LeConte." *Transactions of the American Entomological Society* 2 (1868–1869), 141–49.

Secondary Sources

Unpublished Monographs and Papers

Aldrich, Michele Alexis LaClergue. "New York Natural History Survey, 1836–1845." Ph.D. diss., University of Texas, Austin, 1974. University Microfilms International no. 75–4311.

American Association of Economic Entomologists. "Annual Meetings and Presidential Addresses: Annual Meetings 1–24 (1889–1911)." University of California Library, Davis, California. Typescript.

Baatz, Simon. "Patronage, Science, and Ideology in an American City: Patrician Philadelphia, 1800–1860." Ph.D. diss., University of Pennsylvania, 1986. University Microfilms International no. 86–14760.

Blaisdell, Muriel Louise. "Darwinism and Its Data: The Adaptive Coloration of Animals." Ph.D. diss., Harvard University, 1976.

Bonnell, Daniel Ernest. "Some Factors in the Development of Northwestern Entomology." Ph.D. diss., Oregon State University, 1942.

Cart, Theodore Whaley. "The Struggle for Wildlife Protection in the United States, 1870–1900: Attitudes and Events Leading to the Lacey Act." Ph.D. diss., University of North Carolina, Chapel Hill, 1971. University Microfilms International no. 71–30, 545.

Clausen, Lucy W. "The Development of the Entomological Exhibit in Museums of the United States." Ph.D. diss., New York University, 1947. University Microfilms International no. 00–00937.

Elliott, Clark A. "The American Scientist, 1800–1863; His Origins, Career, and Interests." Ph.D. diss., Case Western Reserve University, 1970. University Microfilms International no. 71–1685.

————. "Sketch of Thaddeus William Harris." Harvard University Archives, Cambridge, Mass. Unpublished paper, 1981.

Gerstner, Patsy Ann. "The 'Philadelphia School' of Paleontology: 1820–1845." Ph.D. diss., Case Western Reserve University, 1967. University Microfilms International no. 68–10,158.

Greenfield, Theodore John. "Variation, Heredity, and Scientific Explanation in the Evolutionary Theories of Four American Neo-Lamarckians, 1867–1897." Ph.D. diss., University of Wisconsin, Madison, 1986. University Microfilms International no. 86–14370.

Gross, Walter Elliott. "The American Philosophical Society and the Growth of Science in the United States, 1835–1850." Ph.D. diss., University of Pennsylvania, 1970. University Microfilms International no. 71–19230.

Henson, Pamela M. "Evolution and Taxonomy: J. H. Comstock's Research School in Evolutionary Entomology at Cornell University, 1874–1930." Ph.D. diss., University of Maryland, 1990. University Microfilms International no. 90–30909.

Houk, Wallace Eugene. "A Study of Some Events in the Development of Entomology and Its Application in Michigan." Ph.D. diss., Michigan State University, Lansing, 1954.

Pfeifer, Edward Justin. "The Reception of Darwinism in the United States, 1859–1880. Ph.D. diss., Brown University, 1957. University Microfilms International no. 23,451.

Sorensen, Willis Conner. "American Women Entomologists, 1840–1880." Paper presented at the Columbia History of Science Society, Friday Harbor, Washington, April 21, 1990.

————. "Brethren of the Net: American Entomology, 1840–1880." Ph.D. diss., University of California, Davis, 1984. University Microfilms International no. 8425056.

Reference Works, Periodicals, and Works in Series

Beattie, John R. *The Rhopalocera Directory: A Comprehensive Index to Butterfly Names Found in the Systematic Index of the* Zoological Record *1864–1971 and in* Berichte über die wissenschaftlichen Leistungen im Gebiete der Entomologie *1834–1863.* Berkeley, Calif.: JB Indexes, 1976.

The Canadian Encyclopedia. Edmonton, Alberta: Hurtig Publishers, 1988.

Dictionary of Scientific Biography. Edited by Charles Coulston Gillispie. New York: Charles Scribner's Sons, 1970.

Elliott, Clark A., ed. *Biographical Dictionary of American Science: The Seventeenth Through the Nineteenth Centuries.* Westport, Conn.: Greenwood Press, 1979.

Gilbert, Pamela. *A Compendium of the Biographical Literature on Deceased Entomologists.* London: British Museum (Natural History), 1977.

Miller, Lee D., and F. Martin Brown. *A Catalogue/Checklist of the Butterflies of America North of Mexico.* Memoir no. 2. Lepidopterists' Society, 1981.

The New Century Cyclopedia of Names. New York: Appleton Century Crofts, 1954.

Rothschild, Walter, and Karl Jordan. "A Revision of the American Papilios." *Novitates Zoologicae* 13 (August 1906), 411–752.

Stoetzel, Manya B., Chair of Committee. *Common Names of Insects and Related Organisms.* Beltsville, Md.: Entomological Society of America, 1989.

U.S. Statutes at Large 19 (1876), 376. (U.S. Entomological Commission.)

Webster's American Biographies. Edited by Charles Van Doren. Springfield, Mass.: G. & C. Merriam, 1974.

Who Was Who in America: A Component Volume of Who's Who in American History. Chicago, Ill.: Marquis Who's Who, 1968.

Other Published Books and Articles

Adicks, Richard, ed. *LeConte's Report on East Florida.* Orlando, Fla.: University Presses of Florida, 1978.

Aldrich, Mark. "Earnings of American Civil Engineers, 1820–1859." *Journal of Economic History* 31 (1971), 407–19.

Aldrich, Michele. "Women in Paleontology in the United States, 1840–1960." *Earth Sciences* 1 (1982–1983), 14–22.

Allan, Mae. *Darwin and His Flowers: The Key to Natural Selection.* New York: Taplinger Publishing, 1977.

Allee, W. C., Orlando Park, Alfred Emerson, and Thomas Park. *Principles of Animal Ecology.* Philadelphia: W. B. Saunders, 1949.

Allen, David Elliston. *The Naturalist in Britain: A Social History.* Worcester: Allen Lane, 1976.

Atkins, Annette. *Harvest of Grief: Grasshopper Plagues and Public Assistance in Minnesota, 1873–78.* St. Paul: Minnesota Historical Society Press, 1984.

Baker, Gladys L. "Women in the U.S. Department of Agriculture." *Agricultural History* 50 (January 1976), 190–201.

Baker, W. A., and O. R. Mathews. "Good Farming Helps Control Insects." In *Insects: The Yearbook of Agriculture,* 437–40. Washington, D.C.: U.S. Government Printing Office, 1952.

Barnes, Jeffrey K. *Asa Fitch and the Emergence of American Entomology: With an Entomological Bibliography and a Catalog of Taxonomic Names and Type Specimens.* New York State Museum Bulletin 461. Albany: State Education Department, University of the State of New York, 1988.

———. "Insects in the New Nation: A Cultural Context for the Emergence of American Entomology." *Bulletin of the Entomological Society of America* 31:1 (1985), 21–30.

Basalla, George. "The Spread of Western Science." *Science* 156 (May 5, 1967), 611–22.

Beaver, Donald. *The American Scientific Community: A Historical-Statistical Study.* New York: Arno Press, 1980.

Beddall, Barbara, ed. *Wallace and Bates in the Tropics: An Introduction to the Theory of Natural Selection. Based on the Writings of Alfred Russel Wallace and Henry Walter Bates.* London: Macmillan, 1969.

Bedini, Silvio A. "The Evolution of Science Museums." *Technology and Culture* 6 (1965), 1–29.

Beier, Max. "The Early Naturalists and Anatomists During the Renaissance and Seventeenth Century." In *History of Entomology,* edited by Ray F. Smith, Thomas E. Mittler, and Carroll N. Smith, 81–94. Palo Alto, Calif.: Annual Reviews, 1973.

Bethune, C. J. S. "William Henry Edwards." *Canadian Entomologist* 41 (August 1909), 245–48.

Bidwell, Percy Wells, and John I. Falconer. *Agriculture in the Northern United States, 1620–1860.* 1925. Repr. Magnolia, Mass.: Peter Smith, 1941.

Blacher, L. I. *The Problem of the Inheritance of Acquired Characteristics: A History of a priori and Empirical Methods Used to Find a Solution.* English translation edited by F. B. Churchill, 1971. Repr. New Delhi, India: Amerind Publishing for the Smithsonian Institution Libraries and the National Science Foundation, 1982.

Blackwelder, Richard. *Taxonomy: A Text and Reference Book.* New York: John Wiley, 1967.

Bocking, Stephen. "Alpheus Spring Packard and Cave Fauna in the Evolution Debate." *Journal of the History of Biology* 21:3 (Fall 1988), 425–56.

Bode, Carl. *The American Lyceum: Town Meeting of the Mind.* 1956. Repr. Carbondale, Ill.: Southern Illinois University Press, Arcturus Books, 1968.

Bogue, Allen. *From Prairie to Corn Belt: Farming on the Illinois and Iowa Prairies in the Nineteenth Century.* Chicago: University of Chicago Press, 1963.

Boller, Paul F., Jr. *American Thought in Transition: The Impact of Evolutionary Naturalism, 1865–1900.* Chicago: Rand McNally, 1969.

Bouvé, Thomas T. *Historical Sketch of the Boston Society of Natural History; with a Notice of the Linnaean Society, which Preceded it.* Anniversary Memoirs of the Boston Society of Natural History. Boston: Boston Society of Natural History, 1880.

Bowler, Peter J. "Edward Drinker Cope and Evolution." *Isis* 68 (June 1977), 249–65.

Brown, W. J. "Taxonomic Problems with Closely Related Species." *Annual Review of Entomology* 4 (1959), 77–98.

Bruce, Robert V. *The Launching of Modern American Science.* New York: Alfred A. Knopf, 1987.

———. "A Statistical Profile of American Scientists, 1846–1876." In *Nineteenth-Century American Science: A Reappraisal,* edited by George H. Daniels, 63–94. Evanston, Ill.: Northwestern University Press, 1972.

Calvert, Monte A. *The Mechanical Engineer in America, 1830–1910.* Baltimore, Md.: Johns Hopkins University Press, 1967.

Cameron, Jenks. *The Bureau of Biological Survey: Its History, Activities, and Organization.* Institute for Government Research, Service Monographs of the U.S. Government 54. Baltimore, Md.: Johns Hopkins University Press, 1929.

Carrier, Lyman. "The United States Agricultural Society, 1852–1860: Its Relation to the Origin of the United States Department of Agriculture and the Land Grant Colleges." *Agricultural History* 11 (October 1937), 278–88.

Casagrande, R. A. "The Colorado Potato Beetle: 125 Years of Mismanagement." *Bulletin of the Entomological Society of America* 33:3 (1987), 142–50.

Cassedy, James H. *American Medicine and Statistical Thinking, 1800–1860.* Cambridge, Mass.: Harvard University Press, 1984.

Churchill, Frederick B. "August Weismann and a Break from Tradition." *Journal of the History of Biology* 1 (1968), 91–112.

———. "Hertwig, Weismann, and the Meaning of Reduction Division Circa 1890." *Isis* 61 (1970), 429–57.

———. "Lepidopteran Research and the Rise of Classical Genetics, 1880–1920." Abstract of paper. In *Proceedings of the Sixteenth International Congress of the History of Science,* 212. August 26–September 3, Bucharest, Romania, 1981.

————. "The Weismann-Spencer Controversy over the Inheritance of Acquired Characters." *Proceedings of the 15th International Congress of the History of Science* (1978), 451–68.

Clarke, James M. *James Hall of Albany, Geologist and Paleontologist*. Albany, N.Y.: E. E. Rankin, 1921.

Cockerell, T. D. A. "Alpheus Spring Packard, Jr." In *Biographical Memoirs of The National Academy of Sciences*, vol. 9, pp. 181–236. Washington, D.C.: National Academy of Sciences, 1920.

Cohen, I. Bernard. "The New World as a Source of Science for Europe." In *IX Congreso Internacional de Historia de la Ciencia*, 95–143. Barcelona-Madrid, 1959.

Coleman, William R. *Biology in the Nineteenth Century: Problems of Form, Function, and Transformation*. 1971. Repr. Cambridge: Cambridge University Press, 1977.

————. *Georges Cuvier, Zoologist: A Study in the History of Evolution Theory*. Cambridge, Mass.: Harvard University Press, 1964.

Comstock, Anna Botsford. *The Comstocks of Cornell: John Henry Comstock and Anna Botsford Comstock*. Edited by Glenn W. Herrick and Ruby Green Smith. New York: Comstock Publishing Associates, 1953.

Comstock, John Henry. "Evolution and Taxonomy: An Essay on the Application of the Theory of Natural Selection in the Classification of Animals and Plants, Illustrated by a Study of the Evolution of the Wings of Insects, and by a Contribution to the Classification of the Lepidoptera." In *The Wilder Quarter Century Book: A Collection of Original Papers dedicated to Professor Burt Green Wilder at the close of his Twenty-fifth Year of Service in Cornell University (1868–1893) by some of his former Students*, 34–114. Ithaca, N.Y.: Comstock Publishing, 1893.

Corgan, James X. "Early American Geological Surveys." In *The Geological Sciences in the Antebellum South*, edited by James X. Corgan. University: University of Alabama Press, 1982.

————, ed. *The Geological Sciences in the Antebellum South*. University: University of Alabama Press, 1982.

Cowan, C. F. "Boisduval and LeConte: *Histoire générale et iconographie des lépidoptères et des chenilles de l'Amérique septentrionale*." *Journal of the Society for the Bibliography of Natural History* 5 (1969), 125–34.

————. "Boisduval's *Species générale des lépidoptères*." *Journal of the Society for the Bibliography of Natural History* 5 (1969), 121–22.

Cresson, Ezra Townsend. *A History of the American Entomological Society, Philadelphia, 1859–1909*. Philadelphia: [American Entomological Society, 1911].

————. "Memoir of Thomas B. Wilson." *Proceedings of the Entomological Society of Philadelphia* 5 (1865), 1–38.

Crosland, Maurice. "The Development of a Professional Career in Science in France." *Minerva* 13:3 (1975), 38–57.

Cutright, Paul Russell, and Michael J. Brodhead. *Elliott Coues: Naturalist and Frontier Historian*. Chicago: University of Chicago Press, 1981.

Dall, William Healey. *Spencer Fullerton Baird: A Biography*. Philadelphia: J. B. Lippincott, 1915.

Danhof, Clarence H. *Agricultural Change: The Northern United States, 1820–1870*. Cambridge, Mass.: Harvard University Press, 1969.

Daniels, George H. *American Science in the Age of Jackson.* New York: Columbia University Press, 1968.

Deiss, William A. "Spencer F. Baird and His Collectors." *Journal of the Society for the Bibliography of Natural History* 9:4 (1980), 635–45.

Desmond, Adrian. "The Making of Institutional Zoology in London, 1822–1836, Part 1." *History of Science* 23 (1985), 153–85.

———. "The Making of Institutional Zoology in London, 1822–1836, Part 2." *History of Science* 23 (1985), 223–50.

Dethier, V. G. *Man's Plague? Insects and Agriculture.* Princeton, N.J.: Darwin Press, 1976.

Dexter, Ralph W. "A. S. Packard's Annual Record of American Entomology, 1871–1873." *Bulletin of the Brooklyn Entomological Society* 59–60 (1964–1965), 35–36.

———. "The Development of A. S. Packard, Jr., as a Naturalist and an Entomologist." Pts. 1, 2. *Bulletin of the Brooklyn Entomological Society* 52 (1957): (June), 57–66; (October), 101–12.

———. "The Impact of Evolutionary Theories on the Salem Group of Agassiz Zoologists (Morse, Hyatt, Packard, Putnam)." *Historical Collections of the Essex Institute* 115 (1979), 144–71.

———. "The Organization and Work of the U.S. Entomological Commission (1877–1882)." *Melsheimer Entomological Series* 26 (1979), 28–32.

Dickinson, Thomas A. "Francis Gregory Sanborn." *Proceedings of the Worcester Society of Antiquities* (1884), 155–65.

Dos Passos, Cyril F. "The Dates of Publication of the *Histoire générale et iconographie des lépidoptères et des chenilles de l'Amérique septentrionale,* by Boisduval and LeConte, 1829–1833 [1834]." *Journal of the Society for the Bibliography of Natural History* 4 (1962), 48–56.

Dow, R. P. "The Greatest Coleopterist." *Journal of the New York Entomological Society* 22 (September 1914), 185–91.

Dunlap, Thomas R. "The Triumph of Chemical Pesticides." *Environmental Review* (May 1978), 39–47.

Dupree, A. Hunter. "National Pattern of Learned Societies." In *The Pursuit of Knowledge in the Early American Republic: American Scientific and Learned Societies from Colonial Times to the Civil War,* edited by Alexandra Oleson and Sanborn C. Brown, 21–32. Baltimore, Md.: Johns Hopkins University Press, 1976.

———. *Science in the Federal Government: A History of Policies and Activities to 1940.* 1957. Repr. New York: Arno Press, 1980.

Dupuis, Claude. "Pierre Andre Latreille (1762–1833): The Foremost Entomologist of His Time." *Annual Review of Entomology* 19 (1974), 1–14.

Egerton, Frank N. III. "A Bibliographical Guide to the History of General Ecology and Population Ecology." *History of Science* 15 (1977), 189–215.

———. "Changing Concepts of the Balance of Nature." *Quarterly Review of Biology* 48 (1973), 322–50.

———. "Ecological Studies and Observations Before 1900." In *History of Ecology,* edited by Frank N. Egerton, 311–51. New York: Arno Press, 1977.

———. "Studies of Animal Populations from Lamarck to Darwin." *Journal of the History of Biology* 1 (Fall 1968), 225–59.

Eiseley, Loren. *Darwin's Century: Evolution and the Men Who Discovered It.* 1958. Repr. New York: Anchor Books, 1961.

Elliott, Clark A. "The American Scientist in Antebellum Society: A Quantitative View." *Social Studies of Science* 5 (1975), 93–108.

———. "Models of the American Scientist: A Look at Collective Biography." *Isis* 73 (March 1982), 77–93.

———. "The *Royal Society Catalogue* as an Index to Nineteenth Century American Science." *Journal of the American Society for Information Science* 21 (November–December 1970), 396–401.

Essig, E. O. *A History of Entomology.* New York: Macmillan, 1931.

Evans, Mary Alice. "Mimicry and the Darwinian Heritage." *Journal of the History of Ideas* 26 (1965), 211–20.

Ewan, Joseph. "George Engelmann." In *Dictionary of Scientific Biography,* edited by Charles Coulston Gillispie, 130–31. New York: Charles Scribner's Sons, 1970.

Farber, Paul L. "The Development of Taxidermy and the History of Ornithology." *Isis* 68 (December 1977), 550–66.

———. "The Type-Concept in Zoology During the First Half of the Nineteenth Century." *Journal of the History of Biology* 9 (Spring 1976), 93–119.

Felt, Ephriam Porter. *Memorial of Life and Entomological Work of Joseph Albert Lintner, Ph.D.* Bulletin of the New York State Museum 5:24. Albany: University of the State of New York, 1899.

Fernald, C. H. "Evolution of Economic Entomology." *Science,* n.s. 4 (October 16, 1896), 541–47.

———. "Presidential Address, 1896." In "Annual Meetings and Presidential Addresses of the American Association of Economic Entomologists, 1889–1911: Annual Meetings 1–24." (Typescript in library of the University of California, Davis.)

Fite, Gilbert C. "Daydreams and Nightmares: The Late Nineteenth Century Frontier." *Agricultural History* 40 (October 1966), 289–91.

———. *The Farmers' Frontier, 1865–1900.* New York: Holt, Rinehart and Winston, 1966.

Fleming, James Roger. *Meteorology in America, 1800–1870.* Baltimore, Md.: Johns Hopkins University Press, 1990.

Foose, Richard M., and John Lancaster. "Edward Hitchcock: New England Geologist, Minister, and Educator." In "History of Geology in the Northeast," edited by William M. Jordan, 13–16. *Northeastern Geology* 3:1 (1981).

Forbes, Stephen A. "Presidential Address, 1908." In "Annual Meetings and Presidential Addresses of the American Association of Economic Entomologists, 1889–1911: Annual Meetings 1–24." (Typescript in library of the University of California, Davis.)

Ford, Alice. *John James Audubon.* Norman: University of Oklahoma Press, 1964.

Foster, Mike. "Ferdinand Vandiveer Hayden as Naturalist." *American Zoologist* 26 (1986), 343–49.

Fraser, S. "Potato." In *Cyclopedia of American Agriculture,* vol. 2, edited by Liberty Hyde Bailey, 519–27. 4 vols. New York: Macmillan, 1907.

Friedlander, C. P. *The Biology of Insects.* New York: Pica Press, 1977.

Gates, Paul Wallace. *Agriculture and the Civil War.* New York: Alfred A. Knopf, 1965.

———. *The Farmer's Age: Agriculture, 1815–1860.* New York: Holt, Rinehart and Winston, 1960.

Gerstner, Patsy A. "The Academy of Natural Sciences of Philadelphia, 1812–1850." In *The Pursuit of Knowledge in the Early American Republic: American Scientific and Learned Societies from Colonial Times to the Civil War,* edited by Alexandra Oleson and Sanborn C. Brown, 174–93. Baltimore, Md.: Johns Hopkins University Press, 1976.

————. "Vertebrate Paleontology: An Early Nineteenth Century Transatlantic Science." *Journal of the History of Biology* 3 (Spring 1970), 137–48.

Ghiselin, Michael T. *The Triumph of the Darwinian Method.* Berkeley: University of California Press, 1969.

Goetzmann, William H. *Exploration and Empire: The Explorer and the Scientist in the Winning of the American West.* New York: Alfred A. Knopf, 1971.

Golding, F. W. "Biographical Sketch of William LeBaron." *Entomologica Americana* 2 (October 1885), 122–25.

Göllner-Scheidung, Ursula. "Zur Geschichte der Entomologie in Berlin mit besonderer Berücksichtigung des 19. Jahrhunderts." *Wissenschaftliche Zeitschrift der Humboldt Universität zu Berlin Mathematisch-Naturwissenschaftliche Reihe* 34:3 (1985), 310–20.

Goode, G. B. "A Memorial Appreciation of Charles Valentine Riley." *Science,* n.s. 3 (February 14, 1896), 217–24.

Graef, Edward L. "Some Early Brooklyn Entomologists." *Bulletin of the Brooklyn Entomological Society* 9 (June 1914), 47–56.

Greene, John C. "American Science Comes of Age, 1780–1820." *Journal of American History* 55 (June 1968), 22–41.

————. *American Science in the Age of Jefferson.* Ames: Iowa State University Press, 1984.

————. "Science and the Public in the Age of Jackson." *Isis* 49 (1958), 13–25.

Gunn, D. L. "The Biological Background of Locust Control." *Annual Review of Entomology* 5 (1960), 279–300.

Guralnick, Stanley M. *Science and the Ante-Bellum American College.* Memoirs of the American Philosophical Society 109. Philadelphia: American Philosophical Society, 1975.

Gurney, Ashley B., and A. R. Brooks. "Grasshoppers of the Mexicanus Group, Genus *Melanoplus* (Orthoptera: Acrididae)." In *Proceedings of the U.S. National Museum* 110:3416, pp. 1–98. Washington, D.C.: Smithsonian Institution, 1959.

Haeussler, G. J., and R. W. Leiby. "Surveys of Insect Pests." In *Insects: The Yearbook of Agriculture, 1952,* 444–49. Washington, D.C.: U.S. Government Printing Office, 1952.

Hagen, Herman A. "Dr. Christian Zimmermann." *Canadian Entomologist* 21 (1889), 53–57.

————. "The History of the Origin and Development of Museums." Pts. 1, 2. *American Naturalist* 10 (1876): (February), 80–89; (March), 135–48.

————. "The Melsheimer Family and the Melsheimer Collection." *Canadian Entomologist* 16 (1884), 191–97.

Haskell, Thomas L. "Power to the Experts." *New York Review of Books* (October 13, 1977), 28–33.

Haygood, Tamara Miner. *Henry William Ravenel, 1814–1887: South Carolina Scientist in the Civil War Era.* History of American Science and Technology Series. Tuscaloosa: University of Alabama Press, 1987.

Headley, J. C. "Economics of Agricultural Pest Control." *Annual Review of Entomology* 17 (1972), 273–86.

Hedrick, Ulysses Prentiss. *A History of Agriculture in the State of New York.* Albany: New York State Agricultural Society, 1933.

———. *History of Horticulture in America to 1860.* New York: Oxford University Press, 1950.

Hendrickson, Walter B. "An Illinois Scientist Defends Darwinism: A Case Study in the Diffusion of Scientific Theory." *Transactions of the Illinois Academy of Sciences* 65 [314] (1972), 25–29.

Hennig, Willi. "Phylogenetic Systematics." *Annual Review of Entomology* 10 (1965), 97–116.

Heppell, David. "The Evolution of the Code of Zoological Nomenclature." In *History in the Service of Systematics: Papers from the Conference to Celebrate the Centenary of the British Museum (Natural History) 13–16 April 1981,* edited by Alwyne Wheeler and James H. Price, 135–41. London: Society for the Bibliography of Natural History, 1981.

Hindle, Brooke. *The Pursuit of Science in Revolutionary America.* 1956. Repr. New York: W. W. Norton, 1974.

Holland, G. P. "Entomology in Canada: Address for the Opening at the Royal Ontario Museum of the Special Exhibit Marking the Centennial of Entomology in Canada, 1863." In *Centennial of Entomology in Canada, 1863–1963: A Tribute to Edmund M. Walker,* edited by Glenn B. Wiggens, 7–8. Royal Ontario Museum, Life Sciences Contribution 69. Toronto: University of Toronto Press, 1966.

Holland, William J. *The Butterfly Book: A Popular Guide to a Knowledge of the Butterflies of North America.* New York: Doubleday, Page, 1913.

Holmfeld, John D. "From Amateurs to Professionals in American Science: The Controversy over the Proceedings of an 1853 Scientific Meeting." *Proceedings of the American Philosophical Society* 114 (February 1970), 22–36.

Horn, G. H. "John Lawrence LeConte." *Proceedings of the American Academy of Arts and Sciences* 19 (1884), 511–16.

Howard, Leland O. "A Brief Account of the Rise and Present Condition of Official Economic Entomology." Presidential Address, August 1894. In "Annual Meetings and Presidential Addresses of the American Association of Economic Entomologists, 1889–1911: Annual Meetings 1–24." (Typescript in library of the University of California, Davis.)

———. *Fighting the Insects: The Story of an Entomologist, telling of the Life and Experiences of the Writer.* New York: Macmillan, 1933.

———. *A History of Applied Entomology (Somewhat Anecdotal).* Smithsonian Miscellaneous Collections 84. Washington, D.C.: Smithsonian Institution, 1930.

———. "Progress in Economic Entomology in the United States." In *U.S. Department of Agriculture Yearbook of Agriculture, 1899,* 135–57. Washington, D.C.: U.S. Government Printing Office, 1900.

———. "The Rise of Applied Entomology in the United States." *Agricultural History* 3 (July 1929), 131–39.

Hull, David L. *Darwin and His Critics: The Reception of Darwin's Theory of Evolution by the Scientific Community.* Cambridge, Mass.: Harvard University Press, 1973.

Huxley, Julian. "Morphism and Evolution." *Heredity* 9, pt. 1 (April 1955), 1–52.

James, M. J. *The New Aurelians: A Centenary History of the British Entomological and Natural History Society, 1873–1973.* London: British Entomological and Natural History Society, 1973.

Jones, F G. W., and Margaret G. Jones. *Pests of Field Crops.* 2d ed. New York: St. Martin's Press, 1974.

Kellogg, Remington. "A Century of Progress in Smithsonian Biology." *Science* 104 (August 9, 1946), 132–41.

Kennedy, John Stoddard. "Continuous Polymorphism in Locusts." In *Insect Polymorphism,* edited by John Stoddard Kennedy, 80–90. London: Royal Entomological Society, 1961.

Kohlstedt, Sally Gregory. *The Formation of the American Scientific Community: The American Association for the Advancement of Science, 1848–1860.* Urbana: University of Illinois Press, 1976.

———. "The Geologists' Model for National Science, 1840–1847." *Proceedings of the American Philosophical Society* 118 (April 1974), 179–94.

———. "In from the Periphery: American Women in Science, 1830–1880." *Signs: A Journal of Women in Culture and Society* 4 (Autumn 1978), 81–96.

———. "The Nineteenth-Century Amateur Tradition: The Case of the Boston Society of Natural History." In *Science and Its Public: The Changing Relationship,* edited by Gerald Holton and William A. Blanpied, 174–86. Boston Studies in the Philosophy of Science 33. Boston: D. Reidel Publishing, 1976.

Laird, William E., and James R. Rinehart. "Deflation, Agriculture, and Southern Development." *Agricultural History* 42 (April 1968), 115–24.

Lanham, Url. *The Insects.* New York: Columbia University Press, 1964.

Lemmer, George Francis. "Early Agricultural Editors and Their Farm Philosophies." *Agricultural History* 31 (October 1957), 3–22.

———. *Norman J. Colman and Colman's Rural World: A Study in Agricultural Leadership.* Columbia, Mo.: University of Missouri Press, 1953.

Lindroth, Carl H. "Systematics Specializes Between Fabricius and Darwin, 1800–1859." In *History of Entomology,* edited by Ray F Smith, Thomas E. Mittler, and Carroll N. Smith, 119–54. Palo Alto, Calif.: Annual Reviews, 1973.

Lindstrom, Diane. *Economic Development in the Philadelphia Region, 1810–1850.* New York: Columbia University Press, 1978.

Lorch, Jacob. "The Discovery of Nectar and Nectaries and Its Relation to Views on Flowers and Insects." *Isis* 69 (December 1978), 514–33.

Lowenberg, Bert James. "The Controversy over Evolution in New England." *New England Quarterly* 8 (June 1935), 233–56.

———. "The Reaction of American Scientists to Darwinism." *American Historical Review* 38 (July 1933), 687–701.

Lurie, Edward. *Louis Agassiz: A Life in Science.* Chicago: University of Chicago Press, 1960.

McKinney, H. Lewis. "Henry Walter Bates." In *Dictionary of Scientific Biography* 1, edited by Charles Coulston Gillispie, 500–504. New York: Charles Scribner's Sons, 1970.

———. *Wallace and Natural Selection.* New Haven: Yale University Press, 1972.

Malin, James C. "The Agricultural Regionalism of the Trans-Mississippi West as

Delineated by Cyrus Thomas." *Agricultural History* 21 (October 1947), 210–15.

Mallis, Arnold. *American Entomologists.* New Brunswick, N.J.: Rutgers University Press, 1971.

———, ed. "The Diaries of Asa Fitch, M.D." *Bulletin of the Entomology Society of America* 9 (1963), 262–65.

Marchant, James. *Alfred Russel Wallace: Letters and Reminiscences.* New York: Harper and Brothers, 1916.

Marlatt, C. L. "A Brief History of Entomology." *Proceedings of the Entomological Society of Washington, D.C.* 4 (February 1897), 83–120.

Matthews, J. R. "History of the Cambridge Entomological Club." *Psyche* 81 (March 1974), 3–37.

Mayor, Alfred Goldsborough. "Samuel Hubbard Scudder, 1837–1911." In *Biographical Memoirs,* vol. 27, 81–104. Washington, D.C.: National Academy of Sciences, 1924.

Mayr, Ernst. "The Nature of the Darwinian Revolution." *Science* 176 (1972), 983–88.

———. *Population, Species, and Evolution: An Abridgment of Animal Species and Evolution.* Cambridge, Mass.: Harvard University Press, 1970.

Mendelsohn, Everett. "The Emergence of Science as a Profession in Nineteenth-Century Europe." In *The Management of Scientists,* edited by Karl B. Hill, 3–48. Boston: Beacon Press, 1964.

Merrill, George P., ed. *Contributions to a History of American State Geological and Natural History Surveys.* U.S. National Museum Bulletin 109. Washington, D.C.: Smithsonian Institution, 1920.

Millbrooke, Anne. "Henry Darwin Rogers and the First State Geological Survey of Pennsylvania." *Northeastern Geology* 3:1 (1981), 71–74.

Miller, Howard S. *Dollars for Research: Science and Its Patrons in Nineteenth Century America.* Seattle: University of Washington Press, 1970.

Mills, Harlow B. "Weather and Climate." In *Insects: The Yearbook of Agriculture, 1952,* 422–28. Washington, D.C.: U.S. Government Printing Office, 1952.

Moore, John A. "Zoology of the Pacific Railroad Surveys." *American Zoologist* 26 (1986), 331–41.

Morge, Günter. "Entomology in the Western World in Antiquity and in Medieval Times." In *History of Entomology,* edited by Ray F. Smith, Thomas E. Mittler, and Carroll N. Smith, 37–80. Palo Alto, Calif.: Annual Reviews, 1973.

Müller, H. J. "Probleme der Insektendiapause." *Zoologischer Anzeiger,* Supplementband 29 (1965), 192–222.

———. "Die Wirkung verschiedener Licht-Dunkel-Relationen auf die Saisonformenbildung von Araschnia levana." *Naturwissenschaften* 43 (1956), 503–504.

Nodyne, Kenneth, R. "The Founding of the Lyceum of Natural History." *Annals of the New York Academy of Sciences* 172 (1970), 141–49.

Numbers, Robert L., and Janet S. Numbers. "Science in the Old South: A Reappraisal." *Journal of Southern History* 48 (May 1982), 163–84.

Ogilvie, William Edward. *Pioneer Agricultural Journalists: Brief Biographical Sketches of Some of the Early Editors in the Field of Agricultural Journalism.* Chicago: Arthur G. Leonard, 1927.

Oleson, Alexandra. "Introduction: To Build a New Intellectual Order." In *The Pursuit of Knowledge in the Early American Republic: American Scientific and Learned Societies from Colonial Times to the Civil War,* edited by Alexandra

Oleson and Sanborn C. Brown, xv–xxv. Baltimore, Md.: Johns Hopkins University Press, 1976.

Ordish, George. *The Constant Pest: A Short History of Pests and Their Control*. New York: Charles Scribner's Sons, 1976.

———. *John Curtis and the Pioneering of Pest Control*. Reading, England: Osprey Press, 1974.

———. "Scientific Pest Control and the Influence of John Curtis." *Journal of the Royal Society of Arts* 116 (1968), 298–309.

Orosz, Joel J. "Disloyalty, Dismissal, and a Deal: The Development of the National Museum at the Smithsonian Institution, 1846–1855." *Museum Studies Journal* 2:2 (1986), 22–33.

Osborn, Herbert. *A Brief History of Entomology*. Columbus, Ohio: Spahr and Glenn, 1952.

———. *Fragments of Entomological History: Some Personal Recollections of Men and Events*. Columbus, Ohio: Author, 1937.

———. *The Hessian Fly in the United States*. U.S. Department of Agriculture, Bureau of Entomology, Bulletin 16. Washington, D.C.: U.S. Government Printing Office, 1898.

Osten-Sacken, Carl Robert. *Record of My Life-Work in Entomology, with an Appreciation and Introductory Preface by K. G. V. Smith*. 1903–1904. Repr. Faringdon, England: E. W. Classey, 1977.

Packard, C. M. "Cereal and Forage Insects." In *Insects: The Yearbook of Agriculture, 1952*, 581–94. Washington, D.C.: U.S. Government Printing Office, 1952.

Packard, C. M., and John H. Martin. "Resistant Crops, the Ideal Way." In *Insects: The Yearbook of Agriculture, 1952*, 429–36. Washington, D.C.: U.S. Government Printing Office, 1952.

Parrott, P. J. "Growth and Organization of Applied Entomology in the United States." *Journal of Economic Entomology* 7 (February 1914), 50–64.

Perkins, John H. *Insects, Experts, and the Insecticide Crisis: The Quest for New Pest Management Strategies*. New York: Plenum Press, 1982.

Pfeifer, Edward J. "The Genesis of American Neo-Lamarckism." *Isis* 56 (1965), 156–67.

———. "United States." In *The Comparative Reception of Darwinism*, edited by Thomas F. Glick. Austin: University of Texas Press, 1974.

Porter, Charlotte M. *The Eagle's Nest: Natural History and American Ideas, 1812–1842*. History of American Science and Technology Series. University: University of Alabama Press, 1986.

Proctor, Michael, and Peter Yeo. *The Pollination of Flowers*. London: Collins, 1973.

Rabbitt, Mary C. *Minerals, Lands, and Geology for the Common Defense and General Welfare, Vol. 1, Before 1879: A History of Public Land, Federal Science and Mapping Policy, and Development of Mineral Resources in the United States*. Washington, D.C.: U.S. Government Printing Office, 1979.

Rasmussen, Wayne. "The Civil War: A Catalyst of Agricultural Revolution." *Agricultural History* 39 (October 1965), 187–95.

Reingold, Nathan. "Definitions and Speculations: The Professionalization of Science in America in the Nineteenth Century." In *The Pursuit of Knowledge in the Early American Republic: American Scientific and Learned Societies from Colonial*

Times to the Civil War, edited by Alexandra Oleson and Sanborn C. Brown, 33–69. Baltimore, Md.: Johns Hopkins University Press, 1976.

——, ed. *Science in Nineteenth Century America: A Documentary History.* American Century Series. New York: Hill and Wang, 1964.

Remington, Jeanne, and Charles L. Remington. "Darwin's Contributions to Entomology." *Annual Review of Entomology* 6 (1961), 1–12.

Riegert, Paul W. *From Arsenic to DDT: A History of Entomology in Western Canada.* Toronto: University of Toronto Press, 1980.

Ritvo, Harriett. "Animal Pleasures: Popular Zoology in Eighteenth- and Nineteenth-Century England." *Harvard Library Bulletin* 33 (1985), 239–79.

Rivinus, E. F, and E. M. Youssef. *Spencer Baird of the Smithsonian.* Washington, D.C.: Smithsonian Institution Press, 1992.

Roberts, Jon H. *Darwinism and the Divine in America: Protestant Intellectuals and Organic Evolution, 1859–1900.* Madison: University of Wisconsin Press, 1988.

Robinson, Gloria. "August Friedrich Leopold Weismann." In *Dictionary of Scientific Biography,* edited by Charles Coulston Gillispie, 232–39. New York: Charles Scribner's Sons, 1970.

——. *A Prelude to Genetics: Theories of a Material Substance of Heredity: Darwin to Weismann.* Lawrence, Kans.: Coronado Press, 1979.

Rosenberg, Charles E. *No Other Gods: On Science and American Social Thought.* Baltimore, Md.: Johns Hopkins University Press, 1976.

Ross, Earle D. *Democracy's College: The Land Grant Movement in the Formative Stage.* Ames: Iowa State College Press, 1942.

——. "The United States Department of Agriculture during the Commissionership." *Agricultural History* 20 (July 1946), 129–43.

Ross, Edward S. "Systematic Entomology." In *A Century of Progress in the Natural Sciences, 1853–1953: Published in Celebration of the Centennial of the California Academy of Sciences,* 485–95. San Francisco: California Academy of Sciences, 1955.

Rossiter, Margaret W. *The Emergence of Agricultural Science: Justus Liebig and the Americans, 1840–1880.* Yale Studies in the History of Science and Medicine 9. New Haven: Yale University Press, 1975.

——. "The Organization of Agricultural Improvement in the United States." In *The Pursuit of Knowledge in the Early American Republic: American Scientific and Learned Societies from Colonial Times to the Civil War,* edited by Alexandra Oleson and Sanborn C. Brown, 279–98. Baltimore, Md.: Johns Hopkins University Press, 1976.

——. "The Organization of the Agricultural Sciences." In *The Organization of Knowledge in Modern America, 1860–1920,* edited by Alexandra Oleson and John Voss, 211–48. Baltimore, Md.: Johns Hopkins University Press, 1979.

——. *Women Scientists in America: Struggles and Strategies to 1940.* Baltimore, Md.: Johns Hopkins University Press, 1982.

Rothstein, Morton. "Antebellum Wheat and Cotton Exports: A Contrast in Marketing Organization and Economic Development." *Agricultural History* 40 (April 1966), 91–100.

Russell, E. S. *Form and Function: A Contribution to the History of Animal Morphology.* London: John Murray, 1916.

Sandved, Kjell B., and Michael G. Emsley. *Insect Magic.* New York: Viking Press, 1978.

Schlebecker, John T. "Grasshoppers in American Agricultural History." *Agricultural History* 27 (July 1953), 85–93.

Schwarz, E. A. "Some Notes of Melsheimer's Catalogue of the Coleoptera of Pennsylvania." *Proceedings of the Entomological Society of Washington, D.C.* 3 (1895), 134–38.

Silverman, Robert, and Mark Beach. "A National University for Upstate New York." *American Quarterly* 22:3 (Fall 1970), 701–13.

Smith, Edward H. "The Comstocks and Cornell: In the People's Service." *Annual Review of Entomology* 21 (1976), 1–25.

———. "The Entomological Society of America: The First Hundred Years, 1889–1989." *Bulletin of the Entomological Society of America* 35:3 (Fall 1989), 10–32.

———. "The Grape Pylloxora: A Celebration of its Own." *American Entomologist* (Winter 1992), 212–21.

Snetsinger, Robert. *Frederick Valentine Melsheimer, Parent of American Entomology.* University Park: Entomological Society of Pennsylvania, 1973.

Sorensen, W. Conner. "The Rise of Government Sponsored Applied Entomology, 1840–1870." *Agricultural History* 62 (Spring 1988), 98–115.

———. "Uses of Weather Data by American Entomologists, 1830–1880." *Agricultural History* 63 (Spring 1989), 162–74.

———. "William Henry Edwards, August Weismann and Polymorphism in Butterflies." In *August Weismann (1834–1914) und die theoretische Biologie des 19. Jahrhunderts: Urkunden, Berichte und Analysen,* special issue edited by Klaus Sander, 157–65. Freiburger *Universitätsblätter,* Heft 87/88 (July 1985).

Stanton, William. *The Great United States Exploring Expedition of 1838–1842.* Berkeley: University of California Press, 1975.

Stephens, Lester D. "The Appointment of the Commissioner of Agriculture in 1877: A Case Study in Political Ambition and Patronage." *Southern Quarterly* 15:4 (1977), 371–86.

Sterling, Keir B. *Last of the Naturalists: The Career of C. Hart Merriam.* Natural Sciences in America. Rev. ed. New York: Arno Press, 1977.

Stern, Vernon M. "Economic Thresholds." *Annual Review of Entomology* 18 (1973), 259–80.

Taylor, George Rogers. "The National Economy Before and After the Civil War." In *Economic Change in the Civil War Era: Proceedings of a Conference on American Economic Institutional Change, 1850–1873, and the Impact of the Civil War, [held] March 12–14, 1964,* edited by David Gilchrist and W. David Lewis. Greenville, Del.: Eleutherian Mills-Hagley Foundation, 1965.

———. *The Transportation Revolution, 1815–1860.* New York: Holt, Rinehart and Winston, 1951.

Teale, Edwin Way. "The Life of the Chinch Bug." In *The Strange Lives of Familiar Insects,* by Edwin Way Teale, 121–29. New York: Dodd, Mead, 1962.

Traub, Hamilton. "Tendencies in the Development of American Horticultural Associations, 1800–1850." *National Horticultural Magazine* 9 (July 1930), 134–40.

True, F. W. "An Account of the U.S. National Museum." In *The Smithsonian Institution, 1846–1896: The History of Its First Half Century,* edited by George Brown Goode. Washington, D.C.: Smithsonian Institution, 1897.

Tuxen, S. L. "Entomology Systematizes and Describes, 1700–1815." In *History of*

Entomology, edited by Ray F Smith, Thomas E. Mittler, and Carroll N. Smith, 95–118. Palo Alto, Calif.: Annual Reviews, 1973.

Usinger, R. L. "Prefatory Chapter: The Role of Linnaeus in the Advancement of Entomology." *Annual Review of Entomology* 9 (1964), 1–16.

Uvarov, Sir Boris. *Grasshoppers and Locusts: A Handbook of General Acridology.* 2 vols. Cambridge, England: Cambridge University Press, 1966.

Verrill, A. E. "Sidney Irving Smith." *Science* 64 (1926), 57–58.

Veysey, Laurence R. *The Emergence of the American University.* Chicago: University of Chicago Press, 1965.

Wakeland, Claude. "The Chinch Bug." In *Insects: The Yearbook of Agriculture, 1952,* 611–13. Washington, D.C.: U.S. Government Printing Office, 1952.

Warner, Deborah Jean. "Science Education for Women in Antebellum America." *Isis* 69 (1978), 58–67.

Warner, Sam Bass, Jr. *The Private City: Philadelphia in Three Periods of Its Growth.* Philadelphia: University of Pennsylvania Press, 1968.

Weber, Gustavus A. *The Bureau of Entomology: Its History, Activities, and Organization.* Institute for Government Research, Science Monographs of the U.S. Government 60. Washington, D.C.: Institute for Government Research, 1930.

Webster, F M. "Presidential Address, 1897." In "Annual Meetings and Presidential Addresses of the American Association of Economic Entomologists, 1889–1911: Annual Meetings 1–24." (Typescript in library of the University of California, Davis.)

Weiss, Harry B. *The Pioneer Century of American Entomology.* New Brunswick, N.J.: Author, 1936.

Weiss, Harry B., and Grace Ziegler. *Thomas Say: Early American Naturalist.* Springfield, Ill.: Charles C. Thomas, 1931.

Welter, Barbara. "The Cult of True Womanhood, 1820–1860." In *Our American Sisters: Women in American Life and Thought,* edited by Jane E. Friedman and William G. Shade, 96–123. Boston: Allyn and Bacon, 1973.

Wheeler, A. G., Jr. "Rev. Daniel Ziegler, D.D.: A Promoter of Entomology in Mid-Nineteenth Century America." *Melsheimer Entomological Series* 28 (1980), 21–26.

———. "The Rev. Solomon Oswald, Forgotten Member of the Original Entomological Society of Pennsylvania." *Melsheimer Entomological Series* 30 (1981), 43–46.

Wheeler, A. G., Jr., and Karl Valley. "A History of the Entomological Society of Pennsylvania, 1842–44 and 1924–Present." *Melsheimer Entomological Series* 24 (1978), 16–26.

Wheeler, William Morton. "Entomology at Harvard University." In *Notes Concerning the History and Contents of the Museum of Comparative Zoology.* By Members of the Staff. Cambridge, Mass.: Harvard University, Museum of Comparative Zoology, 1936.

Whitnah, Donald R. *A History of the United States Weather Bureau.* Urbana: University of Illinois Press, 1961.

Whorton, James. *Before Silent Spring: Pesticides and Public Health in Pre-DDT America.* Princeton, N.J.: Princeton University Press, 1974.

Williams, C. B. "Insect Migration." *Annual Review of Entomology* 2 (1957), 163–80.

Wilson, H. F., and M. H. Doner. *The Historical Development of Insect Classification*. St. Louis, Mo.: John S. Swift, 1937.

Wilson, Joan Hoff. "Dancing Dogs of the Colonial Period: Women Scientists." *Early American Literature* 7 (Winter 1973), 225–35.

Winsor, Mary P. "The Development of Linnaean Insect Classification." *Taxon* 25 (February 1976), 57–67.

———. "Louis Agassiz and the Species Question." *Studies in the History of Biology* 3 (1979), 89–117.

———. *Starfish, Jellyfish, and the Order of Life: Issues in Nineteenth Century Science*. Yale Studies in the History of Science and Medicine 10. New Haven: Yale University Press, 1976.

Woodcock, George. *Henry Walter Bates: Naturalist of the Amazons*. New York: Barnes and Noble, 1969.

Index

Abbe, Cleveland, 146–47
Abbot, John, 14, 228; biography, 7–8; *The Rarer Lepidopterous Insects of Georgia,* 8
Abraupen, 66
Academy of Natural Sciences of Philadelphia: and early American natural history, 5; and the Entomological Society of Philadelphia, 50–51, 54; response to evolution of, 208; William Maclure a sponsor of, 27, 42
Ackhurst, John. *See* Akhurst, John
Adalbert, Prince, 307 (n. 6)
Adams, John, 5–6
Agassiz, Louis: biography, 54; encourages American entomologists, 29; influences Cambridge Entomological Club, 207; instruction in entomology, 81; on the relationship of the Museum of Comparative Zoology to other museums, 280 (n. 108); opposes evolution, 199–202, 211; recommends appointments at Cornell University, 83
Agricultural change: and funding for entomology, 86, 90–91; and insect pests, 65–66; during the Civil War, 74; produces leadership among agriculturists, 92–93
Agricultural Department. *See* U.S. Department of Agriculture
Agricultural revolution, 65–66
Agriculture, mechanization of, 66
Agriculturists, 61, 94–95, 115–16
Akhurst, John, 166, 216

Albany, New York, 66
Allen, Joel A., 102
Amateurs, entomological: and entomological nomenclature, 244, 246; as characteristic of American entomology, 175; contributions of, 259; examples of, 187–88; numbers of, 187, 301 (n. 34); replaced by professionals, 60–61
American Academy of Arts and Sciences, 5–6
American Association for the Advancement of Science, 15, 30, 275 (n. 116). *See also* Committee on Nomenclature of the American Association for the Advancement of Science
American Association of Geologists and Naturalists, 30, 243
American Conchology (Say), 191
American Entomological Society, 78; membership list as source of data, 184–86. *See also* Entomological Society of Philadelphia
American Entomologist, 68, 75
American Entomology; or, Descriptions of the Insects of North America (Say), 10, 191
American Journal of Science, 12, 254
American Naturalist, 88–89
American Ornithology (Wilson), 8
American Philosophical Society, 5
American Pomological Society, 67
"America's Entomological Declaration of Independence," 23
Araschnia prorsa, 217, 223–26

345

Argument from design, 211, 235–36, 238–40. *See also* Natural theology; Teleology

Audubon, John James, 23

Aughey, Samuel, 102

Austin, W. P., 208

Authors, entomological: numbers of, 185–87

Baird, Spencer F.: and reform of American zoology, 22–23; and Smithsonian insects, 46–47, 51–52, 56–58; assistant secretary, Smithsonian Institution, 43; sends Lepidoptera to William Henry Edwards, 216

Balance of nature: acceptance by agriculturists, 103–106; and agricultural entomology, 92–94; and evolutionary theory, 241; and insect-bird relationships, 100–103; as basis of control, 106; as employed by entomologists, 96–100; concept of, 95–97; weather a factor in, 104

Barnard, F. A. P., 211

Barton, Benjamin S., 5

Bartram, John, 5, 14, 95

Bartram, William, 5, 10, 14, 95

Basnett, Thomas, 190

Bates, Henry Walter: and William Henry Edwards, 215, 218, 223, 310 (n. 66); biography, 229; president, Entomological Society of London, 210

Behrens, James, 246

Belfrage, Gustav Wilhelm, 35–37, 281 (n. 131)

Belt, Thomas, 203

Bethune, Charles James, 77, 122, 242–43

Bigelow, Jacob, 96

Biological control, 96–103, 106, 116

Bird question, 100–103

Boisduval, J. B. A., 4, 10

Bonaparte, Charles L., 5

Bordeaux mixture, 125

Boston, Massachusetts, 5–6

Boston Society of Natural History, 6, 88, 179

Bowen, Lydia, 191, 217

Bowles, George John, 209

Böll, Jacob, 239, 276 (n. 7)

Bradley, Richard, 96–97

British Association for the Advancement of Science, 243, 245

British Museum of Natural History, 216

Brooklyn Entomological Society, 209, 219

Brown University, 88

Bruce, Robert V.: analysis of American scientists cited, 152–54

Buel, Jesse, 66

Buffon, Count George Louis Leclerk, 4–5, 212

Bulletin of the Brooklyn Entomological Society, 209

Burns, J. H., 99

Butterflies: as objects of study, 218, 308 (n. 17). *See also* Edwards, William Henry: method of rearing butterflies, and on polymorphism

Butterflies of North America (Edwards), 216–17, 222, 228

Butterfly, monarch, 203

Caloptenus spretus (Rocky Mountain locust), 141

Calverley, Stephen H., 166

Cambridge, Massachusetts, 5–6

Cambridge Entomological Club, 56, 207–208

Canadian Entomologist, 78, 208–209

Catalogue of the Described Coleoptera of the United States (Frederick Ernst Melsheimer), 25

Catalogue of the Insects of Pennsylvania (Frederick Valentine Melsheimer), 11

Catalogue of the Lepidoptera of America North of Mexico (Edwards), 250

Catesby, Mark, 14

Cecidomyia destructor (Say) (Hessian fly), 111

Centennial Exposition, Philadelphia, 58, 83, 132

Chambers, Robert: *Vestiges of the Natural History of Creation,* 211

Chambers, Vactor T., 239

Chapters on Ants (Treat), 189

Charlatanism, in science, 274 (n. 88)

"Checklist of North American Coleoptera" (Crotch), 244

Chemistry, agricultural, 105–106

Chickies, Pennsylvania, 17

Chinch bug, 116–20

Classification, insect: and evolution, 241, 244, 247–50; origins of, 2–4. *See also* Committee on Nomenclature of the American Association for the Advancement of Science; Committee on Nomenclature of the Entomological Club; Nomenclature, entomological; Nomenclature, rules of; Priority of description; Systematics

Classification, zoological, 3
Clay, Henry, 108
Clifford, E. A., 100–101
Coalburgh, West Virginia, 216
Collecting, insect, 34, 38, 43–44
Collection, insect: at the Centennial Exposition, 58, 83, 132; in New York State, 71; of Charles V. Riley, 58; of Frederick E. Melsheimer, 18, 24; of the Academy of Natural Sciences of Philadelphia, 50; of the Entomological Society of Philadelphia, 51–54; of the Museum of Comparative Zoology, 31–32, 55; of the Smithsonian Institution, 43–47, 56–59, 281 (n. 118); of the U.S. Department of Agriculture, 57–58, 281 (n. 118)
Collections, insect: agricultural, 41–42; American compared to European, 59; as standard references, 33–34; care of, 40–41; growth of in America, 35–38, 59, 254–55; support for, 41–43
Collector, insect: as explorer, 35–37
Colleges, land grant, 81–82
Colman, Norman J., 75
Colman's Rural World, 75
Colorado potato beetle, 120–24, 294 (n. 85)
Committee on Nomenclature of the American Association for the Advancement of Science, 243–44, 251
Committee on Nomenclature of the Entomological Club, 243, 248–49
Community, agricultural, 92
Community, American entomological: advances during forty years, 253–55; affiliations of members, 179; agricultural entomologists in, 64; analysis of, 150–54; and insect outbreaks, 125–27, 149; and revisions of nomenclature, 251; and William Henry Edwards, 233; as practicing ecologists, 96–103, 256; British origins of, 155–57; characteristics of, 60–61, 103, 193–94, 210–11, 241–42, 255–59; education of members, 161–63, 168; elevation of status, 198, 212–13, 256; European influences on, 155–57, 171, 182–83, 193–94; geographic location of members, 154–59, 185; leadership of, 234; maturity of, 252; members typical of, 182–84; numbers of, 38, 150–51, 184–87, 253–54, 276 (n. 12); occupations of members, 172–75; origins of, 160; parental occupations

of members, 159–61; professional status of members, 256; professional titles of members, 181–82; professionalization of occupations, 176–82; reception of evolutionary theory, 210–11, 234; solidarity of, 64, 194–96, 259; teachers of members, 170–71; training of members, 164–70; women members of, 188–93. *See also* Entomologists, American: socioeconomic status of; Entomology, American: compared with entomology in Europe
Compendium of the Biographical Literature on Deceased Entomologists, A (Gilbert), 150
Comstock, Anna Botsford, 84
Comstock, John Henry, 62, 99, 314 (n. 25); biography, 83–84; chief, Division of Entomology, 140; evolutionary studies of, 308 (n. 17)
Control methods. *See* Insects, injurious: control (various subheads)
Cook, Albert John, 61–62, 82–83
Cope, Edward Drinker, 204, 211, 280 (n. 94)
Coquillett, Daniel W., 190
Cornell University, 83
Coues, Elliott, 290 (n. 33)
Council of Agriculture and Arts Association of Ontario, 78
Country Gentleman: previously *Cultivator and Country Gentleman*, 86
Cresson, Esra Townsend: and the Entomological Society of Philadelphia, 47; and John L. LeConte, 51–52; and William Henry Edwards, 250; biography, 48–49
Critical review, 30, 275 (n. 116)
Crotch, George R.: "Checklist of North American Coleoptera," 244
Curculio, 108–10
Curtis, John, 86
Cuvier, Georges, 3, 201, 207; *Règne Animal*, 3

Dall, William H., 251
Dana, James D., 199–202
Danaus archippus Fabr., 203
Danaus berenice Cram., 203
Darwin, Charles, 189; and Benjamin D. Walsh, 199–200; and classification, 247; and Henry Walter Bates, 229; and Lamarckism, 206; as an entomologist, 197–98; attacks the argument from design, 235–36; *On the Origin of Species*, 197, 206, 229; *On the Various Contrivances*

(Darwin, Charles, *continued*)
by which British and Foreign Orchids are Fertilized by Insects, 235–36; on insectivorous plants, 189; on mimicry, 203; on polymorphism, 228; on ecological relationships, 235; *The Descent of Man,* 198, 211
Darwinism. *See* Evolutionary theory
Dechéniller, 66
Dejean, Compte P. F. M. A., 4, 10–11
DeKay, James, 70
Department of Agriculture. *See* U.S. Department of Agriculture
Depredations, insect. *See* Insects, injurious: estimates of damage
Descent of Man, The, (Darwin), 198, 211, 229
Descriptions, insect: by Americans, 13, 23–25, 47; by Entomological Society of Pennsylvania members, 23–25; by LeConte, 31; by Say, 10. *See also* Priority of description
Dewey, John, 236
Dimorphism, 217, 232, 248. *See also* Polymorphism
Disease, insect, 99–100
Dodge, Charles, 57–58
Doubleday, Edward, 250
Downing, Andrew J., 67, 72, 108
Drury, Dru, 7, 222
Dupree, A. Hunter: cited on "problem centered" approach, 140–41; defines scientific society, 28

East London Entomological Society, 48
Ecological relationships, 235, 256
Economic ornithology. *See* Ornithology
Economy of nature. *See* Balance of nature: concept of
Education, reform of, 80–82
Edwards, Amory, 215
Edwards, Helen Ann Mann, 215
Edwards, Jonathan, 215
Edwards, William, 215
Edwards, William Henry: *A Voyage up the River Amazon,* 215; and Alfred Russel Wallace, 223, 229; and August Weismann, 223–26, 229–32; and entomological nomenclature, 243–52; and Henry Walter Bates, 223, 229; and insect classification, 222–23; and Joseph Albert Lintner, 86; and Jacob Hübner, 245, 250; and Philipp C. Zeller, 233, 250; as authority on the Lepidoptera, 242; biography,

214–16, 232–33; *Catalogue of the Lepidoptera of America North of Mexico,* 250; entomological correspondence of, 186–87; evolutionary studies of, 213, 228–29, 255; method of rearing butterflies, 219–21; network of supporters, 233; on polymorphism, 209, 217–19, 221–23, 226–28, 234; on speciation, 227–28; on women as entomological collectors, 192; *The Butterflies of North America,* 216–17, 222, 228
Edwards, William W., 215
Elliott, Clark A.: analysis of American scientists cited, 152–54
Embargo on insects, 23
Emmons, Ebenezer, 26, 70–71, 80
Engelmann, George, 212, 236–37
Enlightenment, European, 63–64
Entomological Club of the American Association for the Advancement of Science, 64, 90, 133, 249, 251
Entomological commission, 133–37. *See also* U.S. Entomological Commission
Entomological Society of Canada (Entomological Society of Ontario), 64, 77–79, 208–209
Entomological Society of London, 197, 210
Entomological Society of Pennsylvania: and agricultural entomology, 26–28; and the Academy of Natural Sciences of Philadelphia, 27; compared with the Entomological Society of Philadelphia, 273–74 (n. 77); demise of, 25; embargo on insects proposed by, 23; founding of, 16–18; goals of, 17, 22–23; membership of, 17, 21–22; publications of members, 23–26; significance of, 26, 30–32
Entomological Society of Philadelphia (American Entomological Society): and William Henry Edwards, 216; compared with the Entomological Society of Pennsylvania, 273–74 (n. 77); dispute over insect specimens, 50–53; founding of, 47–48; membership list of, 184–86; reorganized as section of the Academy of Natural Sciences of Philadelphia, 54; response to evolutionary theory, 208; Thomas B. Wilson's patronage of, 49–50
Entomologists, agricultural: and insect-bird relationships, 101–102; and the balance of nature, 105–106; as government Entomologists, 93–94; as leaders, 61–65; as lobbyists, 67–68; careers of, 86–89;

contrasted with systematists, 138; in Canada, 78; relationship to farmers, 92–93; salaries of, 137; use of data from amateurs, 105–106; use of weather data, 114, 119, 142–47, 298–99 (n. 96). *See also* Entomology, American: compared with entomology in Europe

Entomologists, American: advantages and disadvantages of, 6–7, 11–13; and animal populations, 97; numbers of, 38, 150–51, 184–87, 253–54, 276 (n. 12); relationship to European entomologists, 25–26; socioeconomic status of, 28, 47–48, 61–62; "type specimen" of about 1870, 182–83

Entomologists, federal. *See* Comstock, John Henry; Glover, Townend; Howard, Leland O.; Riley, Charles Valentine

Entomologists, official, 78–80, 94

Entomologists, professional, 79–80. *See also* Professionalization of entomology

Entomologists, state and provincial. *See* Fitch, Asa; Forbes, Stephen Alfred; Lintner, Joseph Albert; Packard, Alpheus Spring, Jr.; Sanborn, Francis G.; Saunders, William; Riley, Charles Valentine; Thomas, Cyrus; Walsh, Benjamin D.

Entomologists, women: as entomological illustrators, 191; as entomological investigators, 188–90; as entomological observers, 192; as entomological popularizers, 190–91; difficulties faced by, 190–93; exclusion from leadership, 257–58

Entomologists: earnings of, 137, 287 (n. 104); relationship to the U.S. Department of Agriculture, 138–40; stereotype as "bug hunter," 34

Entomology, academic, 81

Entomology, agricultural: a distinctive American institution, 91; American leadership in, 72, 77, 107, 125–27, 149; and related institutions, 86–89; and the U.S. Department of Agriculture, 130; funding of, 85–86, 137, 287 (n. 104); in Canada, 77–79; in colleges and universities, 80–85; in Georgia, 80; in Maine, 80; in Michigan, 79–80; in Minnesota, 84; in Pennsylvania, 79; institutional changes proposed for, 69; origins of, 61; support for, 110, 128

Entomology, American: advances in theory noted, 254–55; compared with entomology in Europe, 6–7, 55–56, 59, 91, 213, 220–21, 228, 230, 233, 252; dark side of, 257–58; development of, 254–59; European influence on, 155–57, 171, 182–83, 193–94, 256; influence of agriculture on, 60–61, 65–66, 89–91, 255–57, 259; societies dedicated to, 254; support for, 255, 257

Entomology, American style of, 103, 256. *See also* Community, American entomological: characteristics of

Entomology, applied, 72, 107. *See also* Entomology, agricultural

Entomology: college degrees in, 85

Entomology, economic, 77. *See also* Entomology, agricultural

Entomology, European, 1–4. *See also* Entomology, American: compared with entomology in Europe

Erichson, Wilhelm F., 4, 20

Espy, James P., 146

Essex Institute, 88

Evans, Gurdon, 101, 103–104

Evolutionary theory: and religion, 205–206; debate among entomologists, 198–200; dissemination by entomologists, 211–12; essentialism vs. population thinking in, 200–202; insect classification based on, 244, 247–48; Lamarckism and neo-Lamarckism in, 199, 204–206, 212, 232, 255, 310 (n. 66); mimicry as evidence of, 202–204; natural selection in, 214, 218, 241; neo-Darwinism in, 224–25; reception by the American entomological community, 210–11, 234; reception of, 204–10, 229, 303 (n. 5), 306 (n. 77); status of entomologists related to, 198, 212–13; species defined according to, 250, 310 (n. 65)

Fabricius, Johann Christian, 3–4

Farm culture methods, 115–16

Farmers. *See* Agriculturists

Fitch, Asa, 12, 21, 42; and John Henry Comstock, 83; as agricultural entomologist, 61–63; at Rensselaer, 82, 85; biography, 71; Colorado potato beetle investigations, 122; contact with other entomologists, 272 (n. 50); Hessian fly investigations, 111–12, 115–16; on natural theology, 211

Flagg, Willard C., 139

Flint, C. L., 74

Florula Bostoniensis (Bigelow), 96

Fly-free dates (for Hessian fly), 114
Forbes, Stephen Alfred, 61–62, 100,
102–103, 118; biography, 88
Franklin, Benjamin, 95
Fraternity, entomological, 64. *See also*
Community, American entomological:
solidarity of
Frelinghuysen, Senator Frederick T., 135
French, George A., 190
Fruit growers. *See* Horticulturists
Fruit Growers Association of Ontario, 79
Fruit growing. *See* Horticulture

Genus: debate over, 202
Germ plasm, 206, 232
Ghiselin, Michael: cited on the interpreta-
tions of Darwin's orchid book, 235–36
Gilbert, Pamela: *A Compendium of the Bio-
graphical Literature on Deceased
Entomologists*, 150
Glover, Townend, 122, 285 (n. 56); and the
insect collection of the federal govern-
ment, 57–58, 83, 132; and the Colorado
potato beetle, 122; as agricultural ento-
mologist, 61–62; biography, 72;
criticism of, 132–34; publications of, 295
(n. 21), resigns as chief of the Division
of Entomology, 140
Goadby, Henry, 81–82, 191
Gould, Augustus A., 273 (n. 56)
Graef, Edward L., 166
Grange, National, 133–34
Grant, Frederick W., 194
Grant, President Ulysses S., 131–32
Grapta interrogationis, 248
Grasshoppers, 128, 297 (n. 76). *See also*
Locust; Locust, Rocky Mountain
Gray, Asa, 131, 189, 238; wins Harvard
appointment, 16; and evolution, 200,
235–36
Gray, Francis C., 42–43, 54–55
Great Chain of Being, 201
Grote, August Radcliffe, 166, 210; and en-
tomological nomenclature, 243, 246; and
evolution, 202, 204, 213
Guide to the Study of Insects (Packard), 85,
89

Hagen, Herman August: and evolution,
202, 255; and William Henry Edwards,
249; biography, 55, 81; monograph on
the Neuroptera, 273 (n. 76); *Synopsis of
the Neuroptera of North America*, 81
Hager, Albert D., 285 (n. 71)

Haldeman, Samuel Stehman: and reform
of American zoology, 22–23; biography,
17–18; corresponds with members of the
Entomological Society of Pennsylvania,
20–21; criticizes Ebenezer Emmons,
27, 71, 284 (n. 49); difficulties of entomologists noted by, 269 (n. 32); edits
Melsheimer catalog, 25–26; facilitates
work of the Entomological Society of
Pennsylvania, 29–31; financial independence of, 28; urges publication of insect
descriptions, 24
Hall, James, 42, 86
Hardin, Governor Charles H., 143
Harlan, Richard, 5
Harrington, Mark W., 80
Harris, Thaddeus William: and P. F. M. A.
Dejean, 269 (n. 36); and John G. Morris,
15; and Thomas Say, 12; and the Entomological Society of Pennsylvania, 17,
20; and the Massachusetts survey, 69–70;
applies for Harvard professorship, 16; at
Harvard, 81; biography, 11; blends entomological traditions, 64–65; compared
with Alpheus Spring Packard, Jr., 88–89;
death of (1856), 74; on priority of de-
scription, 13; role model for agricultural
entomologists, 62; uses binomial nomenclature, 14
Harvard College, 5–6
Havens, Jonathan N., 111
Hayatt, Alpheus, 204
Hayden, Ferdinand Vandiveer, 45, 137
Hayden survey: and the entomological
commission, 136; entomologists associated with, 45, 297 (n. 75); locust
investigations of, 134, 141–42
Hayes, President Rutherford B., 138
Hayhurst, L. K., 187–88
Henry, Joseph, 25, 45–46, 52, 139
Henry Institute, 48, 51
Hentz, Nicholas Marcellus: biography, 17,
20–21; *The Spiders of the United States*,
272 (n. 41)
Herrick, Edward Claudius, 112
Hessian fly: control of, 114–16; debate over
origin of, 291 (n. 13); life history and
outbreaks of, 110–14; outbreaks related
to weather, 114
Hilgard, Eugene W., 131, 139
Hilgard, Julius Erasmus, 138
*Histoire générale et iconographie des lépidop-
tères et des chenilles de l'Amérique
septentrionale* (Boisduval and LeConte),
10

Hitchcock, Edward, 69, 211
Hooker, C. M., 110
Horticulture, 108
Horticulturists, 61, 67, 74–75, 110
Howard, Leland O., 62, 287 (n. 99)
Hoy, Philo R., 101
Hübner, Jacob, 243–46, 250; *Tentamen*, 245; *Verzeichniss bekannter Schmetterlinge*, 245
Hulst, George D., 209
Humphreys, J. T., 80

Ichneumon wasp, 115
Illinois, University of. *See* Southern Illinois Normal University
Illinois State Agricultural Society, 75
Illinois State Horticultural Society, 67, 74–75
Illinois State Laboratory of Natural History, 88
Illinois State Natural History Survey, 88
Illustrators, entomological: women engaged as, 191. *See also* Abbot, John; Bowen, Lydia; Peart, Mary; Say, Lucy Way Sistare
Ingalls, Senator John J., 134
Inheritance: Weismann's theory of, 230
Inheritance of acquired characteristics, 232
Injurious Insects of the Farm and Garden (Treat), 189
Insect collecting, 34, 38, 43–44
Insect disease, 99–100
Insecticide, kerosene-soap emulsion, 83
Insecticides, chemical, 106, 125
Insect-plant relationships, 235–38
Insects: as objects of study, 2, 198, 214, 218, 308 (n. 17); beneficial, 97–99, 115; definition of, 4; introduced to North America, 97; raising of, 219–21
Insects, injurious: and appointment of state Entomologists, 110; biological control of, 106; control methods practiced by farmers, 115–16, 120; control methods proposed by entomologists, 94; control of locusts and grasshoppers, 142, 147–48, 297 (n. 76); control of the chinch bug, 119; control of the Colorado potato beetle, 122, 124–25; control of the curculio, 108–10; control of the Hessian fly, 115–16; control through chemical insecticides, 120; control through importation of predators, 99; control through insect disease, 99–100, 118–19; estimates of damage, 68–69, 75, 130; monoculture as contributing factor

to, 97; reported in the agricultural press, 93
Insects, noxious. *See* Insects, injurious: estimates of damage
International Zoological Congress, 13
Introduction to Entomology (Kirby and Spence), 4, 95–96, 156

Jefferson, Thomas, 4–5, 95, 108, 212
Jewett, Charles Coffin, 46
Johnson, Samuel William, 63, 131
Journals, entomological. *See Bulletin of the Brooklyn Entomological Society; Canadian Entomologist; Papilio; Psyche*

Kalm, Peter, 14, 95
Kanawha Valley, West Virginia, 216
Kelvin, Lord, 205
Kendall, E. O., 193
Kennicott, Robert, 47
Kirby, William, 8, 168; *Introduction to Entomology* (with William Spence), 4, 95–96, 156
Klug, Johann C., 20
Knoch, August Wilhelm, 10
Kohlstedt, Sally Gregory: cited, 30, 152–54
Köppen, Fedora Petrovich, 144, 298 (n. 92)

Lamarck, Jean Baptiste, 3–4, 201, 204
Latreille, Pierre André, 3–4, 8
LeBaron, William, 21, 138–39; as agricultural entomologist, 61–62; biography, 87; chinch bug investigations, 117–18; on the balance of nature, 104–105; recommends Paris green, 110
Lecanium acericorticis Fitch (bark louse), 190
LeConte, John Eatton, 8, 12, 14, 19; biography, 10–11; on the insect embargo, 23; physical collapse, 16; social status of, 279 (n. 65)
LeConte, John Lawrence: and the Entomological Society of Pennsylvania, 17, 24–25, 30–31; and entomological nomenclature, 243–44, 251; and reform of American zoology, 22–23; and reform of the department of agriculture, 90, 130–33; and the Entomological Society of Philadelphia, 49, 51–53; and the Smithsonian insect collection, 46–47; bid to become commissioner of agriculture, 138–40; biography, 19–20; collection of, 55, 280 (n. 108); criticizes Ebenezer Emmons, 71, 132, 284 (n. 49); death of, 31; dispute with Thomas B. Wilson, 50–51;

(LeConte, John Lawrence, *continued*)
entomological exploration of, 19, 37–38, 271 (n. 26); income of, 28, 139; monograph on the Coleoptera, 273 (n. 76); on evolution, 202, 205–206, 208; opposes use of Paris green, 125; proposes insect disease as control measure, 99; social status of, 279 (n. 65); suspects theft of specimens from cabinet, 279 (n. 81)
LeDuc, William Gates, 140
Legislation: protection of birds, 101. *See also* U.S. Entomological Commission: establishment of
Leidy, Joseph, 179
Leptinotarsa decemlinata (Say) (Colorado potato beetle), 121
Lesueur, Charles A., 5
Limenitis arthemus-prosperpina, 227
Linnaean Society of London, 202
Linnaean Society of New England, 6
Linnaeus, Carl, 2–4, 95, 245, 267 (n. 8)
Lintner, Joseph Albert: and Ebenezer Emmons, 284 (n. 49); and the New York State Museum of Natural History, 42; as agricultural entomologist, 61–62; biography, 86–87; notes advances in American entomology, 253; publishes calendar of butterflies, 96; raises Lepidoptera, 219; seeks appointment at Cornell University, 83; separates *Grapta interrogationis* and *G. fabricii* as species, 221; succeeds Asa Fitch as New York State entomologist, 287 (n. 109)
Little Turk. *See* Curculio
Lobbies, 283 (n. 32). *See also* Entomologists, agricultural: as lobbyists
Locust: control of, 148; definition of, 128; outbreaks in North America, 129–30
Locust, Asiatic, 148, 298 (n. 92)
Locust, Rocky Mountain: accounts of, 127; and establishment of entomological commission, 133–37; and reform of agricultural institutions, 127, 130–33, 138–40; and scientific organization in the federal government, 140–41, 149; barrier to frontier settlement, 129–30, 142; biological limits of, 144–45; breeding ground of, 143–45; control of, 142, 147–49; defined as species, 145, 298 (n. 87); estimates of damage, 130; extent of plague, 132, 136; extinction of, 147–48; swarms related to weather, 142–47, 298–99 (n. 96)
Locust investigations, American leadership in, 127–28, 143, 148–49

Loew, F. Hermann, 55–56
Logan, Senator John A., 87, 134
London purple, 125
Long, Major Stephen H., 14
Lyceum of Natural History of New York, 6
Lygaeus leucopterus (chinch bug), 117

Maclure, William, 5, 10, 27, 42
Maine Board of Agriculture, 89
Maine Farmer, 89
Maine Geological Survey, 89
"Malayan Papilionidae, The" (Wallace), 218
Marsh, Othaniel Charles, 211, 280 (n. 94)
Massachusetts Agricultural Repository and Journal, 8
Massachusetts Natural History Survey, 69–70
Massachusetts Society for Promoting Agriculture, 6, 8
Maxey, Senator Samuel B., 135
Mayr, Ernst: cited on essentialist-population debate, 200
McBride, Sara J., 188–89
Mead, Theodore L., 209, 216–17, 222
Mediterranean wheat variety, 115
Meehan, Thomas, 240
Melanoplus Mexicanus (Sussure) (Rocky Mountain locust), 145
Meldola, Raphael, 230
Melsheimer, Frederick Ernst: and agricultural entomology, 26; biography, 17–18; calls for insect descriptions by Americans, 23–24; *Catalogue of the Described Coleoptera of the United States,* 25, 29; collection of, 31
Melsheimer, Frederick Valentine, 11, 13; *Catalogue of the Insects of Pennsylvania,* 11
Melsheimer, John, 10
Mémoires pour servir à l'histoire des Insectes (Réaumur), 2
Merriam, Clinton Hart, 141
Meske, Otto von, 187
Meteorology, 146. *See also* Balance of nature: weather a factor in; Entomologists, agricultural: use of weather data; Locust, Rocky Mountain: swarms related to weather
Michigan Agricultural College, 82
Migration, insect, 145–46. *See also* Colorado potato beetle; Locust, Rocky Mountain: swarms related to weather
Miles, Manley, 61–62, 82
Mimicry, 202–204, 211–12, 214

"Mimicry and other Protective Resemblances among Animals" (Wallace), 218
Minnesota State Natural History Survey, 84
Minot, Charles S., 179
Missouri Botanical Garden, 236
Missouri Fruit Growers Association, 67
Missouri Geological Survey, 285 (n. 71)
Missouri State Board of Agriculture, 75
Missouri State Horticultural Society, 67, 75
Mitchell, Maria, 188
Monoculture, 65–66, 97, 125–26
Morrill, Senator Justin, 135–36
Morrill Land Grant Act, 82
Morris, John G., 20; and entomological nomenclature, 243; and European insect collections, 18, 33; and the Entomological Club of the American Association for the Advancement of Science, 31; and the Entomological Society of Pennsylvania, 15–18, 21–24, 29–30; lectures on natural history, 274 (n. 80); monograph on the Lepidoptera, 273 (n. 76); on agricultural entomology, 26
Morris, Margaretta, 112, 188, 257
Morris, Miss [?], 193
Morrison, Herbert K., 37, 246–47
Morse, Edward S., 213, 255, 328
Motschulsky, Victor, 144
Müller, Hermann, 236, 238
Murtfeldt, Mary, 189–90
Museum National d'Histoire Naturelle, 3
Museum of Comparative Zoology, 31–32, 55–56, 59, 280 (n. 108)
Museum of Natural History, St. Petersburg, 216
Myer, Colonel Albert J., 146

Nation, 136
National Academy of Sciences, 131, 133
National Agricultural Society, 81
Nationalism, American, 17
Nationalism, scientific, 23–24. See also Patriotism, American
National Museum, U.S. (Smithsonian Museum), 58
National university, 82
Natural history: American contributions to, 4–5
Natural History of Carolina, Florida, and the Bahama Islands (Catesby), 7
Naturalist on the River Amazons, A (Bates), 229
Naturalists, Pennsylvania, 28

Naturalist's Directory, 151, 184
Natural theology, 238–39, 241, 282 (n. 14). See also Balance of nature, concept of; Argument from design
New England Butterflies (Scudder), 251
New Hampshire Geological Survey, 80
New Harmony, Indiana, 10
Newman, George, 47, 217
New York City, 6
New York College of Physicians and Surgeons, 6
New York Natural History Survey, 70
New York State Agricultural Society, 66, 70–71
New York State Cabinet of Natural History, 21, 42, 71
New York State Museum, 287–88 (n. 110)
Nomenclature, binomial, 2, 14
Nomenclature, code of, 243–44
Nomenclature, entomological: changes proposed in, 244–48; crisis in, 242; implementation of changes in, 249–52. See also Classification, insect; Committee on Nomenclature of the American Association for the Advancement of Science; Committee on Nomenclature of the Entomological Club; Priority of description
Nomenclature, rules of, 249
Nomenclature, zoological, 2–3
North West Mounted Police, 148
Norton, Edward, 184
Norton, John Pitkin, 105
Norton, John Treadwell, 184
Numbers, Janet S.: cited on scientists in the South, 157–58
Numbers, Ronald L.: cited on scientists in the South, 157–58
Nuttall, Thomas, 14
Nymphalis disippus Godt., 203

Observers, entomological. See Amateurs, entomological: examples of; Entomologists, women: as entomological observers
Ohio Pomological Society, 67
Olivier, Guillaume Antoine, 4
"On Certain Entomological Speculations of the New England School of Naturalists" (Walsh), 199
On the Effect of Isolation in the Formation of Species (Weismann), 223
On the Various Contrivances by which British and Foreign Orchids are Fertilized by Insects (Darwin), 235–36
Orders, insect, 2–3, 267 (n. 8)

Organization, scientific, 15. *See also* Entomology, American: development of
Origin of Species (Darwin), 197, 206, 229, 235
Ornithologists, 101, 136–37
Ornithology, economic, 102–103, 141
Osborn, Herbert, 116
Osten-Sacken, Baron Charles Robert, 39, 56, 132, 187, 273 (n. 76)
Oswald, Solomon, 17, 19
Outbreaks, insect, 107–108. *See also* Colorado potato beetle; Curculio; Hessian fly; Insects, injurious: estimates of damage; Locust, Rocky Mountain
Owen, Richard, 201
Owen, Robert, 10

Packard, Alpheus Spring, Jr.: and evolution, 199–202, 206, 213, 255; as agricultural entomologist, 61–62, 80; assistant at the Museum of Comparative Zoology, 55; biography, 88–89; chinch bug investigations, 119; college instructor, 84–85; denounces slaughter of birds, 101; Hessian fly investigations, 112–16; locust investigations, 139, 143–47, 298–99 (n. 96); on the work of the entomological commission, 148–49
Papilio, 209
Papilio ajax, 219, 221, 225–26, 230, 308 (n. 32)
Papilio turnus-glaucus, 217, 227
Parasites, insect, 99
Paris green, 124
Parry, C. C., 130–31
Patricians, Boston, 54–55
Patricians, Philadelphia, 27
Patriotism, American, 4, 270 (n. 9); and American science, 273 (n. 60). *See also* Nationalism, scientific
Peabody Academy of Science, 88
Peale, Charles W., 5
Peale's Museum, 5
Pearsall, John, 51
Peart, Mary, 191, 217
Peck, George W., 183
Peck, William Dandridge, 6, 64–65, 81, 268 (n. 21); biography, 8–11
Pennsylvania, Society for the Promotion of Agriculture, 111
Pennsylvania Geological Survey, 79
Pennsylvania Horticultural Society, 28
Periodicals, scientific, 12
Pesticides, 125. *See also* Insecticides, chemical

Pests, agricultural. *See* Insects, injurious
Pests, imported, 99
Pests, insect: and agricultural change, 65–66
Phenology, 96
Philadelphia, Pennsylvania, 5
Philadelphia Society for Promoting Agriculture, 27
Phylloxera, 99, 289 (n. 20)
Pieris napi-napeae-bryoniae, 224
Piety. *See* Religion
Pillsbury, Governor John S., 130
Plum curculio. *See* Curculio
Political system, U.S., 86–89
Pollination of yucca, 237–38
Polymorphism, 214, 218, 306–307 (n. 2). *See also* Edwards, William Henry: on polymorphism
Pomologists. *See* Horticulturists
Popularizers: women entomologists as, 190–91
Populationist, 201
Populations, animal, 97. *See also* Balance of nature: concept of
Potato, 120–22
Potato blight, 121–22
Powell, Major John Wesley, 45
Practical Entomologist, 68, 75
Prairie Farmer, 87
Predator-parasite-prey relationships, 97
Press, agricultural, 92–93. *See also* Entomologists, agricultural: as lobbyists
Priority of description, 12–13, 23–26, 30, 244–47, 251–52, 270 (n. 39). *See also* Classification, insect; Committee on Nomenclature of the Entomological Club; Nomenclature, entomological; Systematics
Prodoxus decipiens (false yucca moth), 239–40
Professionalization of entomology: affiliations of entomologists and, 179–82; and insect outbreaks, 107, 125–27; call for, 89–90; professionals replace amateurs, 60–61; role of agricultural funding in, 85–86; scientific income contributes to, 175–79
Professionalization of scientists, 60–61, 69
Pronuba moth, 236–41
Pronuba yuccasella, 237, 239
Psyche, 56
Publication, scientific: critical review of, 30, 275 (n. 116)
Pugh, Evan, 63
Purpose in nature. *See* Natural theology

Putnam, Frederick Ward, 205, 213

Rafinesque, Constantine S., 5
Rarer Lepidopterous Insects of Georgia, The
(Abbot), 8
Reakirt, Tryon, 34, 222
Réaumur, Réné Antoine, 2
Record of American Entomology, 150, 184
Reed, Edmund B., 246
Règne Animal (Cuvier), 3
Religion, 63–64, 205–206, 210–12. *See also*
Natural theology; Argument from de-
sign
Renaissance, 1
*Report on the Insects of Massachusetts, Injuri-
ous to Vegetation* (Harris), 11, 62, 70, 88
Resistant varieties (wheat), 114–15
Reynolds, John P., 74
Rhyparochromus devastator (chinch bug), 118
Ridings, James, 47, 217
Riley, Charles Valentine, 189, 289 (n. 20);
and entomological nomenclature, 243,
251; and the entomological commission,
133–34, 136–38; as agricultural entomologist, 61–62; biography, 75–77; calls
for reform of the agricultural department, 131–32; calls for study of cotton
insects, 257; chief, Division of Entomology, 140; chinch bug investigations, 118;
collection of, 58–59; college instruction
of, 83–85; Colorado potato beetle investigations, 122–25; control methods
proposed by, 99; evolution and religion
reconciled by, 205; evolution as basis of
insect classification, 247–48; evolution
exhibited in mimicry, 203–204; evolution exhibited in moth larvae, 303
(n. 12); evolutionary studies cited, 255;
evolutionary synthesis of, 206; evolutionary theory disseminated by, 212–13;
Hessian fly investigations, 112–14;
locust investigations, 141–43, 298
(n. 87); locust plague described by, 129;
locust swarms related to weather, 104,
145–47; on insect-bird relationships,
101–102; on insects as subjects for evolutionary studies, 197–98; phylloxera
investigations cited, 289 (n. 20); pronuba
moth investigations, 237–40; resigns as
Missouri state entomologist, 80
Robinson, Solon, 81
Rogers, Henry D., 79
Rogers, William K., 140
Ross, Barnard, 192
Ross, Christina, 192

Royal Geographic Society (London), 229
Royal Society Catalogue of Scientific Papers,
153

St. Louis Academy of Science, 212, 236–37
Sanborn, Francis G., 61–63, 74, 89–90
Sargent, Senator Aaron A., 135
Saunders, Sidney Smith, 210
Saunders, William: as agricultural entomologist, 61–62; and entomological
nomenclature, 245–46, 250–51; and the
Colorado potato beetle, 122; biography,
77; experiments with pesticides, 110;
on the sex of butterflies, 189; raises
Lepidoptera, 219
Say, Lucy Way Sistare, 191
Say, Thomas: *American Entomology; or, Descriptions of the Insects of North America*,
10, 191; *American Conchology*, 191; biography, 10; chinch bug (scientific name
given), 117; Colorado potato beetle (scientific name given), 121; death of, 16;
entomological work of, 11–14; Hessian
fly (scientific name given), 111
Schaum, Hermann, 20, 24–25, 29
Schurz, Carl, 138
Science, American: advancement of, 25;
early center of, 5–6
Science in America: compared with science
in Europe, 14, 213. *See also* Entomology,
American: compared with Entomology
in Europe
Scudder, Samuel Hubbard, 55–56, 80, 216;
entomological nomenclature changes
proposed by, 242–46; entomological nomenclature rules debated by, 249–52;
evolutionary contributions cited, 213,
255; evolutionary theory debated by,
199–204, 206–207; evolutionary views
challenged, 247; locust investigations of,
141, 145; raises Lepidoptera, 219; *The
Butterflies of New England*, 242, 251
Shimer, Henry, 104, 118
Signal Service, U.S., 148
Smith, Emily A., 189–90
Smith, James E., 7–8
Smith, John B., 209
Smith, Sidney Irving, 80
Smithson, James, 43
Smithsonian Institution, 15, 25, 43. *See
also* Collection, insect: of the Smithso-
nian Institution
Smithsonian meteorological project, 146
Societies, agricultural, 66–67, 100–101. *See
also* Illinois State Agricultural Society;

(Societies, agricultural, *continued*)
  Massachusetts Society for Promoting
  Agriculture; Missouri State Board of
  Agriculture; National Agricultural So-
  ciety; New York State Agricultural
  Society; Pennsylvania, Society for the
  Promotion of Agriculture; Philadelphia
  Society for Promoting Agriculture
Societies, entomological, 254. *See also*
  Cambridge Entomological Club; East
  London Entomological Society; Ento-
  mological Society of Canada;
  Entomological Society of London; Ento-
  mological Society of Pennsylvania;
  Entomological Society of Philadelphia
Societies, horticultural, 74–75, 100–101.
  See also American Pomological Society;
  Fruit Growers Association of Ontario;
  Illinois State Horticultural Society; Mis-
  souri Fruit Growers Association;
  Missouri State Horticultural Society;
  Ohio Pomological Society; Pennsylvania
  Horticultural Society; State Pomological
  Society of Michigan; Western New York
  Horticultural Society
Society, scientific: defined, 28
Soil analysis, 105
South, the: as home for science, 272 (n.
  42). *See also* Community, American en-
  tomological: geographic location of
  members
Southern Illinois Normal University, 88
Speciation: and polymorphism, 227–28;
  defined, 310 (n. 65)
Spécies Général des Coléoptères (Dejean),
  10–11
Spence, William, 4; *Introduction to Entomol-
  ogy* (with William Kirby), 4, 95–96, 156
Speyer, Adolph, 250
Spiders of the United States, The (Hentz), 272
  (n. 41)
Spoils system, 131, 139–40, 295 (n. 17)
Stanford University, 84
State Natural History Society of Normal,
  Illinois, 88
State Pomological Society of Michigan, 83
Statistical analysis, 103
Stauffer, Jacob, 79
Studien zur Descendenztheorie (Weismann),
  223, 232
Studies in the Theory of Descent (Weismann),
  232
Survey, annual insect pest, 114, 120
Survey, state: Illinois, 88; Maine, 80, 89;
  Massachusetts, 27, 69–70, 72–74; Michi-
gan, 79–80; Minnesota, 84; Missouri,
  285 (n. 71); New Hampshire, 80; New
  York, 26–27, 70–71, 86; North Caro-
  lina, 80; Pennsylvania, 79
Swallow, G. C., 285 (n. 71)
Synonyms, taxonomic, 24
Synonymy, 30, 242. *See also* Nomen-
  clature, entomological: changes
  proposed in; Nomenclature, rules of;
  Priority of description
Synopsis of the Neuroptera of North America
  (Hagen), 81
Systema Naturae (Linnaeus), 2
"Systematic Revision of American But-
  terflies" (Scudder), 242
Systematics, insect: and Darwinism, 242;
  as stimulus to entomology, 4; of Thomas
  Say, 10. *See also* Classification, insect;
  Evolutionary theory: insect classification
  based on; Nomenclature, entomological;
  Priority of description

Taylor, Alexander S., 144
Taylor, Charlotte DeBernier Scarbrough,
  190–91
Teleology, 95–96, 207. *See also* Natural
  theology
Tentamen (Hübner), 245
Theft: of insect specimens, 34, 51–52, 279
  (n. 81)
Thomas, Cyrus, 190; and the entomological commission, 134–35, 137–38; as
  agricultural entomologist, 61–62; biography, 87–88; calls for reform of the
  agricultural department, 132; chinch bug
  investigations, 119–20; Colorado potato
  beetle investigations, 122; Hessian
  fly investigations, 112–14; locust investigations, 139, 142–44; locust swarms
  related to weather, 145–47, 298–99
  (n. 96); on insect-bird relationships, 101;
  opposes evolutionary theory, 201–202
Torrey, John, 131
Treat, Mary, 188–89, 257
Trimen, Roland, 203
Tucker, Luther, 66
Turner, Jonathan, 82
Type concept, 39–40, 200–202, 245, 276
  (n. 15)
Type specimens, 39–40
Typological concepts, 200–202, 207,
  210–11, 227–28, 256

"Über den Saison-Dimorphismus der
  Schmetterlinge" (Weismann), 225

Uhler, Philip Reese, 55, 139, 273 (n. 76)
U.S. Department of Agriculture, 127, 130–33, 138–41, 149
U.S. Entomological Commission: and scientific organization, 140–41; establishment of, 137; history of, 297 (n. 70); locust investigations of, 141–49; staffing of, 137–38
U.S. Exploring Expedition, 215
U.S. Signal Service, 148
Utility of butterfly wing patterns, 214
Uvarov, Sir Boris, 128

Variation (biological), 224–25, 228
Variety, Underhill (wheat), 111
Verrill, Addison E., 83
Verzeichnis bekannter Schmetterlinge (Hübner), 245
Vestiges of the Natural History of Creation (Chambers), 211
Voyage up the River Amazon, A (Edwards), 215

Walker, Francis, 209
Wallace, Alfred Russel: and William Henry Edwards, 215, 223, 229; biography, 229; on butterflies and orchids, 308 (n. 17); on mimicry, 203; on polymorphism, 218; on utility in organisms, 214; president, Entomological Society of London, 210
Walsh, Benjamin D.: as agricultural entomologist, 61–62; biography, 74–75; calls for evaluation of control methods, 93; calls for study of cotton insects, 257; chinch bug investigations, 118; Colorado potato beetle investigations, 121–24; edits the *American Entomologist,* 68, 75; edits the *Practical Entomology,* 68, 75; evolution demonstrated through gall insects, 200, 303 (n. 12); evolution demonstrated through mimicry, 203; evolution as basis of insect classification, 247; evolutionary theory advocated by, 199–200; evolutionary investigations cited, 213, 255; evolutionary theory as basis of the genus, 202; evolutionary theory disseminated by, 212; locust investigations, 141, 144; "On Certain Entomological Speculations of the New England School of Naturalists," 199; on beef and dairy farmers, 283 (n. 32);

on dimorphism in butterflies, 217; on insect-bird relationships, 102; on solidarity among entomologists, 196; proposes importation of predators, 99
Washington, George, 108
Watts, Frederick, 130–33
Weather. *See* Balance of nature: weather a factor in; Chinch bug; Hessian fly: outbreaks related to weather; Locust, Rocky Mountain: swarms related to weather; Meteorology
Webster, Francis M., 213
Weismann, August: and William Henry Edwards, 223, 225–26, 228–32; biography, 223–24, 232; *On the Effect of Isolation in the Formation of Species,* 223; *Studien zur Descendenz Theorie,* 223, 232; *Studies in the Theory of Descent,* 232; theory of germ plasm, 206, 232; "Über den Saison-Dimorphismus der Schmetterlinge," 225
Westcott, Oliver Spink, 183–84
Western New York Horticultural Society, 83
Westwood, John O., 210
Wheat, 65
White, Andrew D., 83–84
Whitman, Alexander, 84
Whorton, James: cited on Entomologists' endorsement of arsenic insecticides, 125
Wilder, Burt G., 83
Wilder, Marshall P., 67, 81–82
Wilson, Alexander, 5
Wilson, Thomas B.: biography, 50; death, 54; dispute with John L. LeConte, 50–51, 53; supports Academy of Natural Sciences of Philadelphia, 42; supports Entomological Society of Philadelphia, 47, 49
Winchell, Alexander, 79–80, 84
Withers, Senator Robert E., 135
Women. *See* Entomologists, women
Woodward, J. S., 110

Yucca, 237
Yucca moth, 237

Zeller, Philipp C., 233, 250
Ziegler, Daniel, 16–19
Zimmermann, Karl (Christian), 17, 29–31, 259; biography, 20
Zoology, American: reform of, 22–23